Nonlinear Analysis of Physiological Data

Springer
Berlin
Heidelberg
New York
Barcelona
Budapest
Hong Kong
London
Milan
Paris
Santa Clara
Singapore
Tokyo

H. Kantz · J. Kurths
G. Mayer-Kress (Eds.)

Nonlinear Analysis of Physiological Data

With 123 Figures

Springer

Priv.-Doz. Dr. rer. nat. H. Kantz
Max-Planck-Institut für Physik Komplexer Systeme
Nöthnitzer Strasse 38
D-01187 Dresden, Germany

Prof. Dr. rer. nat. habil. J. Kurths
Arbeitsgruppe „Nichtlineare Dynamik",
Universität Potsdam
Am Neuen Palais, Gebäude 19
Postfach 60 15 53
D-14415 Potsdam, Germany

Prof. Dr. G. Mayer-Kress
Ross Institute
East Hampton
NY 11937, USA

Library of Congress Cataloging-in-Publication Data applied for.

Die Deutsche Bibliothek – CIP-Einheitsaufnahme

Nonlinear analysis of physiological data: with 17 tables / H. Kantz... (ed.). Berlin; Heidelberg; New York; Barcelona; Budapest; Hong Kong; London; Milan; Paris; Santa Clara; Singapore; Tokyo: Springer, 1998

ISBN-13:978-3-642-71951-6 e-ISBN-13:978-3-642-71949-3

DOI.10/1007:978-3-642-71949-3

This work is subject to copyright. All rights are reserved, whether the whole or part of the material is concerned, specifically the rights of translation, reprinting, reuse of illustrations, recitation, broadcasting, reproduction on microfilm or in any other way, and storage in data banks. Duplication of this publication or parts thereof is permitted only under the provisions of the German Copyright Law of September 9, 1965, in its current version, and permission for use must always be obtained from Springer-Verlag. Violations are liable for prosecution under the German Copyright Law.

© Springer-Verlag Berlin Heidelberg 1998
Softcover reprint of the hardcover 1st edition 1998

The use of general descriptive names, registered names, trademarks, etc. in this publication does not imply, even in the absence of a specific statement, that such names are exempt from the relevant protective laws and regulations and therefore free for general use.

Typesetting: Camera-ready copy by editors
Cover design: Erich Kirchner, Heidelberg

SPIN 10643606 55/3144 – 5 4 3 2 1 0 – Printed on acid-free paper

Preface

This book is more than a standard proceedings volume, although it is an almost direct result of the workshop on "Nonlinear Analysis of Physiological Time Series" held in Freital near Dresden, Germany, in October 1995. The idea of the meeting was, as for previous meetings devoted to related topics, such as the conference on dynamical diseases held near Montreal in February 1994 (see CHAOS Vol. 5(1), 1995), to bring together experts on the techniques of nonlinear analysis and the theory of chaos and applicants from the most fascinating field where such methods could potentially be useful: the life sciences. The former group consisted mainly of physicists and mathematicians, the latter was represented by physiologists and medical researchers and practitioners.

Many aspects of this workshop were unusual and not previously experienced. Also, the hosting institution, the Max Planck Institute for Physics of Complex Systems (MPIPKS), at this time was brand new. The organizers' rather unconventional intention was to bring specialists of both groups together to really work together. Therefore, there was an excessive availability of computers and the possibility to numerically study time series data sets practitioners had supplied from their own fields, e.g. electrocardiogram (ECG) data, electroencephalogram (EEG) data, data from the respiratory system, from human voice, human posture control, and several others. These data formed a much stronger link between theoreticians and applicants than any of the common ideas. The results of collaborations initiated in the course of this workshop are the main content of this volume: The application of the various techniques presented in the first part of the book to data sets presented on the workshop. Many interesting and fruitful collaborations resulted from this meeting and are reflected by the groups of authors contributing to this volume; a fact that we organizers are proud of.

The general result of this meeting and of the work documented in this book is that this new approach to time series analysis contains tremendous potentials, but has to be further developed. It is a truism that present-day data acquisition techniques in medicine produce much more data than can be analysed by human visual inspection. Furthermore, a continuous stream of data contains orders of magnitude more information than a single measurement extracted from it. Therefore, there is an urgent need for refined and computer-aided diagnostics. Traditional (linear) statistical methods, which regard irregular data predominantly as the outcome of stochastic processes, are only of limited use as diagnostic tools. The methods called *nonlinear* differ from linear methods not just by allowing for nonlinear feedback. More important is the different philosophy behind them: the idea that irregular data can contain deterministic structure, the aperiodicity being related to the phenomenon of deterministic chaos. In every single case where this is found to be the case at least to some extent, the possibility for a much deeper understanding of the phenomena and therefore a much more powerful diagnosis arises. But even in cases where the process is more complicated than low-dimensional determinism, these methods supply new concepts for data classification, such as predictability, divergence rates, geometrical properties in embedding spaces, and statistics of symbolic representations, to mention a few. When sufficient experience with these methods is collected, they might serve for many diagnostic purposes, such as recognition of particular diseases (e.g. high risk for sudden cardiac death, voice disorder such as papillon and laryngitis), or understanding physiological mechanisms (e.g. brain action by localization of activities).

It is the common impression of the contributors to this volume that nonlinear techniques open a new window to view physiological data. A lot of work has still to be done, to adopt the rigorous mathematical concepts to real-world data, to test for clinical and statistical relevance of the outcome, to convince medical doctors and the manufacturers of medical diagnostic instruments and software that they can rely in their daily work on the new methods. For the future, one can speculate about the availability of small devices, carried by all of us on our wrist, continuously recording relevant parameters such as heart rate or blood pressure; the data being transmitted to a computer once a day and being analysed, e.g. with respect to diseases of the cardio-vascular system. Such a preventive medicine, based on time series analysis to detect disease in its early stage, will be much more powerful and much more economic than today's mainly curative medicine.

We still have a long way to go. We are confident that this volume will be a step in the right direction, and we thank all contributors and all participants of the workshop as well as all reviewers of the papers. We also gratefully acknowledge the financial support and hospitality of the MPIPKS which was the precondition for the workshop to take place.

Dresden, Potsdam, February 1998 H. Kantz, J. Kurths, G. Mayer-Kress

Table of Contents

Introduction

Holger Kantz and Jürgen Kurths

The restriction of both the workshop and this book to physiological time series has its clear reason in the particularities of data from exactly this field. In principle, one can think of applications of nonlinear methods to aperiodic, irregular time sequences from various fields, such as economy, and climatology, just to mention some examples. However, in contrast to linear statistical tools, the essential aspect is that nonlinear methods cannot be used as black box routines. Many of them do convert a time series into a number, but the relevance of this number to characterize the data sensitively depends on the general understanding of the system under study. Therefore, these methods, when applied to data from field measurements, have to be seen within the particular environment, which justifies the restriction to physiological data. Nevertheless, many problems occurring in the treatment of physiological data will be encountered also in other fields, such as the problem of nonstationarity.

1 Complex and Complicated

Complexity is not a well-defined concept, but intuitively we have the impression that physiological processes are complex. Nevertheless, the systems we are speaking about are also complicated. In order to approach a clearer notion of complexity, we can start to exclude processes that are certainly not complex. Periodic motion clearly belongs to them. But despite its irregular and unpredictable character, also a white noise random process cannot be assumed to be complex, since it does not contain any nontrivial structure. Thus complexity has to do with intricate structures hidden in the dynamics, emerging from a system which itself is much simpler than its dynamics. Spatiotemporal patterns in a homogeneous tissue and low-dimensional chaos in coupled nonlinear oscillators, as a simplistic model of the heart, are good examples for complex processes. Thus complexity is characterized by the paradoxical situation of complicated dynamics of simple systems. In contrast to that, when the system itself is already complicated, since it is composed of many different parts, it is obvious that it may support very complicated dynamics, but perhaps without the emergence of clear and characteristic patterns.

The reason why we emphasize this interplay of complex and complicated behaviour is that nonlinear and geometrical methods of data analysis display their full power when applied to complex dynamics in simple systems. But of course a system's dynamics may be both complicated and complex. This is surely the case for living beings, at least if we consider them as entities. Many different physiological processes are going on simultaneously, with their different locations in the body, different time scales and amplitudes, and all are interacting in a more direct or indirect, a stronger or weaker way. This fact introduces severe difficulties in data analysis and makes physiological data perhaps the most difficult to treat.

A problem closely linked to these considerations lies in the number of active degrees of freedom. It is extremely hard to reconstruct deterministic processes from scalar time series data if more than a couple of degrees of freedom are active. One way to cope with this is to concentrate on subsystems and to prepare them in an isolated way as much as possible. The opposite would be to take into account as many different but interacting components as possible. A prototype for this ambivalence are ECG data which reflect the variability of the heart rate. Surely, it has its origin in the self-sustained oscillations of the pacemaker section on the heart, but it is also closely linked to the whole activity of the cardio-vascular and the nervous systems. Thus one can concentrate on the ECG and try to keep as many other parameters of the body as constant as possible, thus hopefully being able to study the probably low-dimensional pacemaker dynamics. Alternatively, one can simultaneously measure the blood pressure, breath rate and other data, and hope to have thus a direct access to other dynamical variables linked to the heart activity. This could give a more global understanding but is also a more demanding task. In particular when one finds that a single observable does not contain sufficient information to understand a given phenomenon, one has to ask whether another observable measured simultaneously adds more information than it enlarges the system under observation. Therefore, one question is how far different observables supplement each other, and, if they do, how to extract the additional information by a combined analysis. Some of the articles of this volume are concerned with this issue.

2 Physiological Time Series Data

Throughout this book we are concerned with the analysis of *time series*, i.e. the time ordering of the data contains a significant part of the information. In particular, this means that the observations should be equally spaced in time. If they are not, it will be hard to apply the methods discussed in the following contributions. Only with equal sampling intervals, there is a realistic chance to learn something about the underlying dynamics of the phenomenon. The only exception is formed by data which are not the recording of a continuously changing variable (like the voltage at an electrode in an ECG) but a sequence of intrinsically distinguished events. These point processes can be thought of as mathematical maps, which are obtained from the flows by the technique of Poincaré surface of section. Physiological data of this kind are R-R intervals extracted from ECGs, or times between the firing of neurons. Generally, time series should be as long as possible. As a rule of thumb, if there are typical oscillations in a signal, the time series should cover several hundreds of them, in the best case several thousands of cycles.

Data obtained from living beings are irreproducible even in a statistical sense. A living being can never be set again in exactly the state it was before, and if it were only because of ageing effects. But along with a change of the being and its environment between successive series of measurements, there are always also more or less slight changes during measurement, thus giving rise to nonstationarity. For all kinds of data analysis, one has to minimise nonstationarity, or to extract stationary phases from a longer segment of measurements.

In addition to nonstationarity, there are also perturbations of the data which are called noise. Noise can be introduced through the measurement device, through the transmission of the signal from the source to the measurement device, and due to perturbations from the environment. To give an example, ECG data are always recorded with finite precision (measurement noise), but imperfect electric conduction between skin and electrodes as well as activities of muscles in between the heart and the electrode introduce transmission noise. It is obvious that all this makes data analysis more difficult, and if these perturbations are too strong, they can even destroy almost all relevant information. In many cases, however, nonlinear noise reduction may be successful.

3 Merits of Dynamical Analysis

In contrast to traditional methods, where a state of a system is characterized by averaged quantities like mean values, variances, or autocorrelations, nonlinear analysis tries to get hold of properties of the underlying dynamics. This supplies new tools for diagnostics and for signal classification. Used this way, the diagnostic power of nonlinear methods can be directly compared to that of traditional methods, and in case it turns out not to be superior, traditional methods are to be favoured, since they are better understood and usually computationally less demanding. It is, however, our conviction and supported by the articles collected below that in many situations a careful application of nonlinear dynamics will indeed yield new insights. Since nonlinear methods allow for a characterization of data by a whole set of quantities (e.g. divergence rates, predictability, scaling exponents, entropies in symbolic representation), one can always expect to find a signature dividing two groups of data sets, for example ECG data from high-risk and low-risk patients. However, whether such a distinction is really beyond a statistical fluctuation has to be confirmed by an *out of sample test*, i.e. by applying the statistics to data sets one has not used for the development of the criterion. This is often a real problem, since the data base can be very poor, and one would be reluctant to renounce part of the data during design of the diagnostic tool.

Apart from signal classification, a detailed dynamical analysis can help to detect physiological mechanisms and to design or affirm/reject either simplified or realistic models for certain processes and disorders. Generally, there exist simple models only for well isolated phenomena. One example is the delayed feedback in human's perception and reaction, as presented in the article by Tass et al. Also Rosenblum et al. and Herzel present such an approach, the first introducing a simple model for posture control, and the second containing a conceptual model of voice disorder. Many systems, however, can be modeled in a *realistic* way only when their full structure is taken into account, and one example towards such a model is the sophisticated continuum model of the vocal folds developed by Herzel.

As mentioned before, the application of nonlinear methods to data from field measurements like physiological data is yet widely unexplored. Since physiological data contain surely complex structure, since nonlinearities and time delay in feedback loops is common to them, since finally both understanding and diagnostics of disorders is related to dynamical properties of systems, nonlinear, i.e. geometrical and dynamical analysis, will be very promising. Therefore, it is a great challenge to adopt these methods and to apply them in this field. The articles collected in this volume give an insight into the state of the art. They allow to estimate the problems one is faced by, but also the potential gains. There is no intention to present a complete overview of the field, nor do the articles give a guide for a complete data analysis. This book intends to supply sufficient motivation for further studies and applications in medicine. The collaboration between physicists, and physiologists and medical doctors can be very fruitful, as demonstrated by the articles in this book. A deep knowledge about the physiological processes and difficulties of certain measurements on the one hand and an expertise in the background of nonlinear analysis on the other hand is the combination needed for a successful approach.

The articles collected in this book will not introduce the basic concepts of nonlinear dynamics and their characterization. Nor will chaotic phenomena be discussed in detail. We strongly recommend those of the readers who want to enter the field of nonlinear and dynamical data analysis to get acquainted with at least the fundamental aspects. Recommendable text books are e.g. Ott, Sauer, & Yorke, *Coping with Chaos* (Wiley, New York 1994), Kaplan & Glass, *Understanding Nonlinear Dynamics* (Springer, New York 1995), Kantz & Schreiber, NONLINEAR TIME SERIES ANALYSIS (Cambridge University Press 1997).

I Directions in Nonlinear Data Analysis

Processing of Physiological Data

Thomas Schreiber

Physics Department, University of Wuppertal, D–42097 Wuppertal, Germany

Abstract. In the context of nonlinear time series analysis, preprocessing of data brings up two questions: how does the data processing step influence the performance of nonlinear algorithms, and how can ideas from nonlinear dynamics be used to improve the data? We will illustrate these issues with some exemplary physiological time series problems.

1 Introduction

The classical field of linear time series analysis has been very successful in modeling real world data which are notoriously short and noisy. If the dominant structure we can detect in a time series is due to linear (two point) autocorrelations, a spectrum based analysis provides the most stable description of the data. Nevertheless, there are many systems, in particular in animated nature, where we are reluctant to accept that the underlying process is linear. In biology it is prominently the way systems react to changes in their environment which requires a nonlinear description. We expect that somehow this nonlinearity will affect the signals emerging from such systems. We are thus often facing a paradoxical situation. Regarding an isolated time series on its own, the most natural description is provided by the linear approach. However, keeping in mind what kind of system is producing the signal, this approach seems inadequate.

In view of this dilemma, we would like to keep the analysis of time series data as free as possible from any prejudice. But even very basic data processing techniques concerning sampling, filtering, and data representation turn out, at a closer look, to be biased towards a subsequent linear analysis of the data. It has been pointed out for example that some linear autoregressive filters which are available and often silently built into data acquisition hardware, may spuriously increase the dimension of the system, see e.g. Badii et al. (1988). Whether this is of concern or not depends on the intended use of the data. In any case it might be useful to know some alternative filtering techniques.

In this paper, we will briefly discuss some data processing techniques, linear, almost linear, and nonlinear. The final choice will largely be guided by the nature of the subsequent analysis, rather than by prior knowledge about the nature of the data. Thus whenever nonlinear precessing turns out to be successful, we will not take this alone as evidence for nonlinearity in the data, or even nonlinear dynamics.

Very often, data is analysed in an iterative way: starting with some hypothesis we try how far we can get. If the results are inconsistent or otherwise unsatisfactory, we will have to add new features to the analysis, thereby increasing the complexity of the approach. For this we might have to go back as far as to the raw data. It seems natural to gradually increase the complexity of the methods until the data is described adequately. A natural ordering would be to start with a purely descriptive approach: can we immediately understand structures in the data, like seasonal variations, sudden changes due to environmental influence etc. For this first data browsing we can use unprocessed data. Some smoothing could be useful with the sole purpose of rendering clearer pictures. More systematically, we should study the autocorrelations in the data and the power spectrum. Linear time series analysis provides its own processing tools, and data requirements and limitations are well understood.

If the linear analysis seems not to be able to explain the data well enough, we might try, as a working hypothesis, to assume that the data is related to a nonlinear deterministic system. In this case we should apply linear filtering only very carefully before we understand the intrinsic time scales of the process. The most useful representation for deterministic systems is in a reconstructed phase space. Consequently, also the data processing should be oriented towards such a representation. We may try to reduce noise by phase space projections, either global or local ones.

If the data quality permits, we may try more complicated scenarios, like high dimensional dynamics, space–time chaos, or Langevin (nonlinear, noise driven) dynamics. Eventually, we will stick to the *simplest description which is consistent with the data.* In physiology, we have to be prepared to find that this is the descriptive or the linear approach. This might sometimes seem frustrating since we expected nonlinearity on the base of our knowledge about the underlying physiological system.

Before we give some examples of different data processing techniques and their interactions with a subsequent data analysis, let us point out that for the above approach of increasing complexity it is essential that the raw, unprocessed data remains available, because at each level different data requirements will apply.

2 Why Data Processing?

Usually, the information contained in time series data is not immediately obvious but has to be extracted by some means. Not every bit of the record will be equally important and our task is to separate the important features from the unimportant ones. By data processing we mean the very first step in this process where we get rid of utterly unimportant aspects of the data, enhancing the important ones.

Sometimes, data reduction can be quite dramatic: an electroencephalogram (EEG) may be recorded during sleep solely to determine the sleep stage, REM sleep or not. Or an electrocardiogram (ECG) may be taken to determine a set of relevant parameters, like the PR or QT intervals. In such cases we often have a clear picture of which information may be thrown away immediately: we are neither interested in the 60 Hz AC contamination of the amplifier, nor in a slow baseline drift due to static electricity etc. In other situations we have to be much more careful since fluctuations may be of clinical importance. We will in the following use the term "noise" for any identified contamination, not necessarily of random nature.

In short, data processing always begins with some assumption about the data which leads to an objective criterion for the separation into relevant and irrelevant aspects. Very often, the irrelevant part will contain some measurement noise which has contaminated the observations but not influenced the system itself. The time series can then be thought of as a linear superposition of signal and noise, $s_n = x_n + \eta_n$, which may allow for an approximate decomposition. When the irrelevant part interacts with the system (dynamical or multiplicative noise) a decomposition is not usually well defined and must be undertaken with great care. Take for example a purely deterministic time evolution given by a nonlinear map acting on some m dimensional space. The scalar observable could be some smooth function acting on that space:

$$s_n = H(\mathbf{x}_n), \quad \mathbf{x}_n = \mathbf{F}(\mathbf{x}_{n-1}). \tag{1}$$

If the system is perturbed by noise at each time, the dynamics is modified to

$$\mathbf{x}_n = \mathbf{F}(\mathbf{x}_{n-1} + \boldsymbol{\eta}_{n-1}) \tag{2}$$

but in general it will not be possible to write the signal $\{s_n\}$ in the form $s_n = H(\mathbf{x}_n) + \tilde{\eta}_n$.[1]

3 Linear Filters

Processing of time series data with a linear filter is based on the assumption that signal and noise can be distinguished on the base of the power spectrum of the time series. This is a particularly useful assumption if the data is dominated by periodic oscillations. Then the power spectrum shows sharp peaks at the harmonics of the oscillation frequency and any continuous component must be due to the noise. If the system is nonlinear deterministic, it

[1] For *hyperbolic* systems, the shadowing theorem, (Bowen 1975), enssures that for small noise there is a nearby sequence $\{\tilde{\mathbf{x}}\}$ such that $\tilde{\mathbf{x}}_n = \mathbf{F}(\tilde{\mathbf{x}}_{n-1})$ and $s_n = H(\tilde{\mathbf{x}}_n) + \eta_n$. The theorem does not include the more typical nonhyperbolic systems. Nor does it say anything about the properties of the sequence η_n and the stability of the solution $\{\tilde{\mathbf{x}}_\mathbf{n}\}$. Shadowing in nonhyperbolic systems is discussed in Grebogi et al. (1990).

can perform chaotic motion which leads to a continuous component in the power spectrum itself. In such a case, we cannot estimate the noise spectrum given the spectrum of the time series. But even if we could identify the noise contribution to the spectrum, a spectral filter could not effectively remove it without also affecting the chaotic component. Only if we know that the signal spectrum extends only to an upper cutoff frequency, we can reduce high frequency noise using a linear low-pass filter.

A linear filter can be constructed as follows. The power spectrum of a continuous signal can be estimated from a discretely sampled time series, $s_n, n = 1, \dots, N$ by the periodogram $S_k^{(s)} = |\tilde{s}_k|^2$ where $\tilde{s}_k = N^{-1/2} \sum_{n=1}^{N} s_n e^{2\pi i k n/N}$ is the discrete Fourier transform of the signal. Suppose we want to interpret the power spectrum $S_k^{(s)}$ as a superposition of signal $S_k^{(x)}$ and noise $S_k^{(\eta)}$, $S_k^{(s)} = S_k^{(x)} + S_k^{(\eta)}$. We want to estimate the signal by filtering the time series with a suitable filter ϕ_k such that $\hat{\tilde{x}}_k = \phi_k \tilde{s}_k$. The filter which minimises the mean squared distance between $\hat{x}_n$ and x_n, the so called *optimal* or *Wiener* filter is given by $\phi_k = S_k^{(x)}/S_k^{(s)}$ (see e.g. Press et al. (1988) for details). Obviously, we cannot directly compute $S_k^{(x)}$ but we can try to estimate it by inspection of the total power spectrum. This is the point where we need a criterion to distinguish signal and noise by their respective properties. As we said, this is often possible for signals with dominant oscillations. For signals with intrinsic broad band spectra, like those from deterministic chaotic systems, this is usually impossible.

In Fig. 1 we see a human electrocardiogram (ECG)[2] which has been contaminated by Gaussian white noise of standard deviation 10 A/D units. (The total peak-to-peak amplitude of the signal is about 450 A/D units.) In the upper panel it is plotted before and in the lower after Wiener filtering. The filter has been constructed from the known spectrum of the fairly clean underlying signal, which is much better than we can do in practice. We see that the Wiener filter is unable to clean the ECG from random noise.

The situation is different e.g. if the undesired part of the signal consists of 60 Hz AC contamination, which is readily distinguishable in the spectrum. See Fig. 2.

Except for cases where the distinction between signal and noise can be made clearly on the base of the power spectrum, linear filtering cannot be recommended in connection with a nonlinear data analysis. First, in the absence of a clear distinction, the filtering will also effect frequencies which are essential for the nonlinear part of the dynamics. In a reconstructed phase space, this will typically lead to undesired distortions. These distortions can be severe if the attempt is made to "bleach" chaotic data, that is, to remove the linear correlations, (Theiler & Eubank 1993). Second, autoregressive filters (e.g. AR(1): $s'_n = s_n + a s'_{n-1}$) introduce additional degrees of freedom into the system. They may alter the dimension of a strange attractor formed by

[2] The data were kindly provided by Petr Saparin, Potsdam.

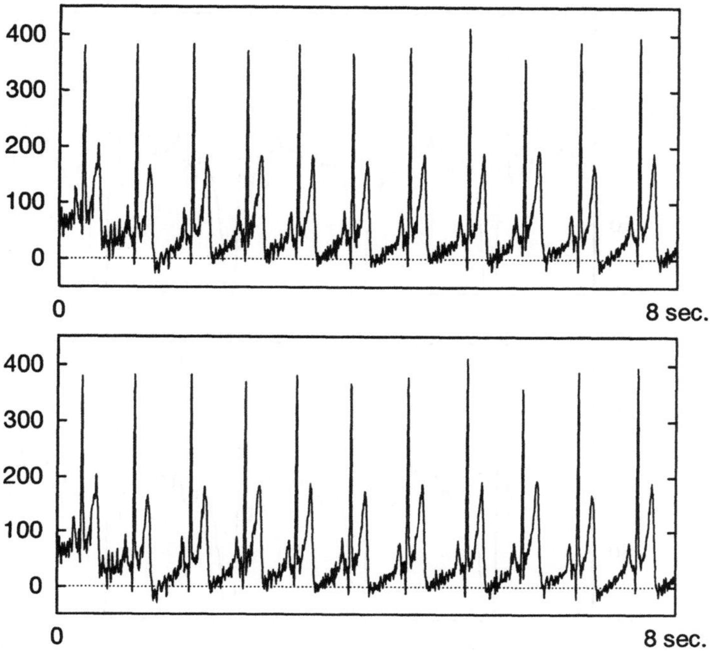

Fig. 1. Noisy human ECG before and after Wiener filtering. We see that the filter has not been able to significantly reduce the noise. (Uncalibrated A/D units.)

the system. Let λ_i be the spectrum of Lyapunov exponents. Then, according to the Kaplan-Yorke formula, the dimension of the attractor is

$$D = k + \frac{\sum_{i=1}^{k} \lambda_i}{|\lambda_{k+1}|} , \tag{3}$$

where $\sum_{i=1}^{k} \lambda_i \geq 0$ and $\sum_{i=1}^{k+1} \lambda_i < 0$. The value of D will change when another direction with small negative Lyapunov exponent is added, which is typical for low-pass AR filters, see Badii et al. (1988).

4 Principal Component Approach

So far, the undesired part of the time series was identified by the frequencies at which it contributes to the power spectrum. In this section we will follow a different route which will eventually lead to a linear filter as well. The construction of the filter will however involve some nonlinearity, hence we could call it "almost" linear. Filtering in the Fourier domain can be understood as a decomposition of the signal with respect to a set of fixed basis functions, sines and cosines. Alternatively, it can be advantageous to construct

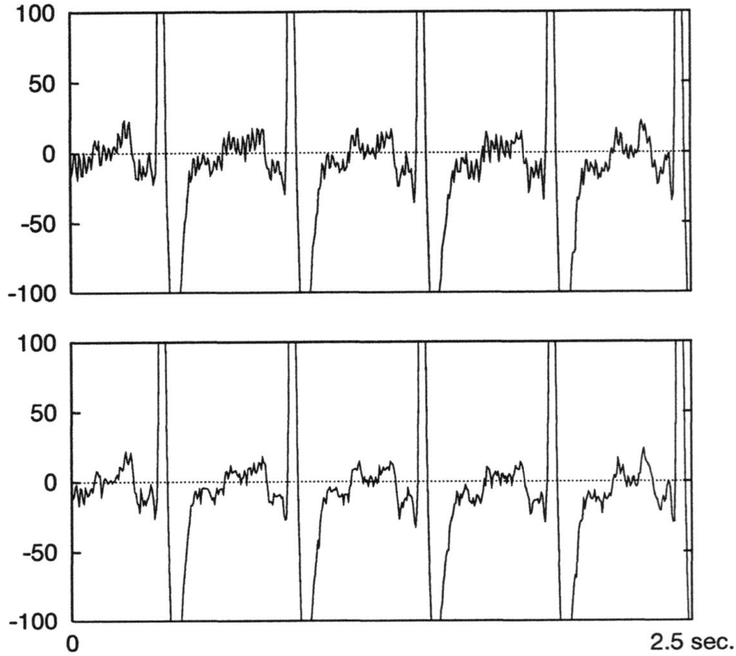

Fig. 2. Here the Wiener filtering was used to remove 60 Hz AC contamination. The corresponding sharp peak in the power spectrum is easy to identify and to remove.

the basis functions in an optimal way based on the data. The methodology (with minor variations mainly concerning the way the autocovariances are estimated from data) is known under several names. We like to use *Principal Component Analysis* (PCA), but it is also referred to as *Karhunen-Löwe* (KL) transformation, *Singular Value Decomposition* (SVD), or *Empirical Orthogonal Functions* (EOF). References in the context of nonlinear signals are e.g. Broomhead & King (1986), Broomhead et al. (1986) and Vautard et al. (1992).

The idea is that if a signal can be thought of as containing a low dimensional component plus some noise, it cannot be embedded in any m dimensional phase space since the noise fills all available directions. However, we could find some embedding dimension m_0 such that the signal can be *effectively* embedded. If we measure the success of this embedding by its correctness in the mean squares sense, we can systematically find the best effective embedding in that sense.

Let us first form m dimensional delay vectors in $\mathcal{R}^m$, where m is large but $\ll N$: $\mathbf{s}_n = (s_{n-m+1}, s_{n-m+2}, \ldots, s_n)$.[3] To obtain a low dimensional embedding in m_0 dimensions we can project onto an m_0 dimensional linear subspace of $\mathcal{R}^m$. To preserve as much power as possible, we will use the projection which maximizes the remaining variance. Thus we are looking for m_0 orthonormal vectors $\mathbf{a}^q$, $q = 1, \ldots, m_0$ such that the average modulus of the projections $\sum_{q=1}^{m_0} \mathbf{a}^q \cdot (\mathbf{a}^q \cdot \mathbf{s}_n)$ is maximal.

If we introduce the constraint that the $\mathbf{a}^q$ have unit length by means of Lagrange multipliers λ^q and use that the $\mathbf{a}^q$ are orthogonal,[4] which means that $\mathbf{a}^q \cdot \mathbf{a}^{q'} = 0, \quad q \neq q'$, we have to minimise the Lagrangian

$$L = \sum_{n=m}^{N} \left[\sum_{q=1}^{m_0} \mathbf{a}^q \cdot (\mathbf{a}^q \cdot \mathbf{s}_n) \right]^2 - \sum_{q=1}^{m_0} \lambda^q (\mathbf{a}^q \cdot \mathbf{a}^q - 1) \tag{4}$$

with respect to $\mathbf{a}^q$ and λ^q. This can be done separately for each q and yields

$$\mathbf{C}\mathbf{a}^q - \lambda^q \mathbf{a}^q = 0, \quad q = 1, \ldots, m_0\,, \tag{5}$$

where $\mathbf{C}$ is the $m \times m$ covariance matrix of the vectors $\mathbf{s}_n$.

$$C_{ij} = \sum_{n=m}^{N} (\mathbf{s}_n)_i (\mathbf{s}_n)_j\,. \tag{6}$$

Of course, the solutions of (5) are nothing but the orthogonal eigenvectors $\mathbf{a}^q$ and eigenvalues λ^q of $\mathbf{C}$. These can be readily determined with standard software. The global maximum is given by the eigenvectors to the m_0 largest eigenvalues. These directions are the *principal components* or *singular vectors* or *empirical orthogonal functions.*

For a noisy harmonic motion e.g., only a few directions would be actually visited by the signal part of the series which could therefore be covered by m_0 principal components. All the directions will contain a noise component which for white noise is isotropic. Thus the finite variance thrown away by projecting onto m_0 dimensions is due to the noise.

The PCA can be seen as an advanced embedding method: One uses the principal directions as a new coordinate system and represents the signal as m_0 dimensional vectors. This approach yields "optimal" coordinates in the least squares sense. It depends on the application whether this is the desired criterion. As an example, let us compare two two dimensional representations of an ECG signal, one using ordinary delay coordinates and the other using the first two principal components of a high dimensional representation (in 50 dimensions, covering 0.25 s at 200 Hz sampling rate). The result is shown in Fig. 3.

[3] If m is of the order of N, estimation of the autocovariances from the time series becomes an issue. This is where the different versions of the method, KL, SVD, EOF, mainly differ.

[4] We could also require orthogonality by additional Lagrange multipliers. This would complicate the algebra but leads to the same result.

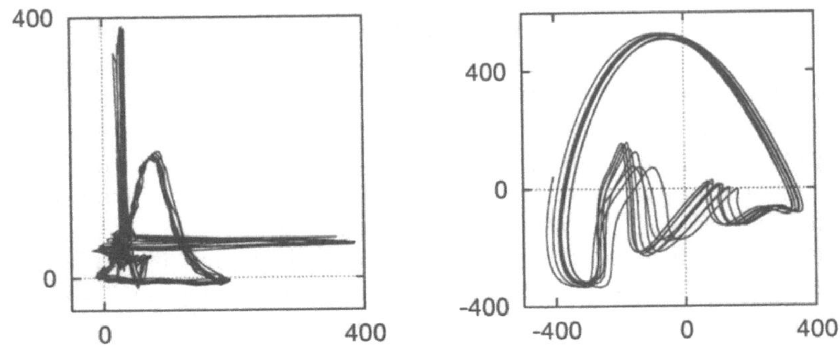

Fig. 3. Two dimensional representations of an ECG signal. Left: usual delay coordinates with delay 0.1 s Right: first two principal components. The dynamical structure is visualized much better in the right panel.

In some cases, the subsequent analysis requires a scalar time series. Note that each scalar measurement occurs as a coordinate in m of the embedding vectors and is therefore affected by m different projections. By averaging all these corrections, the projected vectors can be transformed back into a scalar time series. This kind of filtering is very similar to the linear filter in the Fourier domain we described in the previous section. The difference is that the noise component is not identified by its frequency content but by its smaller rms amplitude. This is very useful for white noise which equally contributes to all the principal components. Additive noise changes the autocovariance matrix by adding a multiple of the unit matrix, corresponding to the δ correlation. If the noise however is autocorrelated, the total autocovariance matrix is the sum of those of the signal and of the noise. While the matrices behave linearly, the eigenvectors and eigenvalues do not. (Except in very special cases, there is no simple relationship between eigenvalues and eigenvectors of the sum of two matrices and those of the single matrices.) Thus the construction of the principal components and thus the filter involves some nonlinearity whence we might call the method "almost linear". The method is linear however in the sense that it does not take any other than two point correlations into account.

Before we pass on to genuine nonlinear techniques, let us make a remark about the effect of linear filters on statistical tests for nonlinearity. It seems obvious that a linear filter acting on the output of a linear stochastic process will not render the time series nonlinear. However, since most signals we are interested in do not follow a Gaussian distribution, most statistical tests, like those using phase randomized surrogate data, Theiler et al. (1992), have been carried out against the null hypothesis of a Gaussian linear random process observed through a static monotonous nonlinear function. The idea is that this particular kind of nonlinearity can be easily removed by rescaling the data back to Gaussian and has nothing to do with nonlinear dynamics. Suppose we have such a time series $\{s_n\}$ where $\{x_n\}$ is the output of a Gaussian linear random process and $s_n = H(x_n)$ where $H(\cdot)$ is some fixed monotonous function. If we now apply a linear filter to the time series s_n (rather than to the Gaussian variable x_n) we spread the nonlinearity in $H(\cdot)$ out over several samples. The filtered signal will not longer be consistent with our null hypothesis although the measured data was. It was pointed out by Prichard (1994) that e.g. taking first differences of non-Gaussian data can lead to spurious detection of nonlinearity. Therefore, *tests for nonlinearity should be always carried out on unfiltered data.*

5 Nonlinear Algorithms

In both of the previous approaches, the Fourier and PCA filters, the convenience of having a linear filter was bought at a price we may not be willing to pay if the data exhibits nonlinearity: The filter had to be constructed under the assumption that the power of the signal is concentrated only in part of the modes, either at certain frequencies in the Fourier spectrum or in certain principal components. If we are interested in the possibility of a nonlinear deterministic component of the signal, we cannot make this assumption since the power of the signal itself is expected to be spread over all modes (frequencies resp. principal components). In order to preserve possible deterministic structure, filtering has to be done very carefully using nonlinear methods.

The main problem arises if we are not altogether certain about the deterministic nature of the data. Then we should be aware of the danger to introduce spurious nonlinearity into the data by filtering. Quite obviously, nonlinear noise reduction is unsuitable for preprocessing before a statistical test for nonlinearity. Below we will give an example were some artifacts due to filtering are of no concern since the goal of the analysis is well defined.

But let us first briefly introduce the main concept of nonlinear (phase space) filtering. A deterministic dynamical system in discrete time (or discretely sampled) can be written as

$$\mathbf{x}_{n+1} = \mathbf{F}(\mathbf{x}_n), \qquad n \in \mathcal{Z} . \tag{7}$$

If $\mathbf{F}$ is at least piecewise continuous, this means that similar states $\mathbf{x}_n$ evolve into similar states $\mathbf{x}_{n+1}$ in the future. This can be used to construct a very simple but effective nonlinear prediction scheme: In order to predict the future state following $\mathbf{x}_n$, find similar states (similar means close in phase space) and average their future states. Switching over to scalar observables $s_n = H(\mathbf{x}_n)$ and delay coordinates, $\mathbf{s}_n = (s_{n-(m-1)\nu}, \ldots, s_n)$, this prediction scheme estimates a future measurement by

$$\hat{s}_{n+\Delta n} = \frac{1}{|\mathcal{U}_\epsilon(\mathbf{s}_n)|} \sum_{\mathbf{s}_{n'} \in \mathcal{U}_\epsilon(\mathbf{s}_n)} s_{n'+\Delta n} \, . \tag{8}$$

$\mathcal{U}_\epsilon(\mathbf{s}_n)$ is a neighbourhood in delay space of radius ϵ around $\mathbf{s}_n$ and $|\mathcal{U}_\epsilon(\mathbf{s}_n)|$ is the number of points in it.

For noise reduction and chaotic data, this scheme would not work simply by replacing measured values s_n by their predictions $\hat{s}_n$ due to the sensitive dependence on initial conditions. It turns out that the scheme can be made to work if not a future value is replaced but a coordinate in the middle of the delay vectors. In order to clean e.g. $s_{n-m/2}$ (where the delay ν is set to one) we replace it by

$$\hat{s}_{n-m/2} = \frac{1}{|\mathcal{U}_\epsilon(\mathbf{s}_n)|} \sum_{\mathbf{s}_{n'} \in \mathcal{U}_\epsilon(\mathbf{s}_n)} s_{n'-m/2} \,. \tag{9}$$

Technical details about the choice of neighbourhoods and embeddings may be found in Schreiber (1993). Here we apply the scheme to a 17 min recording of the air flow through the nose of a human during sleep, sampled twice a second. The data is part of a larger, nonstationary multichannel data set provided by A. Goldberger from Beth Israel Hospital in Boston for the Santa Fe Institute time series competition in 1991/92, Rigney et al. (1993). Using nonlinear predictions and surrogate data, Theiler et al. (1992), for this data set the null hypothesis of a rescaled Gaussian linear stochastic process could be rejected at the 99% level of significance, Schreiber & Schmitz (1996). Nevertheless, no positive evidence for low dimensional determinism has been established.

Figure 4 shows the result of nonlinear noise reduction on this data set. We used an embedding in seven dimensions with neighbourhoods of size 4000 A/D units (the total peak-to-peak amplitude of the series is about 15000 A/D units). The result pretty much looks like a limit cycle. However, since the "noise" is correlated with the signal (the correction is smaller and the outcome less clear for the points in the lower right of each panel in Fig. 4), we cannot assume that all we have is a limit cycle plus measurement noise. This becomes evident if we add to the cleaned data a randomized surrogate, made with the algorithm described in Schreiber & Schmitz (1996), of the subtracted "noise". The result, lower panel in Fig. 4, differs significantly from the original, noisy data. Thus a more likely interpretation is that noise interacts with a limit cycle dynamics.

This simple noise reduction scheme based on locally constant approximations to the dynamics is quite effective even for short data sets, in particular at high noise levels. In cases where more subtle details of the signal have to be preserved by the filtering it is advisable to use a more refined algorithm using local phase space projections (Grassberger et al. 1993; Kantz et al. 1993). This algorithm can be seen as a modified version of the principal component filter of the previous section which is however applied *locally in phase space.* The locality is essential in order to allow for nonlinearity in the data. The other modification is only important for coarsely sampled ("map like") data. Due to the sensitive dependence on the initial conditions in chaotic systems,

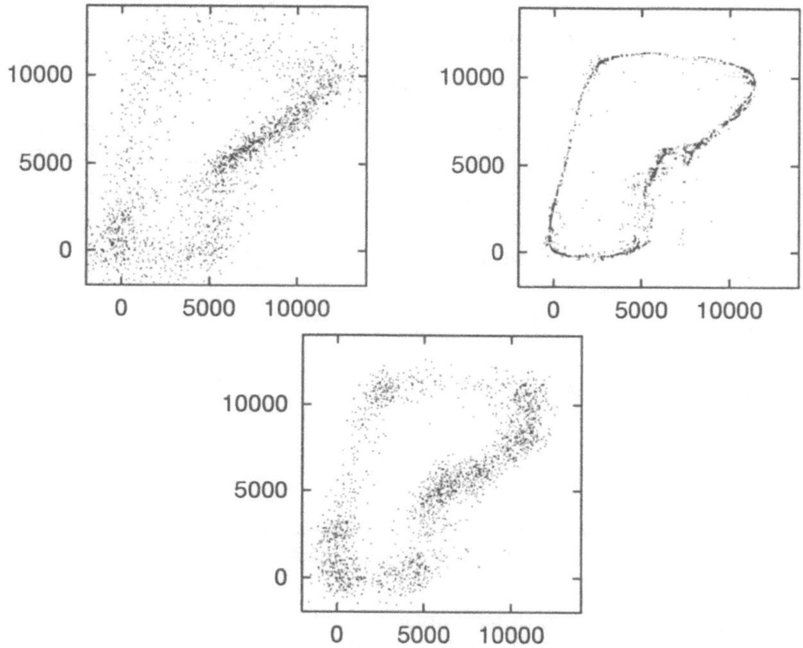

Fig. 4. Air flow through the nose of a human. Upper left: two dimensional delay representation of the original data. Upper right: the same after nonlinear noise reduction. The result suggests a noisy limit cycle. Lower: A random surrogate of the subtracted "noise" is added back on the cleaned data of the upper right panel. Deviations from the original data are evident.

the first and last coordinates within a delay vector are not well under control. Thus, when computing the corrections, changes to these coordinates are discouraged by choosing a suitable metric in phase space.

If we assume that the dynamics is given by an $(m-1)$-dimensional, at least piecewise smooth, map $x_n = F(x_{n-m+1}, \ldots, x_{n-1})$, we can also write it implicitly as $\tilde{F}(x_{n-m+1}, \ldots, x_n) \equiv \tilde{F}(\mathbf{x}_n) = 0$, or linearised as

$$\mathbf{a}^{(n)} \cdot \mathbf{R}(\mathbf{x}_n - \overline{\mathbf{x}}^{(n)}) = 0 \quad + O(\|\mathbf{x}_n - \overline{\mathbf{x}}^{(n)}\|^2)\,. \tag{10}$$

Here, $\overline{\mathbf{x}}^{(n)} = |\mathcal{U}_\epsilon(\mathbf{x}_n)|^{-1} \sum_{\mathbf{x}_{n'} \in \mathcal{U}_\epsilon(\mathbf{x}_n)} \mathbf{x}_{n'}$ is the centre of mass of the delay vectors in a small neighbourhood $\mathcal{U}_n$ of $\mathbf{x}_n$. Further, we took the occasion to introduce a diagonal weight matrix $\mathbf{R}$ which will allow us to focus the noise reduction on the most stable middle coordinates of the delay vectors. This is achieved by choosing R_{11} and R_{mm} large and all other diagonal entries $R_{ii} = 1$. For $\mathbf{R} = \mathbf{1}$ we would obtain orthogonal projections.[5] Of course, for

[5] Local orthogonal projections have been proposed for nonlinear noise reduction independently by Cawley & Hsu (1992) and Sauer (1992).

a noisy sequence s_n, the above relationship will not be valid exactly but only up to some error related to the noise:

$$\mathbf{a}^{(n)} \cdot \mathbf{z}_n = \tilde{\eta}_n \,. \tag{11}$$

Here we introduced the notation $\mathbf{z}_n = \mathbf{R}(\mathbf{s}_n - \bar{\mathbf{s}})$. If the dynamics is actually m_0 dimensional but we work in m dimensions, we will have $m - m_0$ such equations. Completely analogous algebra as in the last section shows that the local projection which induces the least loss is given by the eigenvectors with the largest eigenvalues, or local principal components, of the local covariance matrix

$$C_{ij} = \sum_{\mathbf{z}_{n'} \in \mathcal{U}_n} (\mathbf{z}_{n'})_i (\mathbf{z}_{n'})_j \,. \tag{12}$$

In terms of the original delay vectors $\mathbf{s}_n$ we find for the correction:

$$\hat{\mathbf{s}}_n = \mathbf{R}^{-1} \sum_{q=1}^{m_0} \mathbf{a}^q \cdot [\mathbf{a}^q \cdot \mathbf{R}(\mathbf{s}_n - \bar{\mathbf{s}})] \,. \tag{13}$$

Thus we have to choose an embedding dimension m, a dimension m_0 we want to project on locally, and the size of the neighbourhoods. These issues are discussed in Grassberger et al. (1993).

An example where nonlinear noise reduction can be carried out successfully on physiological data are ECG signals (Schreiber & Kaplan 1996). This is a nontrivial finding since the normal ECG is neither strictly periodic nor has it been possible so far to account for the fluctuations in the interbeat intervals by a deterministic model. Nevertheless, the ECG is found to spend most of each cycle close to a low dimensional manifold. Projections onto this manifold then result in noise reduction while clinically relevant features are preserved. Here we want to show an application of this technique to the problem of *fetal* ECG subtraction.

It is desirable to measure the electrochemical activity of a fetus noninvasively. Depending on the location of the electrodes on the skin of the mother, it is possible to measure an ECG signal where the fetal component is not completely hidden by the maternal ECG. But even in an abdominal ECG, the maternal component will be dominant. Current methods to extract the fetal component from such an ECG are usually either based on multiple lead measurements which allow for an approximate cancellation of the maternal ECG or on the subtraction of a time averaged maternal cycle from a single lead ECG. Both approaches are however unable to cancel exogenous noise which is always present in recordings of such small amplitudes. Here we want to treat the maternal ECG as the signal and the fetal component as the noise. This approach has been proposed in Schreiber & Kaplan (1996a). On this base we apply nonlinear noise reduction in order to separate both parts. The fetal part will however also contain some artifacts and random fluctuations from the recording. Thus we subject it to an additional noise reduction step.

Figure 5 shows a recording[6] of a 16-week-old fetus. Electrodes were put on the abdomen and on the cervix of the mother, Hofmeister at al. (1994). We resampled the recording from 2500 Hz to 250 Hz and removed 60 Hz contamination using the Wiener filter described above. Now first the mother signal was separated by projecting from ten onto two dimensions locally. Neighborhood size was 10 μV. The resulting signal (second line) was further cleaned using embedding dimension 10 and neighborhoods of size 3 μV (bottom line). With this technique we can reconstruct the fetal ECG using only a single ECG lead.

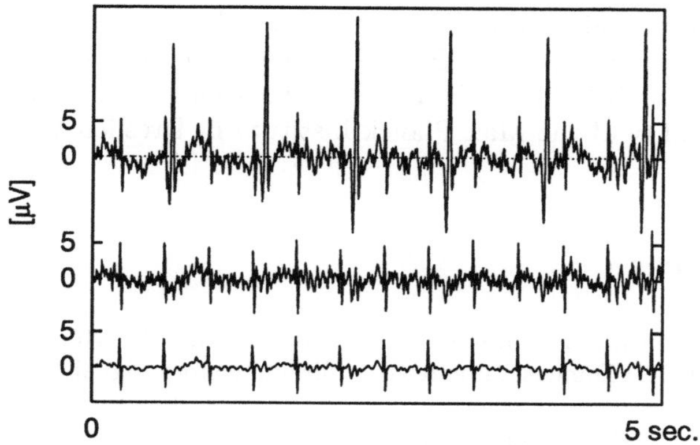

Fig. 5. Noninvasive fetal ECG recording. Upper: original data. Middle: result of projection onto mother ECG. Lower: result of further noise suppression.

6 Conclusions

We discussed a collection of general data processing techniques, spectral linear filters, the almost linear principal component technique, and a phase space approach. These methods are no exception from the rule that one seldom gets anything for nothing: Signal enhancement has to be bought with knowledge or assumptions about the data. If we want to use the data in a statistical test for nonlinearity, we do not want to make any assumptions at all in order not to bias the test. Thus we refrain from any filtering.

If we want to analyse dominant (linear) oscillations of the system, we may assume that the time series is a superposition of an oscillatory signal and broad band noise. We can then separate both components with a Wiener filter. If, as a working hypothesis, we assume that the data can be embedded

[6] The data were kindly provided by John F. Hofmeister, Denver.

at least effectively in a low dimensional phase space, we can do so in an optimal (in the mean squares sense) way using principal components. In the (suspected) presence of strong nonlinearity, phase space filtering may be useful to reduce noise present in the time series. Finally we have seen in the fetal ECG example that nonlinear filtering can be successful even if the formal justification is lacking as long as we know how to use the result.

Acknowledgment

I owe thanks to Petr Saparin, Ari Goldberger, and John F. Hofmeister for generously providing time series data. Further I thank Daniel T. Kaplan, Steven Boker, and Holger Kantz for useful discussions. This work was supported by the SFB 237 of the Deutsche Forschungsgemeinschaft. I appreciate the hospitality of the Max Planck Institute in Dresden and the work of the organizers.

References

Badii, R., Broggi, G., Derighetti, B., Ravani, M., Ciliberto, S., Politi, A.,Rubio, M. A. (1988): Dimension increase in filtered chaotic signals. Phys. Rev. Lett. **60**, 979

Bowen, R. (1975): Equilibrium States and the Ergodic Theory of Anosov-Diffeomorphisms. Springer Lecture Notes in Mathematics, Vol. 470. Springer, Berlin, Heidelberg

Broomhead, D., Jones, R., King, G. P. (1986): Topological dimension and local coordinates from time series data. J. Phys. A **20**, L563

Broomhead, D., King, G. P. (1986): Extracting qualitative dynamics from experimental data. Physica D **20**, 217

Cawley, R., Hsu, G.-H. (1992): Local-geometric-projection method for noise reduction in chaotic maps and flows. Phys. Rev. A **46**, 3057

Grassberger, P., Hegger, R., Kantz, H., Schaffrath, C., Schreiber, T. (1993): On noise reduction methods for chaotic data. Chaos **3**, 127

Grebogi, C., Hammel, S. M., Yorke, J. A., Sauer, T. (1990): Shadowing of physical trajectories in chaotic dynamics: containment and refinement. Phys. Rev. Lett. **65**, 1528

Hofmeister, J. F., Slocumb, J. C., Kottmann, L. M., Picchiottino, J. B., Ellis, D. G. (1994): A noninvasive method for recording the electrical activity of the human uterus in vivo. Biomed. Instr. Technol., Sept. 391

Kantz, H., Schreiber, T., Hoffmann, I., Buzug, T., Pfister, G., Flepp, L. G., Simonet, J., Badii, R., Brun, E. (1993): Nonlinear noise reduction: a case study on experimental data. Phys. Rev. E **48**, 1529

Press, W. H., Flannery, B. P., Teukolsky, S. A., Vetterling, W. T. (1988): Numerical Recipes. Cambridge University Press, Cambridge

Prichard, D. (1994): The correlation dimension of differenced data. Phys. Lett. A **191**, 245

Rigney, D. R., Goldberger, A. L., Ocasio, W., Ichimaru, Y., Moody, G. B., Mark, R. (1993): Multi-Channel Physiological Data: Description and Analysis. in Weigend, A. S., Gershenfeld, N. A., Eds: *Predicting the Future and Understanding the Past: a Comparison of Approaches.* Santa Fe Institute Studies in the Science of Complexity, Proc. Vol. XV, Addison-Wesley, Reading, MA

Sauer, T. (1992): A noise reduction method for signals from nonlinear systems. Physica D **58**, 193

Schreiber, T. (1993): Extremely simple nonlinear noise reduction method. Phys. Rev. E **47**, 2401

Schreiber, T., Kaplan, D. T. (1996): Nonlinear noise reduction for electrocardiograms. CHAOS **6**, 87

Schreiber, T., Kaplan, D. T. (1996): Signal separation by nonlinear projections: the fetal electrocardiogram. Phys. Rev. E **53**, 4326

Schreiber, T., Schmitz, A. (1996): Improved surrogate data for nonlinearity tests. Phys. Rev. Lett. **77**, 635

Theiler, J., Eubank, S. (1993): Don't bleach chaotic data. Chaos **3**, 771

Theiler, J., Eubank, S., Longtin, A., Galdrikian, B., Farmer, J. D. (1992): Testing for nonlinearity in time series: the method of surrogate data. Physica D **58**, 77

Vautard, R., Yiou, P., Ghil, M. (1992): Singular-spectrum analysis: a toolkit for short, noisy chaotic signals. Physica D **58**, 95

Problems in the Reconstruction of High-dimensional Deterministic Dynamics from Time Series

Rainer Hegger, Holger Kantz, and Eckehard Olbrich

Max-Planck-Institut für Physik komplexer Systeme, Nöthnitzer Straße 38, D-01187 Dresden, Germany

Abstract. The reconstruction of a meaningful state space from a scalar time series is the more difficult the more complex the signal. On large scales, a complex deterministic signal is indistinguishable from a random process. Determinism becomes visible only below a critical length scale. We analyse the dependence of this scale on the entropy of the signal and the minimal embedding dimension for state space reconstruction. The extent to which the structures on the larger scales contain system specific information is discussed for two classes of high-dimensional signals: local and global observables in systems with large attractor dimension. Comments on the diagnostic power of nonlinear methods outside their mathematical environment are added.

1 Introduction

One asks for chaos as the origin of irregular time dependence, because chaotic systems may have a relatively simple structure which can be fully understood. A fractal dimension is most often not computed to characterize the data by just another number, but since a finite dimension is the first hint that it might be possible to explain the complicated observations by a simple model which in the best case even gives hints to a deeper understanding of the phenomenon. It is obvious that this is the main difference with respect to stochastic models. For the purpose of diagnostics one may be interested in a mere signal classification, for which ideas from deterministic chaos offer new concepts.

What we call nowadays *nonlinear methods* in time series analysis are concepts which were originally developed in a (from the point of view of us applicants) hostile mathematical environment. To mention some examples, the definition of the Hausdorff-dimension, Oseledec's multiplicative ergodic theorem establishing the existence of Lyapunov exponents, Kolmogorov's idea of merging information theory and ergodic theory, and even the embedding theorems for the reconstruction of a state space from scalar data are all formulated in a highly mathematical language which at first sight seems to be more concealing than elucidating. Unfortunately, mathematicians are right: All these concepts gain their power only due to properties which rely on mathematical intricacy.

One such outstanding feature is invariance. Lyapunov exponents and the dimension of the underlying attractor are meaningful and interesting quantities to characterize the dynamics underlying a time series (and not only the particular time series itself), since they are invariant under most manipulations of the measurements and the data you can imagine. Therefore, rescaling the data, applying nonlinear transformations to the time series, exchanging the measurement apparatus, or even changing from one observable to another should not influence their values, as long as the system under study remains the same. So assume that the state of a patient's heart can be really characterized by Lyapunov exponents, and that the heart were in the same state every time the person comes to see the doctor. Then the Lyapunov exponents would come out independently of the precise position of the electrodes and the bias and amplification the doctor adjusts at the ECG measurement device. This is different from most classical tools in time series analysis as e.g. the mean and variance of the data distribution, but has to be paid by some price. Generally, we have to accept that we can never obtain an estimate of a nonlinear quantity in the statistical sense, i.e. one which is "only" subject to *statistical* errors. Instead, our results are always endangered by *systematic* errors, since we have to make guesses about asymptotic behaviours which we can never observe directly.

The crucial point of nonlinear quantities is that they test for scaling laws and exponential behaviour on the infinitesimal scales, where some kind of universality appears. Theory tells us that on the microscopic level all observables are in some sense equivalent, and every global transformation affects only what we see on the macroscopic level. This establishes the desired invariance. The drawback of this concept is that we always have to extrapolate: no finite data set allows to proceed towards the relevant limits, but in favourable situations it only enables us to guess the asymptotic scaling laws. And here we are right at the issue of this article: Under what circumstances and for what kind of signals can we expect to have access to length scales, where the correct scaling behaviour becomes visible? In fact, as we will see below, a more appropriate question is, in which situations the relevant length scales of a given system are such that they are accessable by a finite amount of data? After a general discussion of the problem we shall present in Sects.4. and 5. quantitative results on a more theoretical ground, and in Sect.6 a resumee under the point of view of time series analysis of physiological data.

2 No Evidence for High-dimensional Chaos in Data?

Attractor dimensions and Lyapunov exponents are quantities which describe state space properties of the data. The state space is spanned by all variables of a system, such that every possible combinaton of values for the different variables corresponds to a point in this space and vice versa. The time evolution of the system can thus be represented by a continuous path in the state space, a trajectory. In deterministic autonomous systems the complete, infinitely long trajectory is unambiguously determined by specifying a single point on it. In this space, Lyapunov exponents are the rate by which infinitesimally nearby trajectories diverge with ongoing time, and the attractor dimension is the dimension of the subset of phase space covered by a typical trajectory asymptotically.

When our system is defined by a set of model equations, the state space obviously is spanned by the variables occuring in the model. There are very robust methods to compute a whole spectrum of Lyapunov exponents (as many exponents as there are directions in space), such that these exponents in fact can be estimated with any desired accuracy. The determination of the attractor dimension and the entropy is already less rigorous, since it involves an extrapolation towards the microscopic scales even in this situation. Nevertheless, this can usually be done quite savely. Do not forget that in numerical simulations data are almost noise free and that data sets can be made extremely large compared to experimental situations. Consequently, one has computed reliable dimensions for systems with attractors of dimensions up to 7 to 10 (see Grassberger and Procaccia 1983c). Along with this go several positive Lyapunov exponents, which are well determined. In so far, there is *no* lack of evidence of high–dimensional chaos.

The first step from numerical simulations towards experiments was performed at the beginning of the eighties: Physicists tried to establish the relevance of the concept of deterministic chaos in physics, and in particular in hydrodynamics. Note that even if one possesses perfect model equations, it is not at all trivial to prove that the model is a good description of what happens in reality, if the dynamics is chaotic. It is not possible to observe the experimental system for some time and integrate simultaneously the model equations for a detailed comparison, simply because the slightest deviation between the initial condition fed into the model equations and the true experimental situation would be magnified exponentially fast in time. Additional approximations in the model and perturbations of the experimental system by noise accelerate this divergence between the observed and the computed solution even more. Therefore, one needs the refined methods of nonlinear time series analysis for this comparison. During the years all the methods we are familiar with today were developed and they have shown to work well, if the signal under investigation really stems from a low-dimensional deterministic system.

Scanning the literature, one will make a remarkable observation: There are meanwhile hundreds of different experimental systems which could be shown to exhibit lowdimensional chaotic motion. But what is the typical dimension of their attractor? It is between two and three. How many positive Lyapunov exponents do they usually have? Exactly one. So, surprisingly, almost all investigations in physical laboratory experiments only brought evidence for the existence of the least complex chaotic motion possible, namely of motion on two plus epsilon dimensional attractors (see Stoop 1988; Stoop et al. 1989; Kruel et al. 1993 for the observation of a second positive Lyapunov exponent).

Nobody can assume that it is a feature of nature that systems have only one positive Lyapunov exponent. As mentioned before, there are several models which exhibit *hyperchaos* (Rössler 1979), i.e. where more than one direction in space is unstable. Of course, one might say that experimental physicists try to simplify their physical problem as much as possible in order to come to conclusive results. This clearly indicates that the analysis of signals with more than one positive Lyapunov exponent becomes so difficult that it is usually avoided to deal with them. In fact there are experiments with clear indications of higher dimensional chaotic motion, e.g. when time delayed feeedbacks are involved, but a quantitative characterization by a value for their dimensionality is almost always impossible.

In this contribution we want to point out the limitations of nonlinear analysis for complex signals. In particular, we shall show that problems arise from the difficulty to reconstruct a meaningful state space from scalar data. Highdimensional chaos and chaos in spatially extended systems can be analysed on the basis of scalar time series by dimensions and Lyapunov exponents only in a limited manner. Apart for exceptional but unrealistic cases, there is currently no convincing concept to relate what can be observed on the large scales and what really happens in the asymptotic regime. The most straightforward way out might be to perform multichannel measurements and to analyse vector-valued time series.

3 The Reconstruction Problem Illustrated

Let us start from an autonomous deterministic dynamical system, defined by an ordinary differential equation

$$\frac{d}{dt}\mathbf{x}(t) = \mathbf{f}(\mathbf{x}(t)), \tag{1}$$

where t is the time, $\mathbf{x} \in \Gamma$ a state vector, and $\mathbf{f}$ some nonlinear vector field. The change of $\mathbf{x}$ in time thus is uniquely determined by $\mathbf{f}$ at the current position. A mathematical theorem guarantees the existence of a unique solution of Eq.(1) for every initial condition $\mathbf{x}_0(t_0)$, as long as f is sufficiently smooth (precisely: Lipschitz-continuous). This property is what we call determinism: in theory, the whole future $\mathbf{x}(t)$ with $t > t_0$ is unambiguously determined as soon as we have specified $\mathbf{x}_0(t_0)$, i.e. a single point in Γ. The measurement records the current value of an observable, i.e. of a scalar function h of the phase space vector $\mathbf{x}$: $s_n = h(\mathbf{x}(n\delta t))$, where δt is the sampling rate.

The original state space is for chaotic motion at least three-dimensional and the signal is scalar, and moreover the function h itself may be very complicated. Altogether, this produces a very complex signal. In general, it is not feasible to reproduce the original phase-space from our scalar observations only. But, due to the invariance of our preferred quantities, this is not required at all. We just need a space, in which the reconstructed attractor, i.e. the relevant subset of the phase space on which the motion takes place asymptotically, is equivalent to the corresponding set in the original space. This is done by the *time delay embedding* first used by Packard et al. 1980 and proven to be useful in Takens 1981 and in a more general form in Sauer, Yorke and Casdagli 1991.

So let us form m-dimensional vectors $\mathbf{s}_n = (s_n, s_{n-\nu}, \ldots, s_{n-(m-1)\nu}) \in \mathbf{R}^m$. Here, ν is an integer and called the time lag, m is the embedding dimension. An *embedding* is a map from the attractor in the original state space into the $\mathbf{R}^m$ which is injective (i.e. one-to-one) and the determinant of its Jacobian is nowhere vanishing (in order to preserve structures in tangent space).

The theorems of Takens and Sauer et al. guarantee that for almost every observable h, almost every sampling rate δt, and almost every[1] time lag ν, the time delay reconstruction by the vectors $\mathbf{s}_n$ forms an embedding, if $m > 2D_f$, where D_f is the box-counting dimension[2] of the attractor.

In time series analysis, we have to cope with several difficulties. Usually, we do not know the box-counting dimension of the attractor and therefore have no idea about the minimal m. However, for many applications this is not a serious problem, since one scans automatically some range of m values (like in dimension or entropy estimates). Using the space of minimal dimension is only important when one wants to determine the underlying dynamics and the whole spectrum of Lyapunov exponents. The correlation dimension D_2 (see Grassberger and Procaccia 1983a) yields a lower bound of the box-counting dimension and therefore allows to estimate m, but it can be sometimes impossible to estimate D_2. In this case one has to try several reasonable m and to see which one performs best. Consistency checks are necessary to prove that one has done something meaningful. We do not want to elaborate this in more detail, since this is outside of our concern.

The time lag ν is not subject of the embedding theorems. In principle, arbitrary ν yields an embedding. The only question is how well we can see the interesting structures, and how far reality, i.e. noise and the lack of data, can veil them. For example, if the sampling rate is high and one uses the lag $\nu = 1$, successive elements of the delay vectors are strongly correlated and all data points lie close to the diagonal in space. All structures in the direction perpendicular to the diagonal are highly compressed, but for mathematically precise data they are present. They can be wiped out completely by noise, but be visible in a representation with larger ν where the attractor is reasonably unfolded. For an illustration, see Fig. 1.

---

[1] There must be no periodic orbit of the system with period $\nu\delta t$ or $2\nu\delta t$.

[2] The box-counting dimension characterises the scale invariance of the attractor and is a real number smaller than or equal to the dimension of the smooth manifold in which the attractor can be embedded locally.

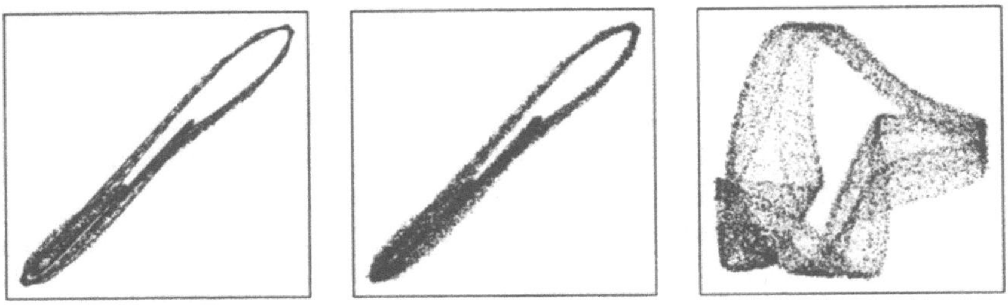

Fig. 1. Two dimensional time delay embedding of a time series from a physical experiment (nonlinear electric circuit) performed by the group of H. Beige in Halle (Beige et al. 1992). The left plot shows the original, extremely clean experimental data embedded with $\nu = 1$. The data underlying the middle plot are corrupted by numerically added 3% measurement noise ($\nu = 1$), the right plot shows the noisy data with $\nu = 4$. The attractor dimension is slightly above two.

This reasoning makes it obvious that the "best" ν cannot be determined by mathematical criteria, but only by physical arguments trying to make quantitative statements about what is called "unfolding of the attractor". In the literature there are many different concepts how to determine the optimal lag, but none of them seems to yield optimal results in *every* situation. However, they all yield values of the same order of magnitude, so that we suggest to start with the simplest and to optimize ν not by a test-statistics but by the application one is interested in. The simplest criterion involves the auto-correlation function: Strong auto-correlations induce the effect of clustering along the diagonal mentioned before, which is resolved at the first zero at the auto–correlation function. Many signals are irregular oscillations, where the times for every single cycle do not differ too strongly. In such a case (as for a strictly periodic signal) one quarter of the average period is best, which coincides with the result of the auto-correlation function criterion.

As stated by the embedding theorems, without noise both attractors in Fig. 1 were characterized by the same invariants: Their dimensions and their Lyapunov-exponents were identical and would agree with the corresponding values of the system in its original state space. However, usually attractors look much smoother in their original state space than in the delay embedding space. The fact that we restore information about a higher-dimensional phase space from a scalar signal introduces always distortions of an attractor which make its appearence on the large scales very complicated. For systems with more space dimensions and a higher degree of chaoticity the distortions are much worse and can become destructive in the sense that they suppress the deterministic structures completely in the range of length scales which are accessible by us.

Let us introduce a simple mathematical model to visualize the problems of state space reconstruction more clearly. It will simulate the situation we are in when we possess a global observable which is a kind of average over many complicated degrees of freedom. In order to facilitate things we start from two uncoupled subsystems, two logistic equations,

$$x_{n+1} = 1 - 2x_n^2, \qquad y_{n+1} = 1 - 2y_n^2, \tag{2}$$

where the initial conditions (x_0, y_0) are chosen at random. Only in the observation we mix the two degrees of freedom by the measurement function $h = h(x,y) = (x+y)/2$. Therefore, although x and y have a completely independent time evolution, our measurements contain information about both of them and Takens's theorem is valid. The dimension of each single attractor is unity (the interval of the real numbers $[-1,1]$), so that the joint attractor has dimension two. Takens's theorem says that for $m > 2D_f$ time delay coordinates form an embedding, in exceptional cases (this depends on the observable!) smaller m down to $m \geq D_f$ may be enough. In this example $m = 2$ is in fact sufficient. We can even write down the dynamics in terms of delay coordinates of our observable,

$$\begin{aligned} s_n &= (x_n + y_n)/2 \\ s_{n+1} &= (x_{n+1} + y_{n+1})/2 = 1 - x_n^2 - y_n^2 \\ s_{n+2} &= (x_{n+2} + y_{n+2})/2 = 1 - 2s_{n+1}^2 + 16s_t^2(s_{n+1} - 1) + 32s_n^4 \end{aligned} \tag{3}$$

Now let us compare the dynamics in the delay embedding space and in the original state space Γ. In Γ, the map $\mathbf{f} : (x_n, y_n) \mapsto (x_{n+1}, y_{n+1})$ is given by Eq.(2), and in delay embedding space it is $\mathbf{g} : \mathbf{s}_n \mapsto \mathbf{s}_{n+1}$, where $\mathbf{s}_n = (s_n, s_{n-1})$ and $\mathbf{s}_{n+1} = (s_{n+1}, s_n)$. Due to the fact that we are dealing with delay vectors, all but the first components of the new vector are just copied from the old vector. The only nontrivial part of $\mathbf{g}$ concerns the first component, which is given by Eq.(3) and is obviously more complicated than any of the two functions which form the components of $\mathbf{f}$ in the original space. In delay embedding space all nonlinearities of the different components of the map $\mathbf{f}$ are shuffled onto only one scalar function which therefore is more complex. In consequence, the reconstructed attractor is much more complicated in the time delay embedding space than in the original state space. Figure 2 shows the attractors of Eqs. (3) in the three dimensional spaces. The right plot is the delay space representation, the left plot shows the attractor in real space. We can easily see that on large scales the right plot looks much less smooth than the left one. We have much more folding effects there, due to the fact that the whole nonlinearity is mapped into one direction as mentioned above. We will see that the consequence of this additional folding is a reduction of the scaling range of the dimension estimation. Instead of the correct attractor dimension 2, we get an overestimation on large length scales. This is indeed a common problem if we deal with folded objects. But the more folding we induce, the more severe this effect becomes.

Fig. 2. The plots show the attractor formed by two uncoupled logistic maps, in real space (left plot) and in delay embedding space (right plot), respectively.

4 The Upper End of the Scaling Range

Measuring the dimension of an object presumes this object being selfsimilar. That means observing the object on different length scales will reveal similar structure. But, if we reach length scales of the order of the overall size of the attractor, selfsimilarity breaks down. The main point is that the critical length scale, at which this violation sets in, depends on the details of the reconstruction of the attractor. On the other hand the amount of data we receive from a measurement is restricted, and additionally the data are usually corrupted by noise. That means that selfsimilarity also breaks down for length scales of the order of the average distance of the data points in the reconstructed state space or for length scales proportional to the noise level.

How do these limitations show up in numerical calculations of quantities characterising the deterministic structures? The most frequently used quantity in analysing time series is the correlation dimension (Grassberger and Procaccia 1983a). Starting from the correlation integral (N denotes the number of vectors, w a time window which is explained below)

$$C(m,\epsilon) = \frac{2}{(N-w+1)(N-w)} \sum_{j=1}^{N} \sum_{i=j+w}^{N} \Theta(\epsilon - ||\mathbf{s}_i - \mathbf{s}_j||) \tag{4}$$

the correlation dimension is defined as

$$D_2(m,\epsilon) = \frac{d \log C(m,\epsilon)}{d \log \epsilon} \qquad D_2 := \lim_{\epsilon \to 0} D_2(m,\epsilon)\ , m > m_0\ . \tag{5}$$

To get the correct dimension, all pairs of neighbors which are also correlated in time have to be cancelled in the correlation sum (Theiler 1986). Therefore the time window w was introduced in Eq. (4). It is also possible to define an entropy based on the correlation integral (Grassberger and Procaccia 1983b):

$$
\begin{aligned}
H(m,\epsilon) &:= -\ln C(m,\epsilon) \\
h(m,\epsilon) &:= H(m+1,\epsilon) - H(m,\epsilon) \qquad (6)\\
h(0,\epsilon) &:= H(1,\epsilon)
\end{aligned}
$$

$$
h_2 := \lim_{\epsilon\to 0}\lim_{m\to\infty} h(m,\epsilon) \qquad (7)
$$

h_2 gives a lower bound for the Kolmogorov-Sinai entropy. For sufficiently large m the correlation integral in the scaling range can be written as

$$
\ln C(m,\epsilon) = -D_2 \ln \epsilon_0 + D_2 \ln \epsilon - m h_2 \qquad (8)
$$

where ϵ_0 is a proportionality constant. The minimal embedding dimension to see this scaling behaviour is given by the smallest integer which is greater then D_2 (Ding et al. 1993), provided a small noise level and a sufficiently large data set.

Now we can give a rough estimate of the limits of the scaling region for finite amount of data: Let us assume, we have reconstructed the attractor from a scalar time series yielding N delay vectors in a m-dimensional embedding space, $m > D_2$. Then the smallest possible value of $C(m,\epsilon)$ is $2/(N-w+1)(N-w)$, i.e. only one pair with a distance smaller then ϵ was found. If we require a minimal number $k_{\min}$ of pairs for reliable statistics, we get for the lower end of the scaling range ϵ_l:

$$
C(\epsilon_l, m) = \frac{2k_{\min}}{(N-w+1)(N-w)} \ . \qquad (9)
$$

The upper end of the scaling range is determined by the edge (or boundary) effects and the folding as described above. The effect of the boundary of the attractor can be understood easily: For each reference point near the global boundary of the attractor there are directions, where no neighbours exist. So, shrinking the size of the neighbourhood, such points lose less neighbours then points inside the attractor, which leads to an underestimation of the dimension (see also sect. 5). The larger ϵ the more points are in the neighbourhood of the boundary until ϵ reaches the overall size of the attractor, where all points are "near the boundary" and the correlation integral gives the dimension $D_2(\epsilon, m) = 0$. A quantitative study of this effect is given e.g. in Nerenberg and Essex 1990.

The folding effect is illustrated in Fig. 3. Suppose we work on a length scale ϵ and suppose further that ϵ is larger than the typical distance between two branches of our attractor. Then the boxes we use for our neighbour statistics do not only contain 'correct' neighbors, but also false ones, belonging to another branch. If we decrease the box size, the loss of the false neighbours results in a too fast decrease of the correlation sum. This leads to the already mentioned overestimation of the dimension until the boxes are small enough. Thus in an extreme situation a 1-d graph with many folds can look m-dimensional. It is obvious that a stronger folding decreases the characteristic distance between branches, since the overall size of the attractor given by the dynamical range of the observable is left unchanged.

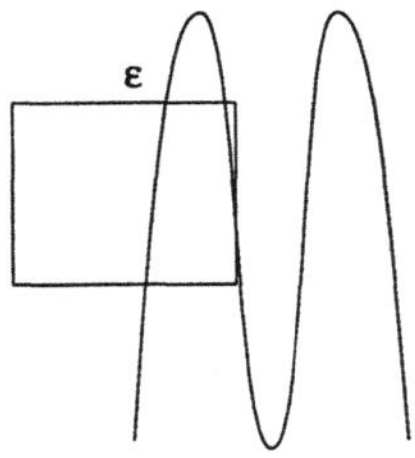

Fig. 3. Schematic representation of the "folding effect".

The following considerations will lead to a quantitative description of this effect. For the information entropy it is known that the following inequality holds:

$$h(m,\epsilon) \geq h(m+1,\epsilon) \quad . \tag{10}$$

If we assume that this is also true for the correlation entropy[3] we get in the scaling range

$$H(m,\epsilon) = \sum_{n=0}^{m-1} h(m,\epsilon) \geq mh_2 \quad , \tag{11}$$

because there

$$h(m,\epsilon) \approx \text{const} \geq h_2 \quad . \tag{12}$$

Since the correlation integral is monotonously increasing with ϵ, at the upper end of the scaling range denoted by ϵ_u the correlation integral has to fulfil

$$\ln C(\epsilon_u, m) \leq -mh_2 \; . \tag{13}$$

Using Eqs. (8),(9) and (13) one can estimate (for $N \gg w$) a lower bound for the minimal amount of data which is necessary to detect the scaling range in a system with a correlation dimension D_2:

$$N \geq e^{\frac{1}{2}mh_2}\sqrt{2k_{\text{min}}}\left(\frac{\epsilon_u}{\epsilon_l}\right)^{D_2/2} \quad . \tag{14}$$

The most important aspect of this result is that in contrast to similar estimations in Holzfuss and Mayer-Kress 1986, Procaccia 1988, Smith 1988, Nerenberg and Essex 1990, Eckmann and Ruelle 1992, Ding et al. 1993 both the

[3] In the framework of Renyi entropies of order q the information entropy corresponds to $q = 1$ and the correlation entropy to $q = 2$. They differ only, if the considered object is multifractal.

entropy and the embedding dimension influences the extent of the scaling range, which reflects the folding effect. A result very similar to Eq. (14) is derived in Malinetskii et al. 1993, though they treated the problem from another point of view.

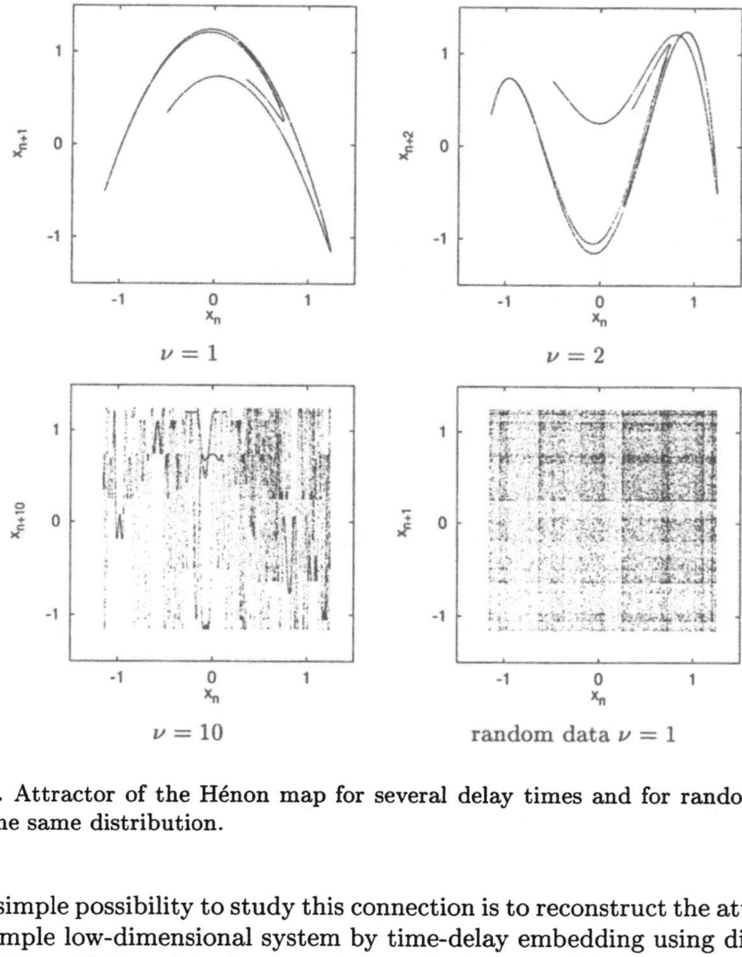

Fig. 4. Attractor of the Hénon map for several delay times and for random data with the same distribution.

A simple possibility to study this connection is to reconstruct the attractor of a simple low-dimensional system by time-delay embedding using different time delays. If the original system is given by a map $F(\mathbf{x})$ then doubling the time delay can be viewed as studying the map $F^2(\mathbf{x})$, which has the twofold entropy. Figure 4 shows the attractor of the Hénon map as an example. For comparison the delay plot of random data with the same scalar distribution (so called surrogate data) is included. The larger the entropy the stronger is the folding of the attractor and the more the picture of the attractor is similar to the picture of the random data.

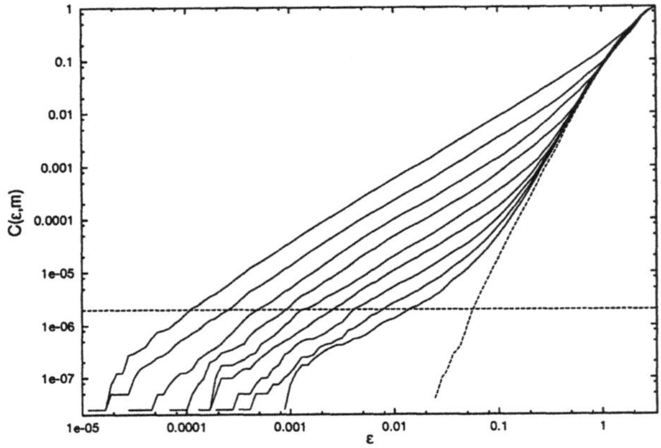

Fig. 5. Correlation integral for the henon map with different delays $\nu = 1, ..., 9$ (solid lines) and surrogate data with $\nu = 1$ (dashed line), $m = 4$.

Let us now have a look at the result of the correlation integral Fig. 5: What we observe is that the upper end of the scaling region is shifted to smaller ϵ, if we increase the delay, i.e. the entropy. Moreover we see that the asymptotic behaviour for $\epsilon > \epsilon_u$ is given by the surrogate data. This means that on these length scales all (nonlinear) correlations between the data are destroyed by the folding effect. What can we learn by the inequality (14) in this case? The horizontal line denotes the value of the correlation integral at the lower end of the scaling range according to Eq. (9) setting $k_{\min} = 100$. We have used 10 000 data points. What would we expect for the extent of the scaling range for say $\nu = 9$ using relation (14)? For the Henon map with $\nu = 1$ we have $h_2 \approx 0.32$ and $D_2 \approx 1.21$, the embedding dimension used in Fig. 5 is $m = 4$. We get $\epsilon_u/\epsilon_l \leq 1.9$. In fact there is no scaling range at all. But there is no contradiction, because relation (14) gives only a upper bound for the length of the scaling region. So we have neglected the fact that $h(m, \epsilon)$ can converge to h_2 very slowly, so that $H(m, \epsilon_u)$ can be much larger then mh_2.

So far we have assumed our data are noise free and of infinite precision. If we have noisy data with noise level ϵ_{noise}, we expect

$$C(m, \epsilon) \propto \epsilon^m \qquad \text{for} \quad \epsilon < \epsilon_{\text{noise}} \,. \tag{15}$$

Therefore the absolute values of ϵ_u and ϵ_l become important. In the argumentation leading to the inequality (14) this was not the case. This gives us an argument to resolve an inconsistency with the statements given in section 3 concerning the problem choosing the optimal time delay ν for flow data. If we consider relation (14), the optimal delay should be $\nu = 0$ because the entropy is proportional to ν. This is obviously nonsense. What happens if we consider the limit $\nu \to 0$ is that the scaling range is shifted to $\epsilon \to -\infty$ due to the linear correlations as discussed in Sect.3 (see also Fig. 1). But because of the finite precision of the given data, determined by ϵ_{noise}, we cannot profit by larger scaling range as soon as ϵ_l is smaller then ϵ_{noise}.

In contrast to the requirements to the amount of data due to the dimensionality of the attractor under study, the restrictions of the scaling region due to the entropy might be reduced, if one uses multichannel measurements instead of a scalar time series. If we measure k values each time and the necessary embedding dimension to reconstruct the attractor properly is greater then k we have to make a delay embedding with this k-dimensional vectors. If we use n time steps, then the embedding dimension is obviously $m = k \cdot n$. Now we can use the same entropy arguments as above, but we have to replace m by $n = m/k$. So if the measurement was taken at N different times relation (14) has to be replaced by

$$N \geq e^{(m/2k)h_2} \sqrt{2k_{\text{min}}} \left(\frac{\epsilon_u}{\epsilon_l} \right)^{D_2/2} . \tag{16}$$

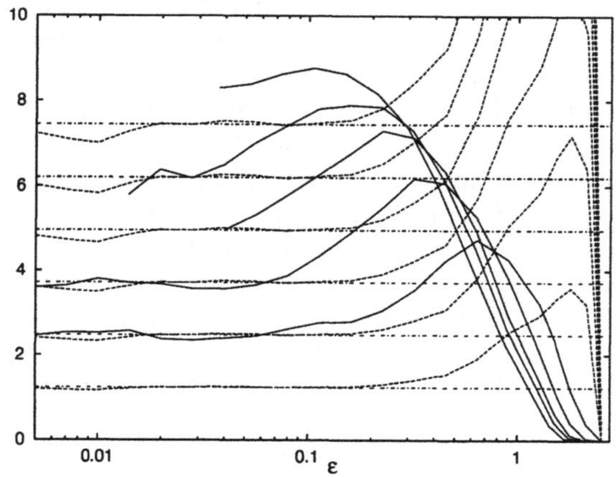

Fig. 6. Correlation dimension estimate for $N = 1$ to $N = 6$ uncoupled Hénon maps. The dashed lines show the results in real space, the solid lines the results in delay space.

Let us compare the time delay embedding of scalar observations to vector valued observations representing the whole state space using a simple example. We consider the mean

$$s_n = \frac{1}{N} \sum_{i=1}^{N} x_n^i \tag{17}$$

of N uncoupled Hénon maps $x_n^i = 1.4-(x_{n-1}^i)^2+0.3x_{n-2}^i$. In Fig. 6 we show the results of the dimension estimates for N ranging from 1 to 6. The solid lines show the estimates in a 10-dimensional delay embedding space of s_n. The dashed lines show the results for vector valued observations $\mathbf{x} = (x_n^1, \ldots, x_n^N)$ (real space). Here, an overembedding was used to demonstrate the effect of folding also in this space. While the maxima of the overestimation of the dimension due to the folding effects remain at fixed length scales in real space, the corresponding maxima in delay space are shifted towards exponentially smaller scales as N is increased. As a consequence, determinism is observable for large N on reasonable length scales only if our measurements represent the full state space, whereas the onset of scaling is shifted towards inaccessibly small scales for delay embeddings of scalar observations.

5 Coupled Map Lattices

In the last section we considered a global observable which contains the information about all degrees of freedom in an equal way. Alternatively, one can think of local observables in a complicated system. To mimic this situation, we first have to construct a model for the high-dimensional dynamics with a reasonable coupling between the different degrees of freedom. Many systems in nature and in particular in physiology are spatially extended. That is, one has some network of almost identical behaving degrees of freedom plus some coupling between them. Think of the heart muscle, on which electrochemical processes create and transport the electrical signals leading to its contraction, or of a network of neurons in the brain. In the cardiac muscle, couplings presumably are local, whereas in a neural net they may be long-range or even global. A simplified model for such systems are coupled map lattices (CML). A CML is a system of maps living on a grid in space. Since we are dealing with maps, time is discrete also. The maps at the different lattice sites have to be coupled. Since a diffusive coupling is quite common in nature, one often introduces a nearest neighbour coupling. Therefore we study the onedimensional chain of maps described by

$$x_i^{t+1} = f\left((1-2\sigma)x_i^t + \sigma(x_{i+1}^t + x_{i-1}^t)\right) , \tag{18}$$

where t and $i = 1, \ldots, N$ are the time and the spatial indices, respectively. N is the number of maps coupled, or in other words, the spatial size of the system. σ is a coupling constant, which determines the interaction strength of next neighbored maps. f is an arbitrary map, which usually is chosen to show chaotic behaviour.

We restrict ourselves here to f being the logistic map in the regime of fully developed chaos. That is,

$$f(x) = 1 - 2x^2 . \tag{19}$$

Since the full phase space of this system is N-dimensional, we find now N different Lyapunov exponents λ_i, characterizing the divergence or the attraction of neighboured trajectories in all possible directions in space. The set of λ_i, ordered by their size usually is called the *Lyapunov spectrum*. There is as well numerical (Paladin and Vulpiani 1986, Kaneko 1993), as analytical (Ruelle 1982) evidence that the Lyapunov spectrum of CML shows a universal behaviour in the limit of the system size N approaching infinity. If we rescale the indices i with respect to the system size, $i \to q = i/N$, the spectrum is independent of the latter. Figure 7 shows the spectra for three different system sizes, namely $N = 100, 200$ and $N = 300$. The coupling is $\sigma = 1/3$. We clearly observe that all curves are more or less identical. The deviations we see are firstly due to numerical fluctuations and secondly to the fact that N is far below infinity.

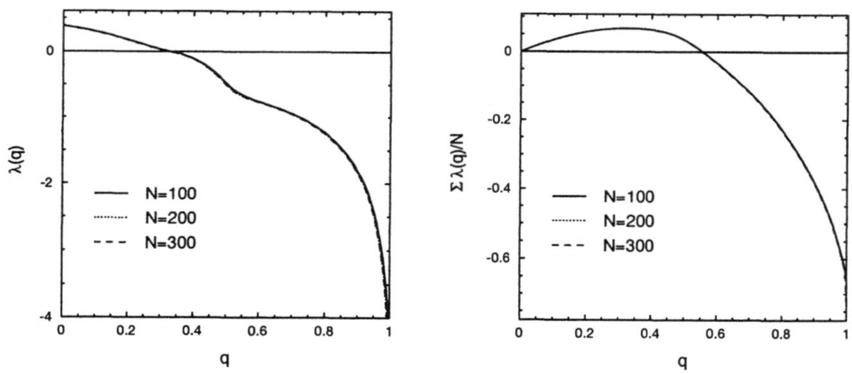

Fig. 7. Lyapunov spectra of three CML systems with sizes $N = 100$, 200 and $N = 300$ (left plot) and the corresponding integrated spectra (right plot), respectively. The coupling constant was set to $\sigma = 1/3$. The entropy density is given by the area created by the positive exponents and the zero line in the left plot, whereas the dimension density is given by the zero of the integrated spectra in the right plot.

A consequence of this scaling behavior of the Lyapunov spectrum is that the entropy and the Kaplan–Yorke dimension of CMLs are extensive quantities. The entropy is related to the Lyapunov spectrum through the Pesin identity (Eckmann and Ruelle 1985). The latter states that

$$h_1 = \sum_{i,\lambda_i>0} \lambda_i \,, \tag{20}$$

where h_1 is the Kolmogorov–Sinai entropy and λ_i are the Lyapunov exponents. Equation (20) together with the scaling behavior of the λ_i in N shows that h_1 is proportional to N. Therefore one can introduce the *entropy density* $\eta = \lim_{N\to\infty} h_1(N)/N$. A similar argumenation holds for the Kaplan–Yorke dimension,

$$D_{KY} = k + \frac{\sum_{i=1}^{k} \lambda_i}{|\lambda_{k+1}|}, \tag{21}$$

where $\sum_{i=1}^{k} \lambda_i \geq 0$ and $\sum_{i=1}^{k+1} \lambda_i < 0$. This yields the *dimension density* $\delta_{KY} = \lim_{N\to\infty} D_{KY}/N$ and in an analog way a δ_2 corresponding to the correlation dimension D_2.

To give an impression, the dimension density of the system underlying Fig. 7 ($\sigma = 1/3$) is $\delta_{KY} \approx 0.55$. This makes obvious that we are never able to estimate the full dimension of a spatially extended system, even for moderate system size N. In order to observe scaling, we had to work with embedding dimensions larger than $\delta_{KY}N$, which would require an unrealistic amount of data. The hope to escape from this dilemma is that some features of the deterministic structure can be reconstructed from data sets we can handle.

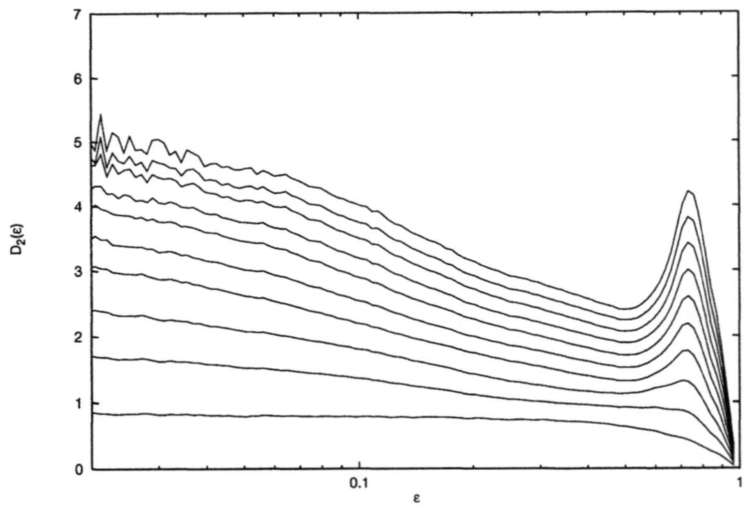

Fig. 8. Estimate of the correlation dimension D_2 for a system of 100 coupled logistic maps. The coupling constant was set to $\sigma = 1/5$.

It was conjectured by several authors (Grassberger 1989, Bauer et al. 1993, Tsimring 1993) that it is possible to see the dimension density for high-dimensional objects even if the embedding dimension is far below the required value. On first view, this sounds unlikely, since one would expect that projecting a highdimensional object to a low dimensional subspace fills the whole volume. Of course, this is true on small length scales. However, as we shall study in the following, short range couplings might create nontrivial structures on the large length scales.

Figure 8 shows the results of the estimation of the correlation dimension D_2 from a scalar time series x_i^t for arbitrary but fixed i of length 10^6, for a system of size $N = 100$ with a coupling constant $\sigma = 1/5$. We observe several regions. For really large ϵ (the attractor is normalized in the interval $[0,1]$) there is a high bump, which is due to folding effects of the attractor. Then we find a valley followed by a more or less linear increase. On even smaller scales we observe statistical fluctuations and saturation effects.

A possible explanation for the linear part is the following: The influence of the neighbours $i+1$ and $i-1$ on the observed variable x_i is of order σ. The influence of the next neighbors $i+2$ and $i-2$ is of order σ^2, and so on. That means, as long as we are considering only length scales larger than σ^n, the influence of the neighbors having a distance larger than n may be neglected. They act more or less like noise, which is only visible on length scales smaller than the 'noise level'.

This effect becomes clearer if we decrease the coupling. Then we can expect to observe plateaus at dimension 1, 3, 5 and so on, above length scales $\sigma, \sigma^2, \sigma^3 \ldots$. Since we usually plot the logarithm of ϵ, these plateaus will have equal length. Figure 9 (upper plot) shows the results for $\sigma = 0.002$. In this case, the attractor of the system looks essentially like a single logistic map (in delay space) which is slightly corrupted by noise. This is exactly what we see in the plot. There is the scaling range, where we see the single map. Then, if the length scale reaches roughly σ, the coarse grained dimension increases, due to the 'noise'. But instead of D_2 becoming equal to m (the embedding dimension), the deterministic origin of the 'noise' becomes visible and we get $D_2 = 3$. The lower plot in Fig. 9 shows the behavior schematically. There are plateaus of length $\log \sigma$ differing 2 in heights. This step function can be described by an average slope s

$$s = \frac{2}{\log \sigma} \, . \tag{22}$$

Of course, we will never see such a picture from a real dimension estimate. The edges of the plateaus will be smeared out and there will be additional structure due to folding effects. What we will see is a superposition of the plateaus and the additional structure. The hope is that this superposition will maintain the average behavior and thus the slope of Eq. (22). This could yield an explanation of the linear part of Fig. 8. In Fig. 10 we show the dimensions estimates for two different couplings, $\sigma = 0.11$ and $\sigma = 0.2$. The straight lines in the plots correspond to Eq. (22). We see that they fit the linear regime reasonably well. For large values of σ we expect a dimension density below unity, which requires a modification of Eq. (22) by a factor δ.

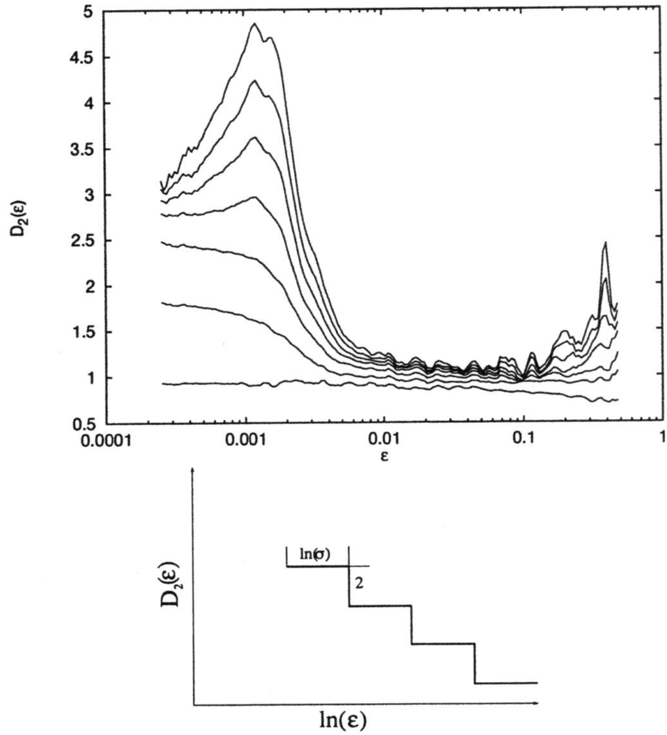

Fig. 9. *Upper Plot*: Results for the dimension estimation of a system of 100 coupled logistic map with coupling constant $\sigma = 0.002$. *Lower Plot*: Schematic plot of the idealized behavior for systems with rather small coupling.

In the right plot, which is identical with Fig. 8, the linear law is violated on small scales. This is related to a special effect of spatiotemporal intermittency and will not be discussed here.

Another way to interpret the linear regime is due to Tsimring 1993. He claimed that the slope is proportional to the dimension density. More precise, he argued that

$$s = 2\delta l \, , \tag{23}$$

where l is the spatial correlation length of the system. This is in obvious contradiction to our results which state that the slope is predominantly related to the coupling strength. Moreover, it is not possible to compute the spatial correlation length l from scalar data.

The last interpretation of plots like Fig.8 is due to Bauer et al. 1993. They claimed that in principle $D_2(m, \epsilon)$ should exhibit a plateau at a value $D_2 = m\delta$, but that this plateau is detroyed by edge effects on large scales. The underestimation of the dimension due to the fact that points on the bor-

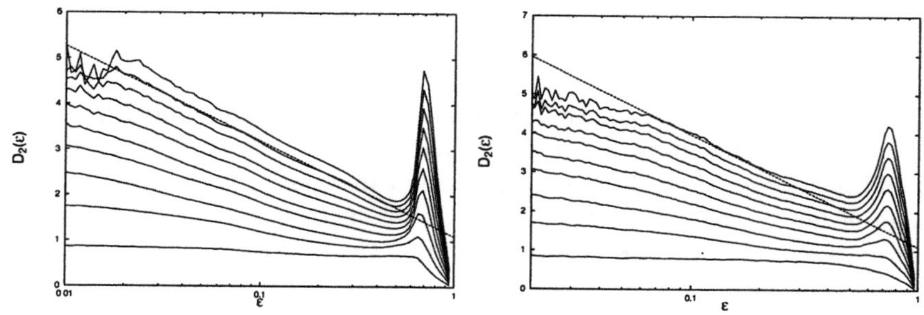

Fig. 10. Dimension estimate for two systems of 100 coupled logistic maps. One with coupling $\sigma = 0.11$ (left plot) and the other with $\sigma = 0.2$ (right plot). The straight lines have slope $2/\ln\sigma$, see Eq. (22)

der of the attractor have neighbours only in certain directions was already mentioned in Sect.4. Bauer et al. suggest a nice procedure to partially eliminate these problems and claim to be able to observe the plateaus. An even simpler and more rigorous way to get rid of edge effect consists in a proper normalization of the correlation sums for embeddding dimension m with respect to the sum in embedding dimension 1. This results in the corrected estimate of the coarse grained dimension

$$\tilde{D}_2(m,\epsilon) = \frac{d \ln C(m,\epsilon)}{d \ln C(1,\epsilon)}. \tag{24}$$

Even with these corrections we were unable to confirm the statements of Bauer et al. 1993. Furthermore, it is most likely that for nonnegligible coupling σ the coarse grained dimensions $D_2(m,\epsilon)$ increase monotonically in $-\log\epsilon$, since eventually they have to saturate at $D_2 = m$.

A possible way out of this trouble was proposed in Grassberger 1989. He suggested to look on the real space of the system, not on the delay space. That means, instead of measuring only one coordinate x_i we now measure a subsystem of say n neighboured coordinates simultaneously. I.e., we measure at every time step a whole vector $(x_1, \ldots, x_n)$.

If n is sufficiently large, larger than the correlation length of the system, we could hope that the extensive character of the system is already observable when comparing such subsystems for different n. If we increase n by one, then the dimension of the new system will not be increased by one, but only by the dimension density δ, since every coordinate does only contain δ degrees of freedom. The rest of the whole system can be treated as a heat bath, which acts only like noise on the subsystem. The influence of this noise is largest at the edges of the subsystem. There it couples in with amplitude σ. The next coordinate is only influenced by σ^2 and so on. This means, the more we approach the center of our subsystem, the less is the influence of the 'external noise'. So we can hope that only few of the coordinates are noisy, while the rest of the system is more or less noise free. If this is true, we can hope to see a clear scaling range in the dimension estimate. The results of such an estimation are shown in Fig. 11. Indeed, there exists a scaling region in contrast to the delay space, where there was none. But our system sizes were not large enough to see the expected dimension density δ. However, we should not forget that all these arguments can again be true only in a finite range of length scales, since in principle this subsystem has to contain the information about the whole system, and on the small scales we again should see either dimension $d = n$ or the full dimension of the attractor.

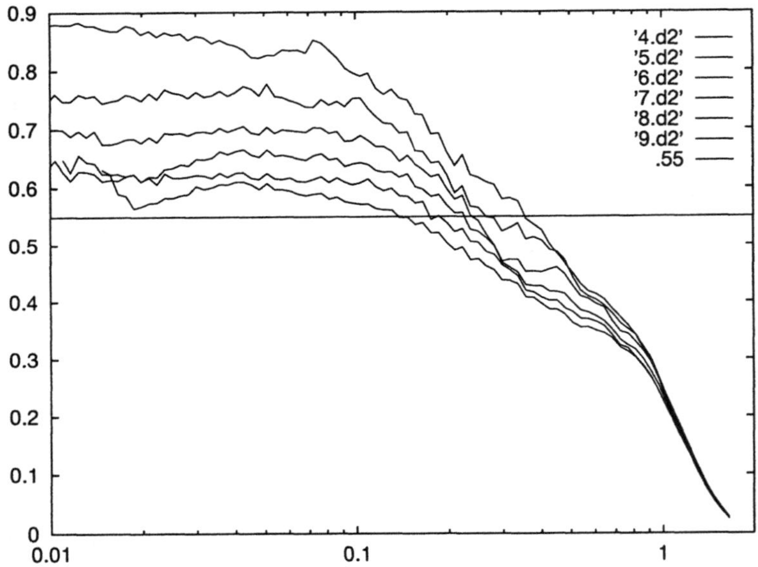

Fig. 11. Estimation of the dimension density in real space for different subsystem sizes $n = 4, \ldots, 9$. The horizontal line represents δ_{KY} computed from the Lyapunov spectrum.

The results shown in the figure are based on 10^6 data points per coordinate. That is roughly the limit of what is possible with present day computers. So, although the results of the real space estimations look more convincing than that from the delay space, we are not able to extract any reasonable δ from that. But if we compare Fig. 8 and Fig. 11 it seems to be obvious that the chance for well founded results is by far larger in real space than in delay embedding space.

6 Potential Meaning of Dimension and Lyapunov Estimates in Real World Data

In the previous two sections we discussed in some detail problems arising in dimension estimates of complicated signals. Let us stress that the problems encountered there are *not restricted to dimension estimates.* The computation of the correlation dimension is only the simplest tool when one is asking for an indication of determinism (provided, one takes care of certain well known pitfalls like linear correlations, nonstationarity, etc., see e.g. Kantz and Schreiber 1995). With all other methods which can be used to prove or characterize determinism one has the same trouble, as with Lyapunov exponents, entropies, or with a fit of the dynamics.

The state space one obtains by a time delay reconstruction from scalar data introduces either complicated folds on the attractor (large time lag) or leaves it badly unfolded (small time lag), and there is no satisfactory compromise in between, when the total entropy of the signal is high and m has to be large, since the attractor dimension is large.

So far there is no evidence that observations on the large scales which are easily accessible *have anything to do* with the true values of the invariants characterizing the limit of infinitesimal scales. If there is system specific information (beyond information about the scalar distribution and linear temporal correlations which can be extracted much easier by linear statistics), then most probably only in local observables from spatially extended systems, and it is a completely different type of information, namely the coupling strength between adjacent degrees of freedom, which potentially could be extracted from the signal.

In this paper we have investigated two classes of signals from high-dimensional systems: In Sect.4. we introduced a signal which is the average over several independent chaotic degrees of freedom. This is the typical situation when one has a global observable. In some sense, in physiology, the ECG signals are of this kind, since they are result of some averaged potential on a large part of the surface of the heart. Depending on the number of sensors, also MEG data are averaged over a large part of the brain. In such signals the different degrees of freedom contribute in a democratic way, and they are therefore very complex already on the large length scales. Nevertheless, what we observe on the large scales when e.g. computing their dimension just reflects their scalar distribution and none of the complicated nonlinear correlations. Therefore, the results of the large scales can be reproduced with surrogate data, and only on scales where the true data and surrogate data start to deviate, we begin to see first hints of additional structure.

In Sect.5 we focussed on another class of signals, on local observables. A local observable in the first place records the dynamics at a certain position inside a spatially extended and therefore complicated system. The farther away another part of the system is, the smaller is the amplitude of its contribution to the local observation. Therefore, on the large scales such a signal appears to be moderately complicated (small dimension, low Lyapunov exponent, low-dimensional dynamics), and the better we resolve the structures by proceeding towards the smaller length scales, the more degrees of freedom appear. This would be a quite encouraging situation, if we only were able to seprarate two sources of increasing complexity: the amplitude by which the next, next to next, etc. degrees of freedom contribute, and their number. In Sect.5 we could predict the slope in Fig. 10 from the knowledge that at each level $\epsilon_n = \sigma^n$ exactly 2 additional degrees of freedom enter, which leads to the factor 2 in the numerator of Eq. (22). If we simulated the maps on a 2-dimensional array instead a 1-d lattice, this would be a factor of 4. In nature, we do not know how the degrees of freedom are distributed in space, therefore the slope s cannot be decomposed into the two interesting quantities coupling strength and number of neighbours. It is our impression that this slope s could nevertheless be some characteristic of the system, but currently there is not enough numerical evidence that such a linear increase is really typical of such a situation to draw any clear conclusions. Only, one problem is again evident: On larger scales, we can observe both larger or smaller dimension values, depending on various details of the data, so that the abolute values of the effective dimension at certain length scales are meaningless.

When dealing with data which do not exhibit reasonable scaling behaviour, people sometimes compute dimensions or Lyapunov-exponent-like quantities on *fixed* large length scales, in the hope that such effective measures of dimension or instability could characterize the data as well. In both situations, local and global observables, *this is almost meaningless*! Quantities computed at large length scales have nothing or not much to do with the characteristic quantities of the system. Therefore they have to be used with extreme care for the purpose of diagnosis. It appears to us reasonable (althoug no systematic investigation was made on model systems) that certain ratios could be useful: e.g. local fixed-scale prediction errors as presented in Kowalik 1995 can possess diagnostic power if one computes their distribution on a given data set and compares this distribution to another one: It might come out that in one situation all these errors are of comparable order, whereas in another situtation there might be parts of the signal with low and other parts with high prediction errors. Or the computation of something like pointwise fixed scale dimensions could give hints on clustering properties of the data in a certain embedding space: Points with a low dimension lie on spiky structures, those with local dimensions close to the embedding dimension lie inside a bulk, and their ratio might be characteristic of a certain signal. The important aspect is that such kinds of statistics must be robust against small changes in the parameters involved in their definition. If one changes the scales on which one computes the prediction errors or the local dimensions, the resulting statitics should remain almost unaffected. The absolute values of pointwise dimensions or local Lyapunov exponents obtained this way, however, are not at all meaningful, since quantities like this are not invariant, and different signals are never taken exactly in the same way, so that one cannot use noninvariant quatities for signal classification. To make our concern more precise: In Figs. 9 and 10 we presented three different plots of dimension estimates of a logistic chain of 100 maps. The Lyapunov-dimension, estimated via the Kaplan Yorke conjecture, is 1.0, 1.0, and 0.7, whereas the effective dimension in an ($m = 5$)-dimensional embedding and at $\epsilon = 0.3$ is about 1.5, 1.2, and 1.5. You can repeat this analysis for any other m or ϵ on the basis of the plots and find that neither is the latter effective dimension proportional to the Lyapunov dimension, nor are the fixed-scale dimensions at different scales compatible among themselves.

7 Conclusions

Time series analysis is a very powerful tool for dealing with low-dimensional dymanic systems. There we can get reliable results for dimension or entropy estimations and for, at least the largest, Lyapunov exponent. This is due to the fact that the average spacing between vectors in the reconstructed phase space is sufficiently small, even in cases where the amount of data is also small. This permits us to investigate relatively small length scales, so that scaling laws can be verified on large regimes.

The situation changes dramatically, if we treat highdimensional objects. On these objects, the density of vectors becomes quite small. As we saw this density decreases exponentially with the dimension. So, we would need an exponentially increasing amount of data to compensate this effect. This seems unfeasible with present day facilities, so that we can no longer hope to reach sufficiently small scales. This means that the chance of observing deterministic structures decreases as well.

At first sight this might be disappointing. Nevertheless, we can draw two positive conclusions. First, a potential way out of the problem consists of suitable multichannel measurements, where the reconstruction of a state space by a time delay embedding is not required. Moreover, in a multichannel measurement one can record a huge amount of data in a much shorter time, which also reduces the ubiquitious problem of nonstationarity. Second, in many situations where currently the only conclusion is that the data are completely described by a stochastic process, we can hope to find more interesting underlying structures and thus a better understanding of the phenomenon by a better reconstruction of the state space. Therefore, the unability to reject the null hypothesis of a linear random process with current methods and data sets can also mean that there are very interesting but high-dimensional and high-entropic structures behind, and it is worth to think about how to make them visible.

Acknowledgements

The data underlying Fig. 1 stem from the nonlinear electronic oscillator constructed and run by the experimental group of Prof. Beige at the University of Halle, and we are very grateful to have access to some of their excellent data. We also acknowledge some very stimulating and encouraging discussions with P. Grassberger, A. Politi, and T. Schreiber.

References

Bauer M., Heng H., Martienssen W. (1993): Characterization of spatiotemporal chaos from time series. Phys. Rev. Lett. **71**(4), 521

Beige H., Diestelhorst M., Forster R., Krietsch T. (1992): Chaos near structural phase transitions. Phase Transitions **37**, 213

Ding M., Grebogi C., Ott E., Sauer T., Yorke J.A. (1993): Plateau onset of correlation dimension: When does it occur? Phys. Rev. Lett. **70**, 3872–3875

Eckmann J.-P., Ruelle D. (1985): Ergodic theory of chaos and strange attractors. Rev. Mod. Phys. **57**, 617

Eckmann J.-P., Ruelle D. (1992): Fundamental limitations for estimating dimensions and Lyapunov exponents in dynamical systems. Physica D **56**, 185–187

Grassberger P., Procaccia I. (1983a): Characterization of strange attractors. Phys. Rev. Lett. **50**, 346–349

Grassberger P., Procaccia I. (1983b): Estimation of the Kolmogorov entropy from a chaotic signal. Phys. Rev. A **28**, 2591–2593

Grassberger P., Procaccia I. (1983c): Measuring the strangeness of strange attractors. Physica D **9**, 189–208

Grassberger P., (1989): Information content and predictability of lumped and distributed dynamical systems. Phys. Scr. **40**, 346

Holzfuss J., Mayer-Kress G. (1986): An approach to error-estimation in the application of dimension algorithms. In G. Mayer-Kress (ed), *Dimensions and Entropies in Chaotic Systems*, Springer, Berlin, Heidelberg 114–122

Kaneko K. (1993) (ed.): *Theory and Applications of Coupled Map Lattices.* Wiley, New York

Kantz H., Schreiber T. (1995): Dimension estimates and physiological data. Chaos **5**, 143–155

Kowalik Z.J. (1995): Relativity of dynamical measures derived from MEG/EEG time series. Talk at the NTPTSA-workshop

Kruel T.M., Eiswirth M., Schneider F.W. (1993): Computation of Lyapunov spectra: effect of interactive noise and application to a chemical oscillator. Physica D **63**, 117–137

Malinetskii G.G., Potapov A.B., Rakhmanov A.I., Rodichev E.B. (1993): Limitations of delay reconstruction for chaotic systems with a broad spectrum. Phys. Lett. A **179**, 15

Nerenberg M.A.H., Essex C. (1990): Correlation dimension and systematic geometric effects. Phys. Rev. A **42**, 7065–7074

Packard N., Crutchfield J., Farmer D., Shaw R. (1980): Geometry from a time series. Phys. Rev. Lett. **45**, 712

Paladin G., Vulpiani A. (1986): Scaling laws and asymptotic distribution of Lyapunov exponents in conservative dynamical systems with many degrees of fereedom. J. Phys. A **19**, 1881–1888

Procaccia I. (1988): Complex or just complicated. Nature **333**, 498–499

Rössler O. E. (1979): An equation for hyperchaos. Phys. Lett. A **71**, 155–159

Ruelle D. (1982): Large volume limit of the distribution of characteristic exponents in turbulence. Commun. Math. Phys. **87**, 137

Sauer T., Yorke J. A., Casdaglii M. (1991): Embedology. J. Stat. Phys. **65**, 579

Smith L. A. (1988): Intrinsic limits on dimension calculations. Phys. Lett. A **133**, 283–287

Stoop R., Meier P. F. (1988): Evaluation of Lyapunov exponents and scaling functions from time series. J. Opt. Soc. Am. B **5**, 1037–1045

Stoop R., Peinke J., Parisi J., Hübener R. P. (1989): A p-Ge semiconductor experiment showing chaos and hyperchaos. Physica D **35**, 425

Takens F. (1981): *Detecting Strange Attractors in Turbulence.* Lecture Notes in Mathematics, Vol. 898. Springer, Berlin, Heidelberg

Theiler J. (1996): Spurious dimension from correlation algorithms applied to limited time-series data. Phys. Rev. A **34**, 2427

Tsimring L.S. (1993): Nested strange attractors in spatiotemporal chaotic systems. Phys. Rev. E **48**, 3421

Chaotic Measures and Real-World Systems: Does the Lyapunov Exponent Always Measure Chaos?

Milan Paluš

Institute of Computer Science, Academy of Sciences of the Czech Republic
Pod vodárenskou věží 2, 182 07 Prague 8, Czech Republic; and
Santa Fe Institute, 1399 Hyde Park Road, Santa Fe, NM 87501, USA
E-mail: `mp@uivt.cas.cz`, `mp@santafe.edu`

Abstract. Direct estimation of the largest Lyapunov exponent as a measure of exponential divergence of nearby trajectories is well established in the case of deterministic dynamical systems. Questions are naturally raised about applicability of Lyapunov exponents and other "chaotic measures" when analyzing data from real-world systems, which are either stochastic or affected by numerous external influences, which cannot be described in any other way than a stochastic component in system dynamics. In a series of numerical experiments, Gaussian random deviates were added to a set of chaotic time series with different Lyapunov exponents. It is demonstrated that the estimated Lyapunov exponents fail to distinguish different noisy chaotic time series when relatively small scales are used. The distinction can be reestablished by using larger scales. Using larger scales, however, the estimated Lyapunov exponent is determined by macroscopic statistical properties of the series and provides the same information as the autocorrelation function and/or coarse-grained mutual information.

1 Introductory Remark

This paper is meant as an informal but, it is hoped, informative contribution to a discussion of problems related to nonlinear techniques used in analysis of physiological or other experimental real-world time series. It is addressed to a broad audience of readers with different educational backgrounds, to both theorists and practitioners, therefore I limit the use of mathematical formulae to the necessary minimum, and try to explain discussed facts verbally and by presenting graphical material. Nevertheless, I suppose that a reader is familiar with basic notions related to time series analysis, both linear theory (spectrum, autocorrelation) and chaos approach (dimensions, Kolmogorov–Sinai entropy, Lyapunov exponents).

2 Introduction

Distinction and classification of different dynamical phenomena or systems, or, distinction and classification of different dynamical states of a system is a common problem in many areas of natural and social sciences. In many cases this problem can be translated into a task of quantitative characterization of observable signals, i.e., into an estimation of a quantitative measure from registered time series.

In physiology and medicine time series are recorded which are related to different physiological and/or pathological states of an organism or its parts. Classification of such data is a challenging task with significance for understanding underlying physiological processes and for medical diagnostics. Traditional, linear techniques for time-series analysis have been applied successfully in various areas of physiology and medicine. However, they may have serious limitations due to the fact that physiological processes are usually nonlinear. Therefore it is not surprising that recently developed methods for nonlinear time series analysis have immediately found their way into physiology and biomedical research.

Many of the current methods in nonlinear signal processing have not arisen as an extension of linear analysis, but have been conceived due to the entirely new idea of deterministic chaos. These innovative techniques have provided experimentalists with new ways of understanding the implications of their data, though the limitations of these new techniques have not always been understood and necessary precautions fully appreciated.

Estimation from time series of descriptive measures such as dimensions, Lyapunov exponents or Kolmogorov entropy, derived from theory of deterministic chaos ("chaotic measures") is well established in the case of data generated by low-dimensional deterministic dynamical systems in numerical and laboratory experiments. Questions are naturally raised about applicability of the chaotic measures when analyzing data from real-world systems, which are either stochastic or affected by numerous external influences, which cannot be described in any other way than a stochastic component in system dynamics. Analyzing time series from physiological systems, many authors have realized that low-dimensional chaos in such systems is improbable, however, they have demonstrated that formal estimates of the chaotic measures may possess some discriminating power with respect to data recorded in different experimental conditions (Layne et al. 1986; Mayer-Kress & Layne 1987; Koukkou et al. 1993; Wackermann et al. 1993). This "relative characterization" of different datasets may surely have its importance in diagnostics, however, the question should be asked: *What do the quantities for measuring chaos actually measure, when in processed data there is no chaos, or a chaotic phenomenon is obscured by noise?* The term "measuring chaos" should be deciphered considering a particular chaos-based method used: What do the small numbers, obtained from dimensional algorithms, actually mean, when the underlying system is high-dimensional or stochastic? What do the estimates of Lyapunov exponents, designed to measure exponential divergence of nearby trajectories, actually characterize, when there is no exponential divergence of trajectories, or even there are no trajectories in the data, or the trajectories are obscured by noise? These questions are important from both practical and theoretical points of view. When the chaotic measures, designed for characterization of low-dimensional dynamics, are applied to analysis of high-dimensional or stochastic systems, precision of their estimates, their robustness with respect to noise, or their sensitivity to changes in underlying dynamics can hardly be established. In theoretical aspect, correct interpretation of obtained results is unclear, while using the original meaning and interpretations of the chaotic measures, i.e., using a "low-dimensional language" for high-dimensional or stochastic systems can be misleading.

This paper does not have the ambition to answer generally these important questions. It is just a demonstration of a particular situation of applying the direct method for estimating the largest Lyapunov exponent (LLE) (Wolf et al. 1985) to noisy chaotic and linear stochastic data. A set of chaotic time series with different positive Lyapunov exponents was generated. It is shown that the LLE algorithm correctly distinguishes and orders the series according to their positive LE's. Then Gaussian noises with zero means and different standard deviations (SD's) are added to the series. The distinction of the series is lost when the LLE algorithm uses scales comparable to, or smaller than the SD of the noise. The noisy chaotic series can be correctly distinguished and ordered, when the scales used in the LLE estimation are larger than the noise's SD.

The requirement to use relatively large scales in practical estimation of the chaotic measures is very typical due to finite precision measurements, limited amounts of data and/or noise in the data, as in this case. Using large scales, however, do the chaotic measures indeed "measure chaos", i.e., does the LLE algorithm measure the exponential divergence of nearby trajectories, or something else? Using isospectral surrogate data approach we show that the LLE in fact distinguishes time series with different autocorrelation functions. This property is probably typical also for dimensional estimates and other measures which explore distributions of distances between points. Thus the chaotic measures, estimations of which usually possess high computational cost and vulnerability to various experimental and numerical factors, in many cases provide the same information as the results obtained by standard linear time series tools such as the spectral or autocovariance analysis. In highly nonlinear systems, when the linear analysis is inadequate, the autocovariance/autocorrelation function should be substituted by (coarse-grained) mutual information and related measures, which can provide reliable relative classification of different dynamical states of nonlinear systems.

3 The Largest Lyapunov Exponent and the Baker Map

Given a scalar time series $x(t)$, an m-dimensional trajectory is reconstructed using the time-delay method Takens 1981 as $\mathbf{x}(t) = \{x(t), x(t+\tau), \ldots, x(t+[m-1]\tau)\}$, where τ is the *delay time* and m is the *embedding dimension.* A neighbour point $\mathbf{x}(t')$ is located so that the initial distance δ_I, $\delta_I = ||\mathbf{x}(t) -$

$\mathbf{x}(t')||$, is $s_{\min} \leq \delta_I \leq s_{\max}$. $||.||$ means the Euclidean distance. The *minimum and maximum scales* $s_{\min}$ and $s_{\max}$, respectively, are chosen so that the points $\mathbf{x}(t)$ and $\mathbf{x}(t')$ are considered to be in a common "infinitesimal" neighborhood. After an *evolution time* $T \in \{1, 2, 3, \ldots\}$, the resulting final distance δ_F is calculated: $\delta_F = ||\mathbf{x}(t+T) - \mathbf{x}(t'+T)||$. Then the local exponential growth rate per time unit is:

$$\lambda_1^{\text{local}} = \frac{1}{T} \log(\delta_F / \delta_I). \tag{1}$$

To estimate the overall growth rate, in the case of deterministic dynamical systems the largest Lyapunov exponent (LLE) λ_1, the local growth rates are averaged along the trajectory:

$$\lambda_1 = \langle \lambda_1^{\text{local}} \rangle = \frac{1}{T}[\langle \log(\delta_F) \rangle - \langle \log(\delta_I) \rangle], \tag{2}$$

where $\langle . \rangle$ denotes averaging over all initial point pairs fulfilling the condition $s_{\min} \leq \delta_I \leq s_{\max}$.

These ideas are applied in the fixed evolution time program for estimating LLE as proposed by Wolf et al. (1985). More details, as well as the code of the program FET1, used in this study, can be found in (Wolf et al. 1985).

The set P of numerical parameters:

$$P = \{m, \tau, T, s_{\min}, s_{\max}\} \tag{3}$$

is chosen by a user.

The data for this study were generated using the well-known chaotic baker transformation:

$$(x_{n+1}, y_{n+1}) = \left(\beta x_n, \frac{1}{\alpha} y_n \right)$$

for $y_n \leq \alpha$, or:

$$(x_{n+1}, y_{n+1}) = \left(0.5 + \beta x_n, \frac{1}{1-\alpha}(y_n - \alpha) \right) \tag{4}$$

for $y_n > \alpha$; $0 \leq x_n, y_n \leq 1$, $0 < \alpha, \beta < 1$, β was set to $\beta = 0.25$. For this system the positive Lyapunov exponent λ_1 can be expressed analytically as the function of the parameter α (Hentschel & Procaccia 1983; Farmer et al. 1983):

$$\lambda_1(\alpha) = \alpha \log \frac{1}{\alpha} + (1-\alpha) \log \frac{1}{1-\alpha}. \tag{5}$$

The second LE of this system is negative and is given as (Schuster 1988):

$$\lambda_2 = \log \beta, \tag{6}$$

in this case $\lambda_2 = -2 \log 2$.

Varying the parameter α from 0.01 to 0.49 with step 0.005, ninety-seven time series with different positive Lyapunov exponents λ_1 were generated.

The component y was recorded,[1] the series length $N = 1024$ samples in each case of this study was used. In addition to the original strictly deterministic series, also noisy data were prepared. The noise considered in this study is the additive "measurement" noise, i.e., the strictly deterministic series y_n, $n = 1, \ldots, N$, were generated according to (4). Then a noisy series ξ_n, $n = 1, \ldots, N$, of Gaussian random deviates with zero mean and unit variance, generated using the GASDEV procedure from Press et al. (1986), were added to the deterministic series:

$$z_n = y_n + c\xi_n, \tag{7}$$

and the noisy series z_n were analyzed. The coefficients c were defined so that the standard deviation (SD) of the noise was equal to a defined portion of the SD of the original noise-free data. That is, the term "10% of noise" means that the SD of the added noise is equal to 0.1SD of the original data.

The set of 97 baker series with different $\lambda_1(\alpha)$ is an ideal material for simulating the task of relative characterization, i.e., the task of distinguishing and ordering the series according to their "chaoticity", i.e., according to their λ_1. The exact dependence of $\lambda_1(\alpha)$ on the parameter α, based on the analytic formula (5), is displayed in Fig. 1a. Figure 1b presents estimates of λ_1 from noise-free data using the following numerical parameters: $m = 2$, $\tau = 2$, $T = 1$, $s_{\text{min}} = 0.01\text{SD}$, i.e., 1% of SD of a particular series, s_{max} is always defined as $s_{\text{max}} = 10 s_{\text{min}}$ in this study. The λ_1 estimates in Fig. 1b agree with the correct $\lambda_1(\alpha)$ values only for small α, while the majority of the results in Fig. 1b are overestimated. It is possible to "tune" the results by changing some parameters from P (3), e.g., the estimates would decrease using larger evolution time T. Trying to simulate a real problem of classifying experimental time series, where the correct values of λ_1 are unknown (or, in strict mathematical sense they do not exist), it may be dangerous to tune the parameters P for each estimate individually.[2] As the methodologically correct approach we consider using the same parameters P for the whole set of time series, i.e., in each plot of the type of Fig. 1b the estimated LLE's were obtained using the same numerical parameters. The only varying parameter is the parameter α from (4), used in generating the series. Then, we are not interested in absolute values of estimated LLE's, but in relative quantification of different series. In this case, the results can be considered as successful, if a similar curve as that in Fig. 1a was obtained, irrespectively of a scale on the ordinate. The principal shape of the theoretical curve $\lambda_1(\alpha)$ is reproduced by the λ_1 estimates in Fig. 1b. However, the curve is not smooth

[1] Thus we concentrate to the chaotic dynamics in the y direction, which is equivalent to a one-dimensional system known as the tilted tent map (Hilborn 1994).

[2] This may lead to a subjective bias and false positive results. Even from white-noise data any positive value of the λ_1 estimate may be obtained by tuning the parameters P (Dämmig & Mitschke 1993).

due to numerical instability[3] of the estimates. Fluctuations of the estimates occur due to a relatively short time series length (1k = 1024 samples) used. For a significant decrease of the fluctuations and obtaining smooth curves resembling the theoretical one (Fig. 1a) the series length must be increased by one or two orders of magnitude. We will, however, continue the study using 1k series and consider the results in Fig. 1b as a "good" classification considering "available" amount of data. It should also be noted that

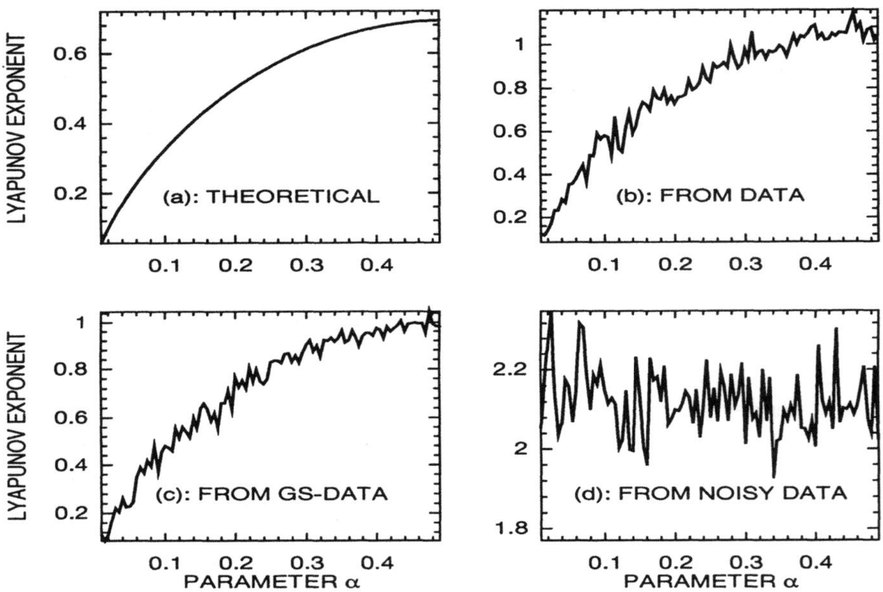

Fig. 1. a–d. Positive Lyapunov exponents of the baker system as functions of the parameter α. (a) Theoretical dependence $\lambda_1(\alpha)$, (b) estimates from noise-free data, (c) estimates from noise-free gaussianized data, (d) estimates from noised data (10% of noise). The parameters, used in estimations (b, c, d) are: $m = 2$, $\tau = 2$, $T = 1$, $s_{\min} = 0.01\mathrm{SD}$, $s_{\max}$ is always $s_{\max} = 10 s_{\min}$.

the results presented below were obtained from (noisy) baker series which underwent so-called gaussianization (Paluš 1995) – a nonlinear transformation which transformed the marginal distribution of the data into a normal distribution. The reason for this transformation is comparison of the results

[3] To decrease fluctuations of the λ_1 estimates due to local properties of time series, as a final estimate we used the mean value from the last third of all iterates (see Fig. 5).

from the "experimental" data with the results from surrogate data, described below. Although the gaussianization has some influence on the estimated λ_1 values, principal dynamical properties and related classification of the series were not changed – cf. Figs. 1b and 1c, the former was obtained from the original baker series, the latter from the same series after the gaussianization, using the same parameters P.

The situation dramatically changed when 10% of noise was added to the baker series and λ_1 estimations were repeated using the same parameters P as in the case of Figs. 1b,c. In this case the LLE algorithm failed to distinguish the series, it yielded random values irrelevant to the actual dynamical properties of the data (Fig. 1d).

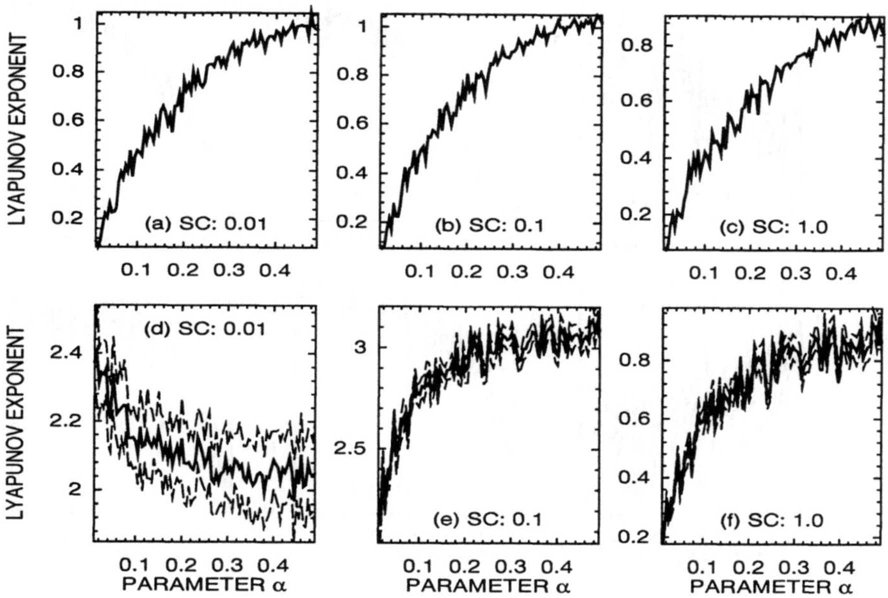

Fig. 2. Estimates of the positive Lyapunov exponent from noise-free baker series (a, b, c) and their surrogate data (d, e, f), plotted as functions of the parameter α. In plots d-e-f solid lines and dashed lines depict mean λ_1 and mean$\pm$SD, respectively, of 15 realizations of the surrogates for each value of α. The scales $s_{\min} = 0.01$SD (a, d), $s_{\min} = 0.1$SD (b, e), and $s_{\min} = 1.0$SD (c, f) were used. The parameters $m = 2$, $\tau = 2$, $T = 1$ were used in all estimations.

In the following we compare the λ_1 estimates obtained from the baker series with different portions of noise, using different scales $s_{\min}$ ($s_{\max} = 10 s_{\min}$). The values of the parameters m, τ, T are the same as above. The results for the noise-free data are presented in Fig. 2, the minimum scales are $s_{min} = 0.01$SD (Fig. 2a), 0.1SD (Fig. 2b) and 1.0SD (Fig. 2c). The largest Lyapunov exponents λ_1, estimated from the noise-free low-dimensional chaotic series, are stable with respect to different scales (cf. Figs. 2a and 2b), only in the case of the largest scales (Fig. 2c) the estimates have lower values and the curve $\lambda_1(\alpha)$ is partially distorted, but still able to classify the series in the relative sense. With 5% of noise in the data the classification is practically impossible for $s_{min} = 0.01$SD (Fig. 3a), possible, though with a higher error rate for $s_{min} = 0.1$SD (Fig. 3b), while for $s_{min} = 1.0$SD (Fig. 3c) the results are almost as good as for the noise-free data. For the data with 10% of noise

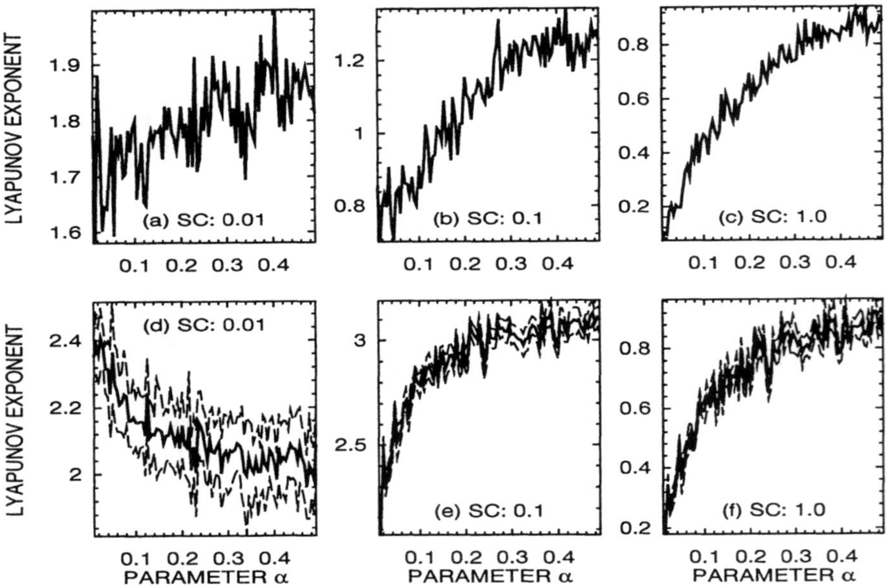

Fig. 3. Estimates of the positive Lyapunov exponent from noisy (5% of noise) baker series (a, b, c) and their surrogate data (d, e, f), plotted as functions of the parameter α. In plots d-e-f solid lines and dashed lines depict mean λ_1 and mean$\pm$SD, respectively, of 15 realizations of the surrogates for each value of α. The scales $s_{min} = 0.01$SD (a, d), $s_{min} = 0.1$SD (b, e), and $s_{min} = 1.0$SD (c, f) were used. The parameters $m = 2$, $\tau = 2$, $T = 1$ were used in all estimations.

(Fig. 4), the classification is impossible for both $s_{min} = 0.01$SD (Fig. 4a) and $s_{min} = 0.1$SD (Fig. 4b), while the classification ability of the algorithm is restored using $s_{min} = 1.0$SD (Fig. 4c). Using $s_{min} = 1.0$SD we obtained the same results as in Fig. 4c also for 30% of noise in the data. (Data with higher portions of noise were not tested.) Thus, the generally known advice that the scales, used in estimating the chaotic measures, should be above the noise level, seems to be valid. Considering, however, that the chaotic measures are

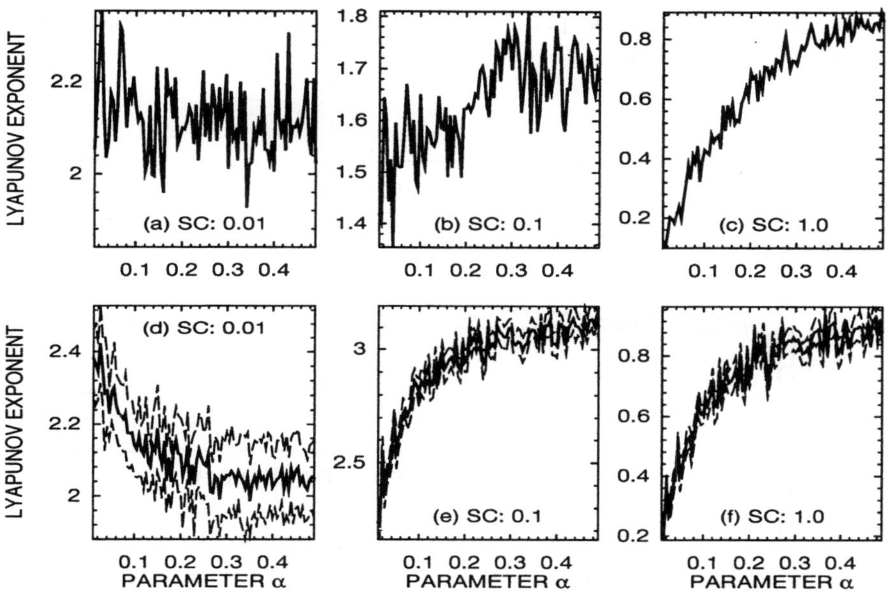

Fig. 4. Estimates of the positive Lyapunov exponent from noisy (10% of noise) baker series (a, b, c) and their surrogate data (d, e, f), plotted as functions of the parameter α. In plots d-e-f solid lines and dashed lines depict mean λ_1 and mean$\pm$SD, respectively, of 15 realizations of the surrogates for each value of α. The scales $s_{\min} = 0.01$SD (a, d), $s_{\min} = 0.1$SD (b, e), and $s_{\min} = 1.0$SD (c, f) were used. The parameters $m = 2$, $\tau = 2$, $T = 1$ were used in all estimations.

defined in terms of vanishing distances between points, one could doubt what is actually measured using the large, macroscopic scales. In this study, is it really the exponential divergence of nearby trajectories, which is reflected in the results in plots (b) and (c) of Figs. 2–4? Searching for an answer, the technique of surrogate data (Theiler et al. 1992; Paluš 1995) was used. The surrogate data to an "observed" series are, in this case, realizations of a Gaussian linear stochastic process with the same spectrum as the "observed" series.

For each time series analyzed above, a set of 15 realizations of the surrogates were constructed and the largest Lyapunov exponents λ_1 were estimated using the same parameters P as for the λ_1 of the relevant "observed" series.

The results from surrogates are presented in plots d, e, f of Figs. 2–4. Solid lines are used for mean λ_1, dashed lines depict mean$\pm$SD of λ_1 estimated from the set of 15 realizations of the surrogates.

Estimating LLE λ_1 from linear stochastic data one could ask whether such estimates converge. The positive answer is illustrated in Fig. 5, where the convergence of λ_1 estimates is presented for a noise-free baker series, generated with $\alpha = 0.3$ (upper panel in Fig. 5) and a realization of its surrogates (lower panel in Fig. 5).

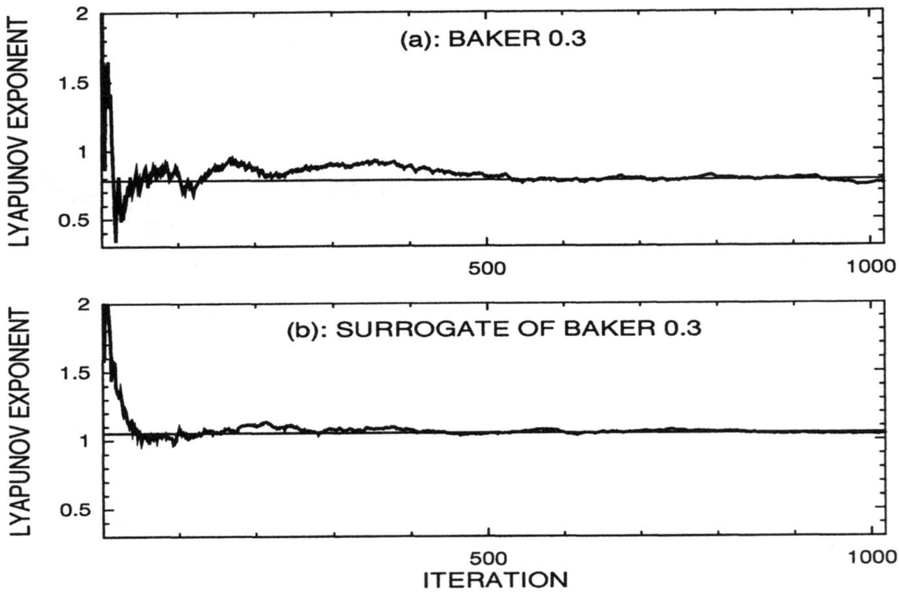

Fig. 5. Convergence of estimates of the positive Lyapunov exponent in the course of averaging along the trajectory ("iteration") for a baker series generated with $\alpha = 0.3$ (a) and a realization of its surrogate data (b). Estimation parameters: $s_{\min} = 0.1$SD, $m = 2$, $\tau = 2$, $T = 1$. The horizontal line presents the final estimate, defined as the average value of the last third of all iterations.

Exploring relatively small scales ($s_{\min} = 0.01$SD, plots d in Figs. 2–4), LLE's λ_1 estimated from the surrogates do not reflect the "chaoticity", i.e., the dependence $\lambda_1(\alpha)$ of the original data. Such a result could be expected as far as the chaotic dynamics and nonlinear properties of the original data were destroyed by phase randomization in the surrogates. Using larger scales $s_{\min} = 0.1$SD and 1.0SD (plots e and f in Figs. 2–4), however, a relative classification, similar to the ordering of the baker series according to their

λ_1, is again observed, though, in the surrogate data, there is no exponential divergence of trajectories, or even no trajectories in the deterministic sense! These time series are realizations of ***Gaussian linear stochastic*** processes, thus their dynamics are fully characterized by their power spectra or, equivalently, by their autocovariance functions. In this situation one can infer that the algorithm for the largest Lyapunov exponent distinguishes time series with different autocorrelation functions.

4 Discussion: From Chaotic to Stochastic Measures and Back

For understanding the results from the previous section we will briefly review relations between two kinds of dynamical measures of chaos – Lyapunov exponents and Kolmogorov–Sinai entropy (KSE), between KSE and mutual information (MI) and between MI and a standard linear statistical measure – the autocorrelation function.

Consider that a time series $x(t)$ started at time t_0 with an initial value $x(t_0)$. If the process underlying the series is not regular, but either stochastic, or chaotic and our knowledge about $x(t)$ is limited by finite precision measurement, after an evolution time $\tau = t_1 - t_0$ it is impossible to find an exact relation between $x(t_0)$ and $x(t_1)$. That is, knowing $x(t_0)$ one cannot exactly[4] predict $x(t_1)$, or, knowing $x(t_1)$ one cannot exactly compute backward the value of $x(t_0)$. In its evolution the underlying system is forgetting the information about its initial condition or, in other words, the system is creating new information, which has to be obtained by new measurement to know the state $x(t)$ of the system. The rate, how quickly the new information is created, is characterized by the *entropy rate* h of the system. When the underlying process is described by a dynamical system, the special case of the entropy rate h can be defined, known as the Kolmogorov–Sinai entropy (KSE). The famous theorem of Pesin (1977) says that the KSE h of a dynamical system is equal to the sum of its positive Lyapunov exponents. In the case of the baker map there is only one positive λ_1 and thus

$$h(\alpha) = \lambda_1(\alpha). \tag{8}$$

The mutual information $I(\tau) \equiv I[x(t); x(t+\tau)]$ (Shannon & Weaver 1964; Cover & Thomas 1991; Paluš 1995; Paluš 1996a; Pompe, this volume) quantifies the average amount of information about $x(t+\tau)$ that is contained in $x(t)$, and vice-versa. The rate of decrease of $I(\tau)$ with increasing τ should be related to the rate of the information creation of the underlying system,

[4] "Exactly" should be understood as "with precision comparable to the precision of measurement".

thus to the system's entropy rate h.[5] In the case of the baker system, where $h = \lambda_1$ increases with the parameter α according to Fig. 1a, the larger α is used for generating the series, the faster should be the decrease of the mutual information $I(\tau)$, estimated from that series. This behaviour is demonstrated

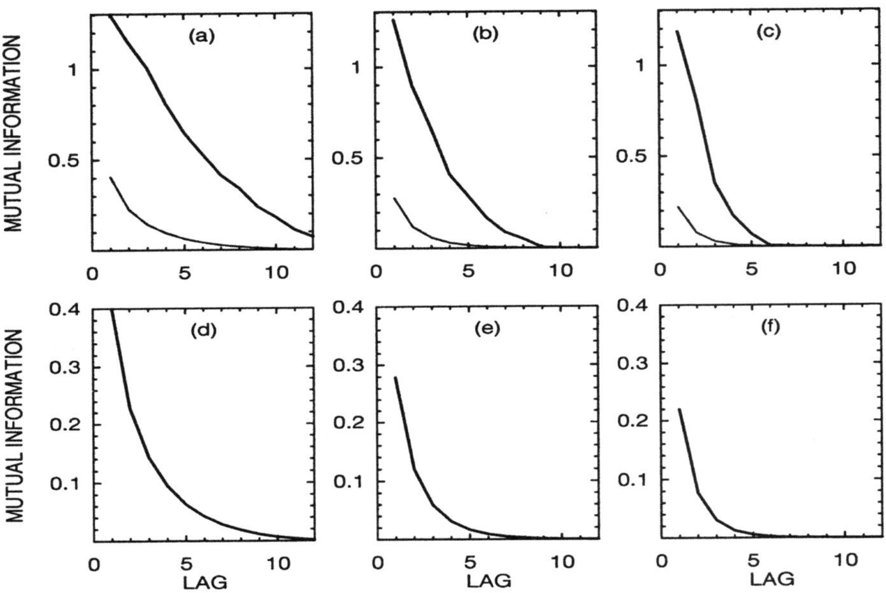

Fig. 6. Mutual information $I(\tau)$ as a function of the lag τ for baker series and their surrogates, generated with $\alpha = 0.1$ (a, d), $\alpha = 0.2$ (b, e), and $\alpha = 0.3$ (c, f). In the plots a-b-c $I(\tau)$ from the baker data is plotted using thick lines, thin lines are used for $I(\tau)$ from the surrogates, which are plotted once more in the plots d-e-f, using the appropriate scale.

in Fig. 6, where $I(\tau)$ is presented, extracted from the baker series generated using $\alpha = 0.1$ (Fig. 6a), $\alpha = 0.2$ (Fig. 6b), and $\alpha = 0.3$ (Fig. 6c). The chaoticity of different baker series is reflected in the character of the τ-dependence of the mutual information $I(\tau)$ estimated from the baker series (thick lines in Figs. 6a–c), but surprisingly also in $I(\tau)$ estimated from related surrogate data (thinner, lower curves in Figs. 6a–c illustrate mean values for the sets of 15 realizations of the surrogates). Thus the chaoticity of the baker series

[5] More details about the relation between higher-order mutual information (called marginal redundancy) and the entropy rates can be found in (Fraser 1989; Paluš 1993; Paluš 1996b) and references therein.

is not only encoded into their "nonlinear properties", characterized by their $I(\tau)$, but also reflected into their "linear properties", which are preserved in the surrogates. $I(\tau)$ from the surrogates are once more displayed using an appropriate scale in Figs. 6d,e,f.

The surrogates are realizations of Gaussian linear stochastic processes, thus their mutual information $I(\tau)$ can be expressed as a function of their autocorrelation functions $C(\tau)$ (Morgera 1985; Paluš 1995) as

$$I = -\frac{1}{2}\log(1 - C^2). \tag{9}$$

Then also the autocorrelation functions $C(\tau)$ (and spectra) of the baker series and their surrogates contain the information about the baker series' chaoticity (dependence $\lambda_1(\alpha)$). This explains why the Gaussian linear stochastic surrogates, related to different baker series, can be distinguished and ordered in the same way (in the relative sense) as the original baker series are classified according to their positive Lyapunov exponents. The question is, however, why this classification was possible to perform by using the Lyapunov exponent algorithm, designed to quantify the exponential divergence of nearby trajectories of chaotic systems.

The LLE algorithm explores changes of initial distances δ_I of pairs of points into final distances δ_F after an evolution time T. Consider a time series generated by white noise (independent identically distributed – IID process). For any initial distance δ_I, the final distance δ_F is a random number independent of δ_I. The averaged $\langle\delta_F\rangle$ is then equal to the overall average distance of the data points. The averaged initial distance $\langle\delta_I\rangle$ is influenced by the choice of the scales $s_{\text{min}}, s_{\text{max}}$. Then, choosing the scales so that $\langle\delta_I\rangle$ is smaller than $\langle\delta_F\rangle$, a positive estimate of λ_1 is obtained. When considered noise is not white but "coloured", i.e., there is some correlation $C(T)$ between $x(t)$ and $x(t+T)$, the increase of distance after the time T is smaller for series with stronger correlations, i.e., the larger $C(T)$, the smaller is the estimated λ_1, and vice versa.

Dämmig and Mitschke (1993) have derived analytic formulae for the λ_1 estimates when applying the considered LLE algorithm to white noise and a very special kind of coloured noise (white noise filtered by a "brickwall filter", the filter function is equal to one for a defined spectral bandwidth, and to zero otherwise). As one could expect, λ_1 estimated from white noise depends exclusively on the parameters P, in the case of the coloured noises λ_1 depends on P and on the spectral bandwidth. Thus for fixed P the estimated Lyapunov exponent λ_1 classifies the series according to their spectra, or, equivalently, according to their autocorrelation functions.

In the case studied here, where the coloured (autocorrelated) noises – the surrogate data – were generated according to given nontrivial spectra, derivation of an analytic formula is probably impossible, but the dependence of λ_1 on autocorrelations has been demonstrated in the presented numerical study.

Consider next the results in Figs. 2b and 2e, where λ_1 was estimated from the baker series and their surrogates using the same scale $s_{\min} = 0.1$SD. We can see that the stronger is the dependence between $x(t)$ and $x(t + T)$, the smaller is the estimate of λ_1. The differences between the strength of the dependence between $x(t)$ and $x(t + T)$, measured by $I(T)$, in the original baker series and their surrogates can be observed in Figs. 6a–c. The estimates of λ_1 for the baker series reach values between 0.1 and 1.1 (Fig. 2b), while in the case of surrogates they are between 2.1 and 3.2 (Fig. 2e). Above it has been found that in linear stochastic series the estimates of λ_1 are determined by the series' autocorrelation function, here it can be inferred that in general nonlinear series the estimates of λ_1 are determined by general (linear + nonlinear) temporal dependences in the series, which can be measured, e.g., by the mutual information $I(\tau)$.

Using the largest scales $s_{\min} = 1.0$SD, however, the λ_1 estimates obtained from the baker data and the surrogates are not significantly different (plots c and f in Figs. 2–4). In these scales dynamical properties of the series and the λ_1 estimates are dominated by linear properties of the series. Considering this result as a formal surrogate-based test for nonlinearity (Theiler et al. 1992; Paluš 1995), the Lyapunov exponent as a statistic fails to distinguish the nonlinear chaotic baker series from their isospectral linear stochastic surrogates. Using smaller scales ($s_{\min} = 0.1$SD, Figs. 2b and 2e), however, the differences between the baker series and their surrogates are statistically significant, though in both cases (the data and the surrogates), equivalent relative classifications of the series were observed.

The direct comparison of the values of λ_1 estimated from the baker series and from the surrogates was possible due to using the methodology of the surrogate-based tests for nonlinearity. The surrogate data had the same linear properties (spectra, autocorrelations) as the original baker series, and also the same marginal histograms, which also influence the estimates of chaotic and other measures. The latter was achieved by the "gaussianization" – a histogram transformation which transformed the marginal distributions of the baker series into the Gaussian distribution. The surrogates were Gaussian by the construction.

An equivalent approach is using the original baker data without transformation, but transforming the surrogates from Gaussian into the distribution of the original data. Using this approach a shift in scales was observed: For $s_{\min} = 1.0$SD, the estimates of λ_1 were negative and irrelevant to the actual chaoticity of the baker series (i.e., the average initial distance, given by $s_{\min}$, was already larger than the overall average distance). Then, the results of this approach for scales $s_{\min} = 0.1$SD and 0.01SD were equivalent to the results from the former approach using $s_{\min} = 1.0$SD and 0.1SD, respectively, while the scale $s_{min} = 0.001SD$ was the first "nonmacroscopic" scale, in which the surrogates were not classified according to the $\lambda_1(\alpha)$ scheme, i.e., this result is equivalent to the result from the scale $s_{min} = 0.01SD$ in the former approach. Using the first or the other approach, the differences in the results are of a technical level, but the main messages of this study, formulated in the conclusion (items 1 and 2), are not changed.

5 Conclusion: From Stochastic to Chaotic Measures and Back

The findings of this study can be summarized as follows.

1. The surrogates of the baker series can be relatively classified according to the $\lambda_1(\alpha)$ scheme (Fig. 1a). Because of the linear stochastic nature of these processes, this classification must be accessible using linear techniques and, consequently, the original chaotic baker series can be distinguished and ordered equivalently to the $\lambda_1(\alpha)$ ordering also by using linear statistical techniques. This result may hold also for other chaotic systems, but *not* generally for all nonlinear systems.

2. The classification of the linear stochastic surrogates was performed by using the algorithm for estimating the largest Lyapunov exponent. This finding may probably be generalized:[6] The chaotic measures may provide meaningful classification (relative characterization) even for linear stochastic data.

The relation between existence of chaos in a system underlying data and the ability of chaotic measures to classify different systems states is not straightforward. Linear techniques may be used successfully for some chaotic systems, while chaotic measures may give meaningful results for linear stochastic data. A successful application of a chaotic measure in relative characterization of system states does not necessarily imply chaos in the system.

About a decade ago the chaotic measures became frequently used in analysis of complex time series as an alternative to stochastic, mostly linear techniques. Deterministic chaos has been usually considered as an opposite alternative to random effects in attempts to explain complicated dynamics. Recent results indicate, however, that low-dimensional chaos may be rather

[6] Especially for those chaotic measures which explore distributions of distances between points like the correlation dimension (Grassberger & Procaccia 1983a; Grassberger & Procaccia 1983b). For example, in an EEG study it was observed that classification of EEG signals, obtained by using the correlation dimension, had been possible to reproduce by linear measures (Paluš et al. 1992).

a rare than ubiquitous phenomenon,[7] or the strict separation between deterministic-chaotic and stochastic dynamics may be impossible (Ellner & Turchin 1995). And even in data generated by a low-dimensional chaotic system, microscopic properties, which are characterized by the chaotic measures, may be unaccessible due to finite precision and measurement noise, as demonstrated in this study. A more comprehensive approach to study real-world systems is emerging, based on mathematical theory of nonlinear stochastic systems. This approach offers data analysis methods that explicitly consider randomness and have a firm basis in statistical theory.

The entropy rate (Cover & Thomas 1991; Paluš 1996b), i.e., the rate of information creation by a system, was the property which made possible to classify the above studied time series. The entropy rate can be defined for both chaotic and stochastic systems. Although the exact entropy rate of a continuous system may be unaccessible from data, there is always a possibility to estimate its "coarse-grained" versions, suitable for classification of system states. An example of such measures, applied to the same baker series, as considered here, is presented in (Paluš 1996b). A comprehensive review of "complexity" measures, related to entropies and entropy rates, can be found in (Wackerbauer et al. 1994).

Creation of information by a system, characterized by its entropy rate, may be caused either by intrinsic dynamical noise, or by a system's "chaoticity" - sensitivity to initial conditions; or by a combination of the two. Detection and characterization of the "stochastic chaos", i.e., of the initial-condition sensitivity of nonlinear stochastic systems, is a task of immense importance in nonlinear time-series analysis. The sources of positive entropy rates can be distinguished neither by the measures of entropy rate, nor by the chaotic measures such as the Lyapunov exponents. Specific techniques, designed for nonlinear stochastic systems, should be used, such as the conditional mean/variance or conditional probability approaches, advocated by Yao & Tong (1994). An interesting overview of nonlinear time series analysis from a chaos perspective has been recently published by Tong (1995).

Acknowledgements

The author was supported by the Grant Agency of the Czech Republic (grant No. 102/96/0183) and by the Ministry of Environment of the Czech Republic (grant No. VaV/520/2/97). Majority of numerical experiments related to this study were performed using computational facilities of the Santa Fe Institute, supported by the core funding from the John D. and Catherine T. MacArthur Foundation, the National Science Foundation (PHY-8714918), and the U.S. Department of Energy (ER-FG05-88ER25054).

[7] Especially when considering open, real-world systems, such as those studied in physiology and medicine. For recent results on nonlinearity/chaos in EEG see (Theiler & Rapp 1996; Paluš 1996c).

References

Cover T.M., Thomas J.A. (1991): *Elements of Information Theory.* Wiley, New York

Dämmig M., Mitschke F. (1993): Estimation of Lyapunov exponents from time series: the stochastic case. Phys. Lett. A **178** 385–394

Ellner S., Turchin P. (1995): Chaos in noisy world: new methods and evidence from time series analysis. The American Naturalist **145(3)** 343–375

Farmer J.D., Ott E., Yorke J.A. (1983): The dimension of chaotic attractors. Physica D **7** 153–180

Fraser A.M. (1989): Information and entropy in strange attractors. IEEE Transactions on Information Theory **35** 245–262

Grassberger P., Procaccia I. (1983a): On the characterization of strange attractors. Phys. Rev. Lett. **50** 346–349

Grassberger P., Procaccia I. (1983b): Measuring the strangeness of strange attractors. Physica D **9** 189–208

Hentschel H.G.E., Procaccia I. (1983): The infinite number of generalized dimensions of fractals and strange attractors. Physica D **8** 435–444

Hilborn R.C. (1994): *Chaos and Nonlinear Dynamics: An Introduction for Scientists and Engineers.* Oxford University Press, Oxford

Koukkou M., Lehmann D., Wackermann J., Dvořák I., Henggeler B. (1993): Dimensional complexity of EEG brain mechanisms in untreated schizophrenia. Biol. Psychiatry **33** 397–407

Mayer-Kress G. (Ed.) (1986): *Dimensions and Entropies in Chaotic Systems* Springer, Berlin, Heidelberg

Mayer-Kress G., Layne S.P. (1987): Dimensionality of the human electroencephalogram. In Mandell A.S., Koslow S. (Eds.): Perspectives in Biological Dynamics and Theoretical Medicine. Ann. N.Y. Acad. Sci. **504** 62–87

Morgera S.D. (1985): Information theoretic covariance complexity and its relation to pattern recognition. IEEE Trans. on Systems, Man, and Cybernetics **SMC-15, 5** 608–619

Layne S.P. , Mayer-Kress G., Holzfuss J. (1986): Problems associated with dimensional analysis of electroencephalogram data. In Mayer-Kress G. (Ed.): *Dimensions and Entropies in Chaotic Systems.* Springer, Berlin, Heidelberg. p. 246.

Paluš M., Dvořák I., David I. (1992): Spatio-temporal dynamics of human EEG. Physica A **185** 433–438

Paluš M., Albrecht V., Dvořák I. (1993): Information-theoretic test for nonlinearity in time series. Phys. Lett. A **175** 203–209

Paluš M. (1993): Identifying and quantifying chaos by using information-theoretic functionals. In Weigend A.S., Gershenfeld N.A. (Eds.): *Time Series Prediction: Forecasting the Future and Understanding the Past.* Santa Fe Institute Studies in the Sciences of Complexity, Proc. Vol. XV. Addison-Wesley, Reading, MA. pp. 387–413

Paluš M. (1995): Testing for nonlinearity using redundancies: Quantitative and qualitative aspects. Physica D **80** 186–205

Paluš M. (1996a): Detecting nonlinearity in multivariate time series. Phys. Lett. A **213** 138–147

Paluš M. (1996b): Coarse-grained entropy rates for characterization of complex time series, Physica D **93** 64–77

Paluš M. (1996c): Nonlinearity in Normal Human EEG: Cycles, temporal asymmetry, nonstationarity and randomness, not chaos. Biol. Cybern. **75** 389–396

Pesin Ya.B. (1977): Characteristic Lyapunov exponents and smooth ergodic theory. Russian Math. Surveys **32** (1977) 55–114

Press W.H., Flannery B.P., Teukolsky S.A., Vetterling W.T. (1986): *Numerical Recipes: The Art of Scientific Computing.* Cambridge University Press, Cambridge

Schuster H.G. (1988): *Deterministic Chaos, An Introduction.* Physik, Weinheim

Shannon C.E., Weaver W. (1964): *The Mathematical Theory of Communication.* University of Illinois Press, Urbana, IL

Takens F. (1981): Detecting strange attractors in turbulence. In Rand D.A., Young L.-S. (Eds.): *Dynamical systems and turbulence.* Springer Lecture Notes in Mathematics Vol. 898. Springer, Berlin, Heidelberg. pp. 366–381

Theiler J., Eubank S., Longtin A., Galdrikian B., Farmer J.D. (1992): Testing for nonlinearity in time series: the method of surrogate data. Physica D **58** 77–94

Theiler J., Rapp P.E. (1996): Re-examination of the evidence for low-dimensional nonlinear structure in the human electroencephalogram. EEG Clin. Neurophysiol. **98** 213–222

Tong H. (1995): A personal overview of nonlinear time series analysis from a chaos perspective, Scandinavian J. Statistics **22(4)** 399–445

Wackerbauer R., Witt A., Atmanspacher H., Kurths J., Scheingraber H. (1994): A comparative classification of complexity measures. Chaos, Solitons and Fractals **4(1)** 133–173

Wackermann J., Lehmann D., Dvořák I., Michel C.M. (1993): Global dimensional complexity of multi-channel EEG indicates change of human brain functional state after a single dose of a nootropic drug. EEG Clin. Neurophysiol. **86** 193–198

Weigend A.S., Gershenfeld N.A. (Eds.) (1993): *Time Series Prediction: Forecasting the Future and Understanding the Past.* Santa Fe Institute Studies in the Sciences of Complexity, Proc. Vol. XV. Addison-Wesley, Reading, MA.

Wolf A., Swift J.B., Swinney H.L., Vastano J.A. (1985): Determining Lyapunov exponents from a time series. Physica D **16** 285–317

Yao Q., Tong H. (1994): On prediction and chaos in stochastic systems. Philos. Trans. Roy. Soc. (Lond.) **A 348** 357–369

Ranking and Entropy Estimation in Nonlinear Time Series Analysis

Bernd Pompe

E.-M.-Arndt-Universität Greifswald, Institut für Physik, F.-L.-Jahn-Straße 16, D-17487 Greifswald, Germany

Abstract. This chapter is concerned with two subjects. The first one is a method of signal preprocessing called *ranking*. It is of special relevance in nonlinear time series analysis and may cause several computational advantages. The second subject is the definition and estimation of a *generalized mutual information* which is useful to analyse statistical dependences in scalar or multivariate time series. A fast algorithm for its estimation is described in detail which essentially profits from ranking of the scalar components of the time series.

1 Ranking

1.1 Definition

Consider a scalar time series

$$\{x_t\}_{t=1}^{T} \tag{1}$$

originating from an experiment. *Ranking* denotes the transformation

$$x_t \longrightarrow R_t = \#\{t^* \ : \ x_{t^*} \leq x_t, \ t^* = 1, \ldots, T\}, \quad r_t \equiv R_t/T \ . \tag{2}$$

R_t is the rank of x_t within the original series (1), and r_t is refered to as its *relative rank.* The smallest value of the series is mapped to $r = 1/T$ and the largest to $r = T/T = 1$. The transformation is unambiguously invertible iff all entries of (1) are pairwise different, which is supposed first of all.[1] Then the new series, the series of rank numbers

$$\{R_t\}_{t=1}^{T} \ , \tag{3}$$

is nothing but a permutation of

$$\{1, 2, \ldots, T\} \ .$$

Hence the relative ranks $\{r_t\}$ are *uniformly* distributed in the unit interval, and we can consider ranking as the transformation

$$F : x_t \rightarrow r_t \ ,$$

[1] The case where some entries of the original time series are equal is refered to as *tied ranks.* They will be considered later in this chapter and in Appendix 3.

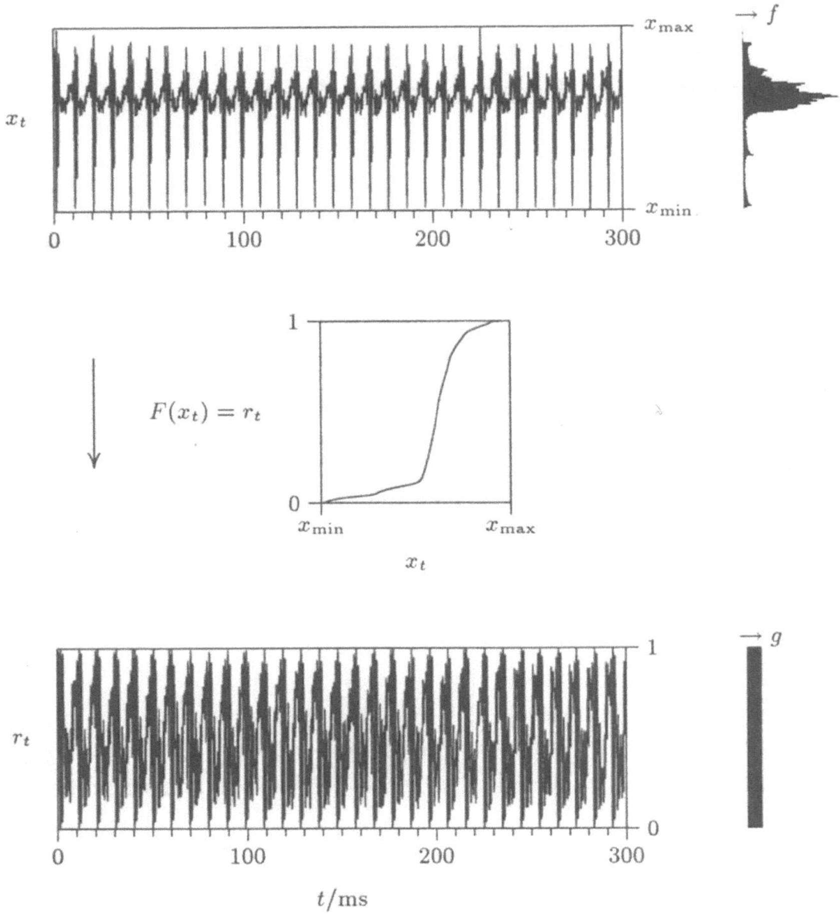

Fig. 1. A speech signal $\{x_t\}$ (top) and its ranked version $\{r_t\}$ (bottom) with the one-dimensional probability density functions f and g, respectively, and the cumulative distribution function F of f

where F is the (empirical) one-dimensional distribution function of the original series (1). In general, this transformation is nonlinear.[2] Figure 1 shows an example. Some consequences of ranking are discussed in the following subsections.

[2] F is linear iff the data $\{x_t\}$ are already uniformly spread over an interval $[0, x_{max}]$.

1.2 Invariance

Suppose we calculate any quantity as a function of the (relative) ranks $\{r_t\}$ instead of the original samples $\{x_t\}$. Then this quantity is invariant with respect to any monotonous increasing transformation $H : x_t \rightarrow y_t$. This is because ranking of $\{x_t\}$ and that of $\{y_t\}$ yield the same series $\{r_t\}$. If F is the distribution function of $\{x_t\}$ then that of $\{y_t\}$ is given by $F \circ H^{-1} : y_t \rightarrow r_t$. The situation is revealed by the diagram

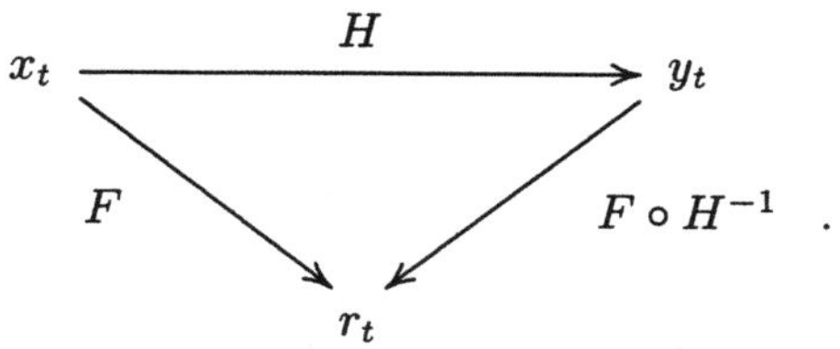

Some practical consequences of this are as follows:

1. Assuming we want to analyse a signal $\{x_t\}$ which cannot be measured directly, or for which the cost of a direct measurement are too high. Then, alternatively, we could try to measure a signal $\{y_t\}$ which is related to $\{x_t\}$ by a transformation H. It does not matter if the explicit expression of H is unknown – if we have reason to assume that an increase of x causes an increase of y, then we get exactly the same series of rank numbers from $\{y_t\}$ as we would have got from the unknown $\{x_t\}$.
2. In general, a signal $\{x_t\}$ is nonlinear distorted along the signal path of the measuring equipment, starting from the sensor and going via several amplifiers to the analog-digital converter. In most cases the experimenter tries to obtain a *linear* signal path. However, linear signal transmission is an ideal construction which can be reached only approximately in practice, often with expensive measuring equipments and time consuming calibration. Ranking can mitigate this problem and hence diminish the measuring cost.

In nonlinear time series analysis we are often interested in estimates of dynamical invariants like Kolmogorov-Sinai entropy, Hausdorff dimension and Lyapunov exponents of supposed underlying ergodic dynamical systems (e.g., Eckmann & Ruelle 1985). Due to ranking these invariants are not changed, where in the case of Lyapunov exponents we have to suppose that F is smooth enough.

1.3 Change of Spectra

The Fourier spectrum of the original signal $\{x_t\}$ may differ drastically from that of its ranked version $\{r_t\}$. Figure 2 reveals this for the signal of Fig. 1. Above 6 kHz the spectral density of $\{x_t\}$ vanishes which is in contrast to that

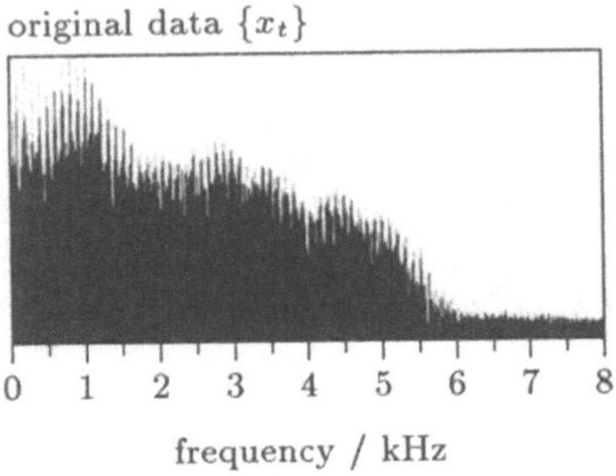

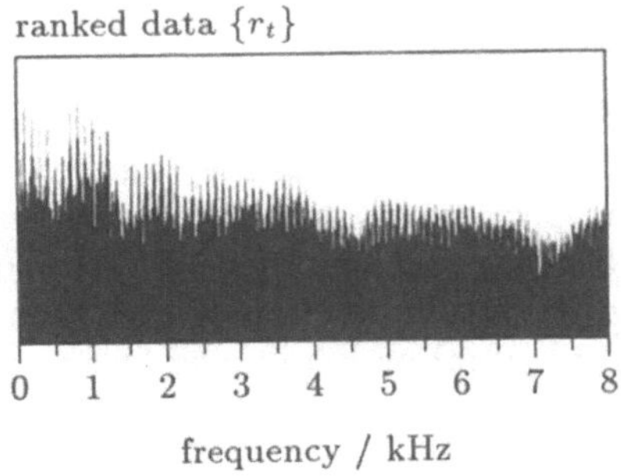

Fig. 2. Power spectra of the signals in Fig. 1

of $\{r_t\}$. This is due to the nonlinearity of the function F. (Nevertheless, in our example some characteristic frequencies of $\{x_t\}$ are preserved in $\{r_t\}$.)

Consider another example, the time-continuous signal $x(t) = \sin 2\pi t/T_\mathrm{p}$. Its distribution function is $F(x) = (2/\pi)\arcsin\sqrt{x}$. In this case the transformation $x(t) \to r(t) \equiv F\big(x(t)\big)$ yields the sawtooth wave which is not band-limited though the original $x(t)$ is. The general consequence is as follows: Suppose $\{x_t = x(t \times T_\mathrm{s})\}_{t=\ldots,-1,0,1,2,\ldots}$ is the time-discrete version of the time-continuous $\{x(t)\}_{t\in\mathbb{R}}$ fulfilling the sampling condition $2T_\mathrm{s} < T_\mathrm{p}$, where $2/T_\mathrm{p}$ is the Nyquist rate of the signal. Then $x(t)$ is completely determined by its samples $\{x_t\}$, using standard interpolation formulas (e.g., Roberts & Mullis 1987). On the other hand, from the ranked samples $\{r_t\}$ we cannot get from direct application of these standard interpolation formulas the original $r(t)$. Now we need, in addition to $\{r_t\}$, the distribution function F,

$$\{r_t\} \xrightarrow{F^{-1}} \{x_t\} \xrightarrow{\qquad\text{standard interpolation}\qquad} x(t) \xrightarrow{F} r(t) \ .$$

In our example the difficulties come from the fact that the slope $\mathrm{d}F(x)/\mathrm{d}x$ goes to infinity for $x \to 0$ and $x \to 1$. However, in practice we have noisy data. Hence the underlying distribution function can be considered to be "bona fide" such that if $x(t)$ is band-limited then $r(t)$ is band-limited as well. Therefore we could obtain $r(t)$ from the samples $\{r_t\}$ directly from standard interpolation formulas, without knowledge of F. However, to do so we must suppose that $x(t)$ was sampled at a sufficiently high rate. In practice this means that we should oversample the signal $x(t)$ or, if necessary, interpolate $\{x_t\}$ *before* ranking.[3]

[3] Of course, if the signal originates from a time-discrete system the consideration of interpolation is irrelevant.

1.4 Some More Consequences of Ranking

The top line of Fig. 3 shows scatter plots of a time series. From such plots we often try to estimate two-dimensional distribution functions in order to investigate statistical dependencies between x_t and $x_{t+\tau}$. In the most simple case we can partition the plane by little bins to get a histogram on them. This is done in line (b) of the figure for 100 × 100 nonoverlapping quadratic bins of equal size. In the lower line a kernel method is used (e.g., Silverman 1986).

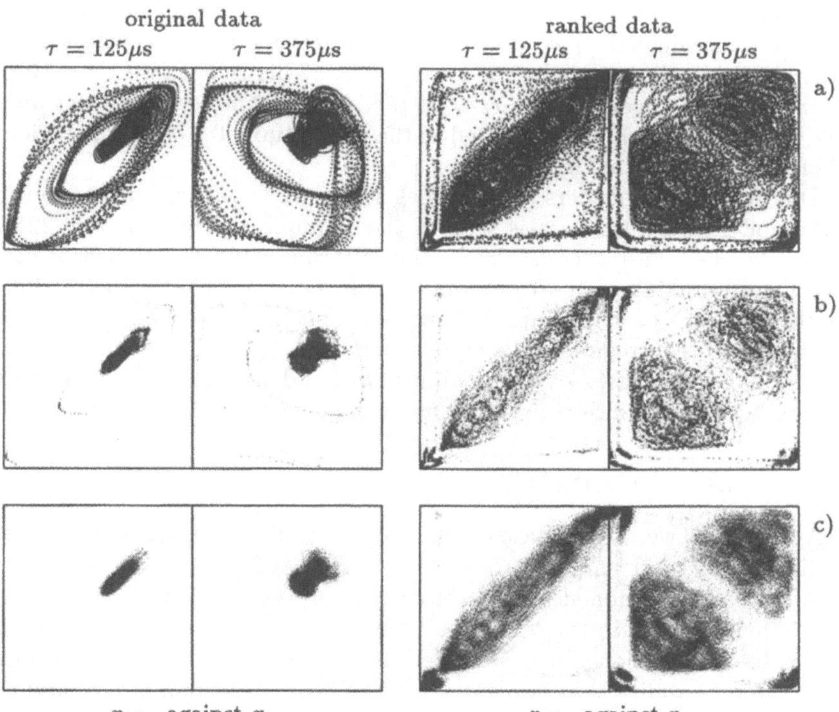

Fig. 3. Delay representations of the time series of Fig. 1: (a) scatter plots; (b) corresponding probability densities functions encoded with gray levels, and using a histogram estimator; (c) the same as (b) but using a Gaussian kernel estimator

A more sophisticated procedure for density estimation is to use a nonuniform partition of the plane such that in regions where probability was dense the bins are smaller than in regions where it was sparse. However, this may lead to rather unhandy algorithms (Fraser & Swinney 1986). Essentially the same is reached much simpler by working with ranked data and a uniform partition.

The drawings reveal that ranking acts like a magnifier of regions visited more probably by the orbit of the scatter plot. This is because on the one hand F spreads out probability where it was dense and on the over hand it crowds probability where it was sparse. But we should be aware that in this way the relative noise level of a noisy signal is changed by the factor

$$\frac{\mathrm{d}F(x)}{\mathrm{d}x} \times (x_{\max} - x_{\min}) = f(x)(x_{\max} - x_{\min}) \ ,$$

exceeding one in regions where probability was dense.

From scatter plots $r_{t+\tau}$ against r_t (or better from plots of the corresponding probability densities) we can immediately get an impression whether there are statistical relations between x_t and $x_{t+\tau}$. This is because there are *no dependencies* between x_t and $x_{t+\tau}$ of the original time series *if and only if* $\{(r_t, r_{t+\tau})\}$ is uniformly distributed in the unit square $[0,1] \times [0,1]$. Obviously this is not the case in Fig. 3.

Due to ranking outliers of the times series are transformed to somewhat more moderate values as it is seen in the example of Fig. 1 at $t \approx 225\,\mathrm{ms}$. Of course, classical methods like median filtering could be applied, if necessary, before ranking to treat outliers (e.g., Rabiner & Schafer 1978). But we should be aware that median filtering implies some smoothing which might be undesireable in regions with no outliers.

Ranking can cause several computational advantages. This will be explained later in the more specific context of some entropy estimation.

1.5 Algorithm for Ranking

In Appendix 3 we give a Pascal code for ranking. The main part of it is a modified "quicksort" making the procedure rather fast.

As already mentioned above, ranking according to (2) is unambiguously invertible iff all data in the original time series are pairwise different. This is a reasonable assumption when the underlying random variables are continuous. Then the probability that two entries of the series are equal is zero and thus so-called *tied ranks* (e.g., Cox & Hinkley 1994) wont occure. However, in practice ties arise quite often. If the time series was recorded, e.g., with an 8-bit analog-digital converter, there are necessarily equal values in the data record if its length T exceeds $2^8 = 256$. In Appendix 3 the problem of tied ranks is discussed in more detail. Here we only note that the Pascal code actually yields the desired uniform distribution because it somewhat arbitrarily discriminates between originally equal values in the series $\{x_t\}$. However, the consequence of this is that the ranked signal $\{r_t\}$ should be considered with a noise level

$$\mathrm{noise}_{\mathrm{rank}} \lesssim \frac{l_{\max}}{T} \ ,$$

where $l_{\max}$ is the maximum number of equal points in $\{x_t\}_{t=1}^T$. Hence, ranking should be applied only for "nearly continuous" time series, such that the data vary in a set of, let's say, 100 different values.

1.6 Ranking Ignores Scales

At a first glance, a conventional experimenter might have some aversion to ranking – usually it is a rather vehement distortion of the signal. However, we should be aware that, in general, measuring means that we map (encode) some objective events by numbers. In any case, this map (random variable) is defined somewhat arbitrarily concerning the scales – nature does not know numbers. This becomes obvious when we consider the reaction of a complex nonlinear system, e.g. in audiology, to the input of any signal. Here we often observe that the same small input deviation causes deviations in the system output of very different magnitude, depending on the operating point of the system. The consequence is that we often use sophisticated *nonlinear* scales revealing the nonlinearity of the system response. (The definition of the linear scale comes from any laboratory of standardization which cannot have in mind our applications.) If we work with ranked time series, the scale becomes irrelevant – here we only take into account that the magnitude (strength, intensity, etc.) of a quantity is larger than another. This is the approach also in order statistics (e.g., Lindgren 1993; Cox & Hinkley 1994).

2 Generalized Mutual Information Function (GMIF)

The estimation of entropies has a long tradition in the analysis of time series originating from chaotic dynamical systems. Instructive overviews are given by Eckmann & Ruelle (1985), and Grassberger et al. (1991). The information theoretical concept of entropies is involved in several dynamical invariants like Kolmogorov-Sinai entropy (metric entropy), Hausdorff dimension (information dimension) and Lyapunov exponents. The importance of these quantities in ergodic theory is striking. However, there are enormous difficulties to estimate them from experimental data. It is desirable to have meaningful quantities which could be estimated more easily. Recently, we have done some attempts in this direction (Bandt & Pompe 1993; Pompe 1993; Pompe & Heilfort 1994; Pompe 1995). Here we first shortly report on the concept of the GMIF describing, on a certain level of coarse graining, statistical dependencies between several time series. Our approach has computational advantages which are first of all due to the two facts:

1. Instead of the original time series we work with its ranked version.
2. Instead of the usual Shannon entropy we work with the second-order Rényi entropy.

Progress in the computational efficiency and robustness of the algorithms is necessary if we want to promote the nice but somewhat academic concepts of nonlinear time series analysis in practice. The GMIF concept is a step in this direction.

2.1 Definition of the GMIF

Consider a multivariate time series

$$\{\mathbf{x}_t\}, \quad \mathbf{x}_t \in \mathbb{R}^{D+1} \ .$$

The components $x_t^{(D)}, \ldots, x_t^{(1)}, x_t^{(0)}$ of $\mathbf{x}_t$ may represent different quantities or they are time shifted versions of an originally scalar series. In the first case *cross*-dependencies between different time series are investigated, and in the second case *auto*-dependencies within one time series are considered. Any mixed version of some delay coordinates of different time series is also possible. We introduce time shifts

$$\boldsymbol{\vartheta} \equiv (\vartheta_D, \ldots, \vartheta_1) \text{ and } \tau \ ,$$

where $\boldsymbol{\vartheta}$ is refered to as the *time comb*, and we write

$$\mathbf{x}_{t,\boldsymbol{\vartheta},\tau} \equiv \left(x_{t-\vartheta_D}^{(D)}, \ldots, x_{t-\vartheta_1}^{(1)}, x_{t+\tau}^{(0)} \right) \ .$$

Usually we let the time comb fixed and ask for statistical dependencies between $\mathbf{x}_{t,\boldsymbol{\vartheta}} \equiv \left(x_{t-\vartheta_D}^{(D)}, \ldots, x_{t-\vartheta_1}^{(1)} \right)$ and $x_{t+\tau}^{(0)}$ for running time lag τ.

There are several quantitities measuring dependencies. One is the mutual information which is defined in our case as follows: Let

$$\mathcal{S} \equiv \{ s_{m_D \ldots m_1 n} \} \tag{4}$$

be the joint probabilities describing the distribution of the time series

$$\{\mathbf{x}_{t,\boldsymbol{\vartheta},\tau}\}_{t=1}^{T} \tag{5}$$

on the level of coarse graining given by the relative quantization levels

$$\boldsymbol{\varepsilon} \equiv (\varepsilon_D, \ldots, \varepsilon_1, \varepsilon_0) \ .$$

Here we assume that the range of each scalar component of the time series is divided into ε_d^{-1}, $d = 0, 1, \ldots, D$, bins of equal size, where ε_d^{-1} is an integer. Then

$$\mathcal{P} \equiv \left\{ p_{m_D \ldots m_1} \equiv \sum_n s_{m_D \ldots m_1 n} \right\}_{m_D \ldots m_1}$$

and

$$\mathcal{Q} \equiv \left\{ q_n \equiv \sum_{m_D \ldots m_1} s_{m_D \ldots m_1 n} \right\}_n$$

are the D- and 1-dimensional marginal distributions assigned to $\mathbf{x}_{t,\vartheta}$ and $x^{(0)}_{t+\tau}$, respectively. Finally, consider the quantity

$$I_\alpha(\mathcal{S}) \equiv H_\alpha(\mathcal{Q}) + H_\alpha(\mathcal{P}) - H_\alpha(\mathcal{S}) \ . \tag{6}$$

H_α denotes the Rényi entropy of order α which is defined for any discrete distibution $\mathcal{P} = \{p_m\}$ as

$$H_\alpha(\mathcal{P}) \equiv \begin{cases} \frac{1}{1-\alpha} \log_2 \sum_m p_m^\alpha & : \ \alpha \geq 0, \ \alpha \neq 1 \\ -\sum_m p_m \log_2 p_m & : \ \alpha = 1 \ . \end{cases} \tag{7}$$

(We have the conventions $0^0 \equiv 0$ and $0 \log_2 0 \equiv 0$.) $H_1(\mathcal{P})$ represents Shannons information, and $H_0(\mathcal{P}) = \log_2 \#\{p_m \neq 0\}$ is Hartleys entropy.

For $\alpha = 1$ we obtain the well-known *mutual information* (MI) (e.g., Rényi 1970) fulfilling the relations

$$0 \leq I_1(\mathcal{S}) \leq \min\{H_1(\mathcal{P}), H_1(\mathcal{Q})\} \ . \tag{8}$$

Assign the random variables $\boldsymbol{\xi}$ and η to $\mathcal{P}$ and $\mathcal{Q}$, respectively. Then we have the following properties:

$$\begin{array}{lcl} I_1(\mathcal{S}) = 0 & \Longleftrightarrow & \boldsymbol{\xi} \text{ and } \eta \text{ are independent,} \\ I_1(\mathcal{S}) = H_1(\mathcal{Q}) & \Longleftrightarrow & \eta \text{ is a function of } \boldsymbol{\xi}, \\ I_1(\mathcal{S}) = H_1(\mathcal{P}) & \Longleftrightarrow & \boldsymbol{\xi} \text{ is a function of } \eta. \end{array} \tag{9}$$

$I_1(\mathcal{S})$ gives the (mean) information we gain on η from the knowledge of $\boldsymbol{\xi}$ and vice versa.

For any distribution $\mathcal{S}$, the quantity $I_\alpha(\mathcal{S})$ satisfies the relation

$$0 \leq I_\alpha(\mathcal{S}) \quad \text{if and only if } \alpha = 0 \text{ or } 1 \tag{10}$$

(e.g., Rényi 1970; Aczél & Daróczy 1975). Hence we cannot get a well-behaved generalized mutual information I_α on the base a generalized entropies H_α, especially not for $\alpha = 2$, by trivial analogy to Shannons case $\alpha = 1$. However, if $\mathcal{Q}$ is a uniform distribution then we also have

$$0 \leq I_2(\mathcal{S}) \leq \min\{H_2(\mathcal{P}), H_2(\mathcal{Q})\} \ ,$$

with the same nice properties (9) of $I_1(\mathcal{S})$. A proof is given in Appendix 3. Thus we are motivated to call $I_2(\mathcal{S})$ *generalized mutual information* (GMI), provided $\mathcal{Q}$ is a uniform distribution.[4]

[4] There could be considered some other criteria a quantity should fulfil to justify the expression "mutual information", coming from coding theory. There is a coding theorem for the Rény entropy H_α, $\alpha > 0$, $\alpha \neq 1$, showing that $H_\alpha(\mathcal{P})$ can be considered as the infimum of the mean length of an exponential coding cost

If we let the time comb $\boldsymbol{\vartheta}$ and the coarse graining levels $\boldsymbol{\varepsilon}$ fixed and vary only the time lag τ we write for clarity

$$I_{2,\boldsymbol{\varepsilon},\boldsymbol{\vartheta}}(\tau) \equiv I_2(\mathcal{S})$$

and call it *generalized mutual information function* (GMIF). In the one-dimensional case we set $\boldsymbol{\vartheta} = \vartheta_1 = 0$ and, for simplicity, omit the paramter $\boldsymbol{\vartheta}$, i.e., $I_{2,\boldsymbol{\varepsilon},\vartheta_1}(\tau) \equiv I_{2,\boldsymbol{\varepsilon}}(\tau)$. Of course, using Shannons information measure we can also consider a mutual information function $I_{1,\boldsymbol{\varepsilon},\boldsymbol{\vartheta}}(\tau)$.

Both mutual informations I_1 and I_2 can be considered as a nonlinear analogon to the well-known correlation function. A relation between correlation and GMI is given in Appendix 3 for a somewhat modified definition of the GMI, working for continuous random variables. From this we learn that the GMI should be compared with the squared correlation ϱ^2 instead of the correlation itself. It reflects the fact that MI do not distinguish between positive and negative correlation.

Table 1 summarizes properties of some "mutual informations" I_α, $\alpha = 0, 1, 2$, and the squared correlation coefficient. (Here the correlation is defined only for the scalar case $D = 1$.) Note that for the uniform distribution I_1 and I_2 have the same properties. Moreover, "uncorrelation" says that there are no linear statistical dependencies. Hence all remaining dependencies detected by I_1 or I_2 are nonlinear. However, a direct comparison of the magnitudes of any I_α with that of the squared correlation ϱ^2 is not possible.

2.2 Estimation of the GMIF

The main problem in estimating "mutual informations" $I_\alpha(\tau)$ is to get estimates for the joint probabilities (4) for varying time lag τ. The direct way is by histograms for which some results are shown in the first line of Fig. 4. Somewhat more sophisticated are kernel methods for density estimation as they were already used in Fig. 3. Some results for a Gaussian kernel with (relative) standard deviation $\varepsilon = 64^{-1}$ in each direction are shown in the second line of Fig. 4. Note that I_1 and I_2 rather well reflect the pitch period of the speech signal which is about 10 ms, and also a formant period of about 2×0.5 ms is clearly visible.[5] The Gaussian kernel smoothes the densities and

function, where the infimum has to be taken over all sets of code word lengths of uniquely decipherable codes (Campbell 1966). However, for our purpose its enough to demand that $I_\alpha(\mathcal{S})$ should clearly distinguish the deterministic case from that of independence (total randomness) revealed by (8) and (9). Nevertheless, to simplify the way of speaking the quantity $I_\alpha(\mathcal{S})$ is refered to as "mutual information", but with quotation marks if it might not have properties which are similar to (8) and (9).

[5] The factor 2 is necessary because negative and positive correlation are not distinguished in our appoach. More information concerning speech signals and its characteristics are given in the chapter by Herzel et al. in this volume.

Table 1. Comparison of properties of some "mutual informations" I_α between two random variables $\boldsymbol{\xi}$ and η and squared correlation ϱ^2 of ξ (scalar) and η

property			I_0	I_1	I_2	ϱ^2
η arbitrarily distributed:						
quantity ≥ 0			+	+	−	+
(*quantity* $= 0$)	$\Longrightarrow$	independence	−	+	−	−
(*quantity* $= 0$)	$\Longrightarrow$	uncorrelation	−	+	−	+
independence	$\Longrightarrow$	(*quantity* $= 0$)	−	+	+	+
uncorrelation	$\Longrightarrow$	(*quantity* $= 0$)	−	−	−	+
η uniformly distributed:						
quantity ≥ 0			+	+	+	+
(*quantity* $= 0$)	$\Longrightarrow$	independence	−	+	+	−
(*quantity* $= 0$)	$\Longrightarrow$	uncorrelation	−	+	+	+
independence	$\Longrightarrow$	(*quantity* $= 0$)	+	+	+	+
uncorrelation	$\Longrightarrow$	(*quantity* $= 0$)	−	−	−	+
			+ : true,		− : false	

Column header: *quantity*

thus also the "mutual information functions". This becomes most obvious for time lags τ around half the pitch ($\tau = 3 - 7$ ms). It means that higher frequencies are masked due to the kernel smoothing which is more obvious for the ranked signal because it has a greater bandwidth (see Fig. 2). The "mutual information" I_0 (thin lines) much more sensitively follows to kernel smoothing because the underlying Hartley entropy regards only the support of the probability density. (Of course, smoothing is also caused by the use of larger coarse graining levels ε.)

For comparison, the lower line shows squared correlations. There are regions on the τ axis where we have almost no correlation and hence all dependencies indicated by positive values of I_1 (left column of the figure) and I_1 or I_2 (right column) are purely nonlinear.

Finally, the single picture in the third line of Fig. 4 shows again I_2, but now obtained from a preferable fast estimation algorithm working in two steps:

1. Step: Transform each (scalar) component of the original series (5) to the series of rank numbers according to (2). Thus obtain

$$\left\{\left(R^{(D)}_{t-\vartheta_D}, \ldots, R^{(1)}_{t-\vartheta_1}, R^{(0)}_{t+\tau}\right)\right\}_{t=1}^{T}.$$

2. Step: Determine

$$C^{(D+1)}_{\boldsymbol{\epsilon},\epsilon_0,\boldsymbol{\vartheta}}(\tau) \equiv \#\left\{(t_1,t_2) \ : \ \left|R^{(d)}_{t_1+\vartheta_d} - R^{(d)}_{t_2+\vartheta_d}\right| < \epsilon_d,\ d = D,\ldots,1,0\right\}$$

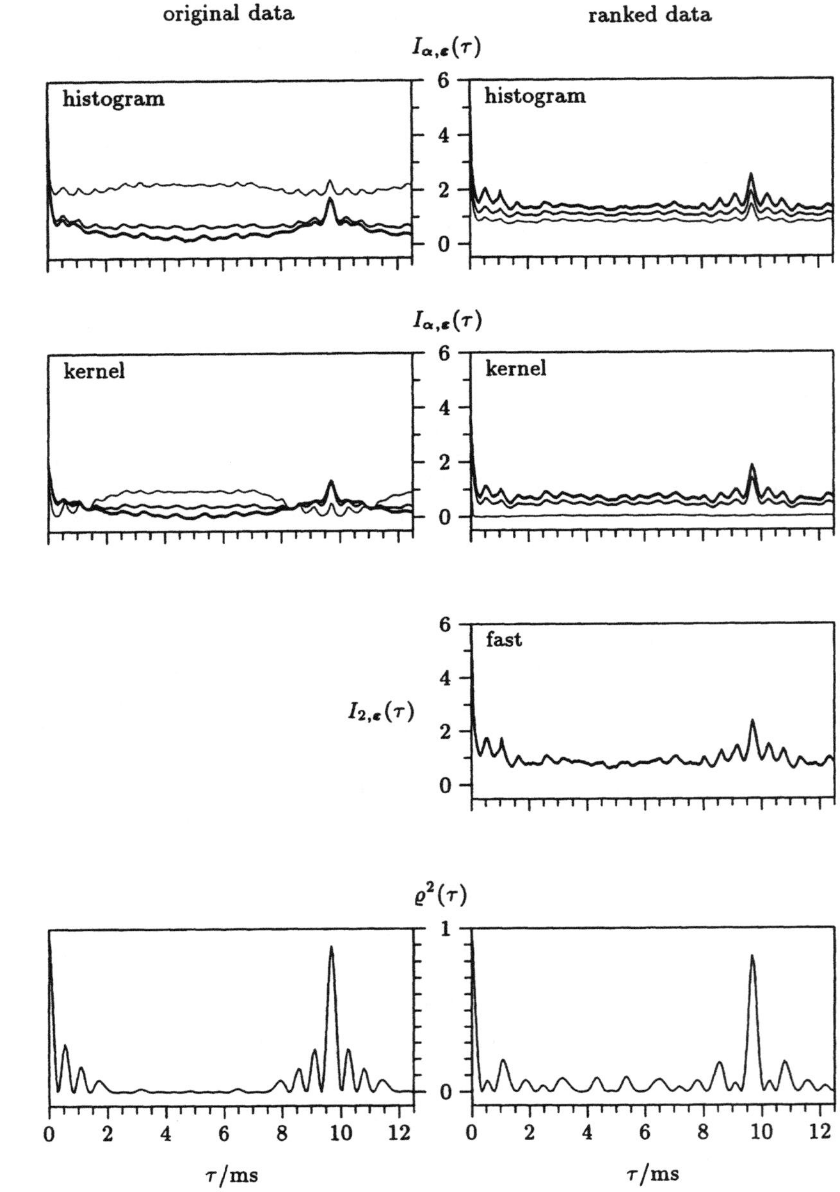

Fig. 4. For the signal of Fig. 1: Comparision of "mutual informations" for different entropy orders ($\alpha = 0, 1, 2$ from thin to thick lines) and several estimation methods; parameters: coarse graining levels $\boldsymbol{\varepsilon} = (64^{-1}, 64^{-1})$, time comb $\boldsymbol{\vartheta} = \vartheta_1 = 0$. The pictures in the lower line represent the squared correlation ϱ^2

where $t_1 = 1, 2, \ldots, T-1$ and $t_2 = t_1 + 1, \ldots, T$. Similarily get $C^{(D)}_{\epsilon,\vartheta}$, with the only difference that $d = D, \ldots, 1$ such that only D instead of $D+1$ tests are performed. Finally calculate

$$C^{(1)}_{\epsilon_0} \equiv \frac{2(\epsilon_0 - 1)}{T-1} - \frac{\epsilon_0(\epsilon_0 - 1)}{T(T-1)} \; .$$

Then

$$\widehat{I}_{2,\varepsilon,\vartheta}(\tau) \equiv -\log_2 C^{(1)}_{\epsilon_0} - \log_2 C^{(D)}_{\epsilon,\vartheta} + \log_2 C^{(D+1)}_{\epsilon,\varepsilon_0,\vartheta}(\tau) \tag{11}$$

is an estimator of the generalized mutual information $I_{2,\varepsilon,\vartheta}(\tau)$ on the relative coarse graining levels

$$\varepsilon = (\varepsilon_D, \ldots, \varepsilon_1, \varepsilon_0) = \frac{2}{T}(\epsilon_D, \ldots, \epsilon_1, \epsilon_0) \; .$$

The algorithm works for ergodic time series and $1 \ll \epsilon_d \lll T$. The expression $C^{(1)}_{\epsilon_0}$ contains finite sample corrections. We have

$$\lim_{\substack{T \to \infty \\ \varepsilon_0 = \text{const.} \ll 1}} C^{(1)}_{\epsilon_0} = \varepsilon_0(1 - \varepsilon_0/4) \simeq \varepsilon_0 \; .$$

The procedure goes back to ideas of Grassberger & Procaccia (1983), and Takens (1983). They considered the so-called correlation integral $C^{(D)}$ in the context of estimation of fractal dimension and metric entropy h_α for ergodic dynamical systems. If the multivariate time series (5) is generated by delay reconstruction from a scalar time series $\{x_t\}$ of a dynamical systems then we would set $\vartheta_D > \ldots > \vartheta_1 = 0$ and $0 \leq \tau$. Hence we have $C^{(D+1)}_{\epsilon,\vartheta}(0) = C^{(D)}_{\epsilon,\vartheta}$ and for the decay of the mutual information we get

$$\begin{aligned} I_{2,\varepsilon,\vartheta}(0) - I_{2,\varepsilon,\vartheta}(\tau) &\approx \log_2 C^{(D+1)}_{\epsilon,\vartheta}(0) - \log_2 C^{(D+1)}_{\epsilon,\vartheta}(\tau) \\ &\longrightarrow h_2/\tau \\ &\quad \text{for } D \to \infty \text{ and } \max\{\varepsilon_d\} \to 0 \; . \end{aligned}$$

In practice τ has to be choosen sufficiently small to detect the decay and the time comb and coarse graining levels must be sophisticated selected. (More numerical hints are given, e.g., by Grassberger et al. (1991) and by the references in there.) The sum over all positive Lyapunov exponents of a dynamical system is an upper bound of h_1. Indeed, we sometimes find that the initial decay of the generalized mutual information resembles properties of h_1 or of the largest Lyapunov exponent. An example is given by Herzel et al. in this volume.

On large scales of the time lag τ the mutual information reflects "folding properties" of the underlying phase flow. This is in contrast to Lyapunov exponents coming from a linear stability theory. However, our point is that the concept of mutual information works equally well for deterministic (chaotic) and noisy data. Some steps in this direction were done also by other authors

(Brock 1991; Scheinkman & LeBaron 1989; Green & Savit 1991; Savit & Green 1991).

A fast code for the GMIF is given in Appendix 3. Indeed, much faster implementations are possible, giving for reasonable parameters reliable estimates of the GMIF in much less than a second using a Pentium/90 MHz processor. Its noteworthy that for fixed coarse graining levels ε and data length T the algorithm works faster if the dimension D of the multivariate time series increases. However, this is at the expense of the reliability of the estimates. Moreover, it should be noted that the kernel routines of the algorithm need no multiplication, which is in contrast, e.g., to the fast Fourier or cosine transform.

2.3 To What Questions Could the GMIF Provide Answers?

Detecting Fundamental Periods. Suppose we have a seasonal scalar signal $\{x_t\}$ with the mean fundamental period $\overline{T_p}$. Voiced speech signals like that in Fig. 1 represent physilogical examples. The variation of T_p is 2 - 3% around the mean $\overline{T_p} \approx 9.7$ ms. Then there are relatively strong statistical dependencies between x_t and $x_{t+\tau}$ for any τ which is an integer multiple of $\overline{T_p}$. This long range dependencies decay at a rate of about 0.103 bit/ms (0.3 bit/$\overline{T_p}$) (see Fig. 5).

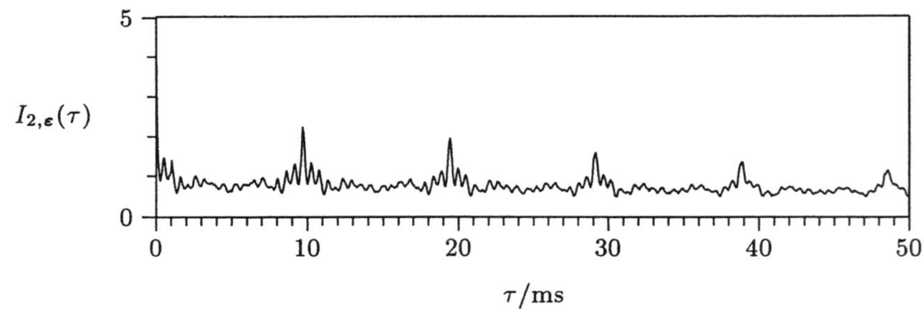

Fig. 5. Long term decay of the generalized mutual information function of the signal of Fig.1 ($\varepsilon_1 = \varepsilon_0 = 2^{-5}$)

Detecting Optimal Time Combs for Forecasting and Modelling. Assuming we want to predict the future $x_{t+\tau}$ from known past values $(x_{t-\vartheta_D}, \ldots, x_{t-\vartheta_1})$. Then the questions are: what time comb $(\vartheta_D, \ldots, \vartheta_1)$ yields a maximum of information on $x_{t+\tau}$, and how many past values (this is the dimension D of the time comb) yield already "almost all" information on it? Of course, for modelling D should be as small as possible. For this optimal time comb we should, in principle, get the "best" predictions, on the considered level of coarse graining.

Ideed, for the signal of Fig. 1 we find that, for a given embedding dimension D, some time combs are better than others (see Fig. 6). To find an optimal time comb we let the time lag $\tau > 0$ fixed and vary the components of $\boldsymbol{\vartheta}$ in the D-dimensional cube $[0, \vartheta_{\max}] \times \ldots \times [0, \vartheta_{\max}]$ where $\vartheta_{\max}$ is reasonably large (for our speech signal $\vartheta_{\max}$ should exceed the pitch period $\overline{T_p}$). Then we get an optimal time comb which is, in general, not uniformly

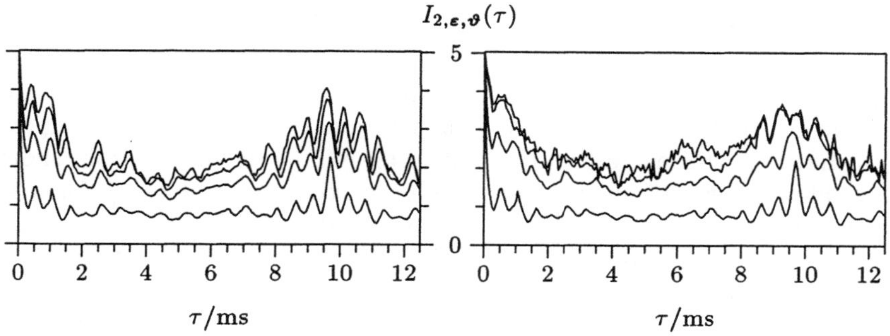

Fig. 6. For the signal of Fig. 1: Higher order generalized mutual informations indicating an increase of information on the future $x_{t+\tau}$ with increase of knowledge on the past $(x_{t-\vartheta_D}, \ldots, x_{t-\vartheta_1})$. Parameter: coarse graining levels $\varepsilon_4 = \ldots = \varepsilon_0 = 2^{-5}$; time combs $\boldsymbol{\vartheta} = (\vartheta_D, \ldots, \vartheta_1)$ from lower to upper lines in the left picture – (0) ms, (0.125, 0) ms, (0.250, 0.125, 0) ms, (0.375, 0.250, 0.125, 0) ms, and in the right picture – (0) ms, (0.25, 0) ms, (0.50, 0.25, 0) ms, (2.50, 0.50, 0.25, 0) ms

spaced. On the right hand side of Fig. 6 the GMIF for optimal time combs up to embedding dimension $D = 4$ are presented. Obviously they provide a little bit more information on the future than the corresponding nonoptimal time combs on the left hand side of Fig. 6. Indeed, another past value $x_{t-\vartheta_5}$, in addition to the optimal time comb for $D = 4$, cannot significantly improve the predictability, whatever we take for ϑ_5. Hence $D = 4$ can be considered as the "degree of freedom" of the signal. With other words: modelling of the signal as a Markov chain of order 4 with the corresponding optimal time comb should be possible. This is most obvious in the figure for small time lags ($\tau \lesssim 0.25$ ms), because our optimization was done for $\tau = 0.125$ ms. The initial decay of the GMIF for $D = 4$ ($\tau = 0 \ldots 0.7$ ms) is nearly linear which is typical for chaotic signals. The rate is given by ≈ 4.8 bit/ms which is much faster than the longterm decay revealed in Fig. 5. Supposing an underlying chaotic dynamical system, this rate could be considered as an estimate of the metric entropy h_2. Indeed, we were able to detect with a very different method (Wolf et al. 1985) for the largest Lyapunov exponent the value ≈ 3.9 bit/ms, which is in rather good agreement.

Consider conditional probability density functions describing the probability to find $x_{t+\tau}$ in an interval (1-dimensional bin) under the condition that

$(x_{t-\vartheta_D}, \ldots, x_{t-\vartheta_1})$ is in a given D-dimensional bin. Typically, for optimal time combs these densities become unimodal and nearly normaly distributed for sufficiently high embedding dimension D. Their mean values should provide "good" predictions.

The numerical detection of optimal time combs may be rather time consuming, but the procedure profits from the fast algorithm in Appendix 3. In the context of dynamical systems we expect that the optimal time comb provides "good" embedding delays for the reconstruction of higher-dimensional orbits from a scalar time series.

Miograms for Sliding Window Analysis of Nonstationary Data. To apply the GMIF analysis technique to nonstationary time series it is important to have an efficient estimator, providing reliable estimates from short data windows. Figure 7 shows some results for the signal of Fig. 1 for a sliding window data analysis. When we encode the GMIF with gray levels and plot it against the starting time of the data window we get the so-called *miogram* which is in analogy to the spectrogram or correlogram. In the figure we used a window length of 50 ms and a window shift of about 3 ms. Obviously the dark horizontal stripes rather well reflect the shape of the GMIFs in Fig. 4. Some more interesting examples of miograms are given in some other chapters of this volume (by Herzel et al. and Hoyer et al.).

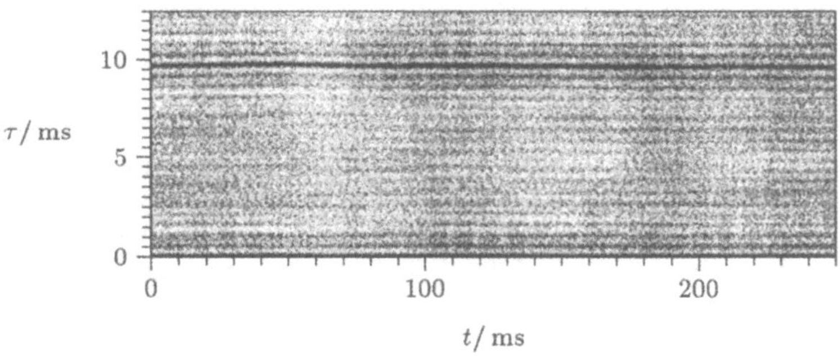

Fig. 7. Miogram of the signal in Fig. 1

Testing Stationarity. Stationarity of the considered time series must be presumed for the GMIF approach as well as for many other procedures in time series analysis. Often stationarity is more a question of belief than of knowledge. Indeed, in practice stationarity cannot be proved rigorously. However, we could ask whether on a certain level of coarse graining the assumption of stationary fails. This can be done using the GMIF in the following way: Take the (multivariate) time series $\mathbf{x}_{t,\boldsymbol{\vartheta}} \equiv \left(x^{(D)}_{t-\vartheta_D}, \ldots, x^{(1)}_{t-\vartheta_1}\right)$, $t = 1, 2, \ldots, T$, for which we want to test stationarity. Create another scalar i.i.d. (independent identically distributed) time series $x^{(0)}_{t+\tau}$ originating, e.g., from a "good" pseudo-random generator. Finally, get $\widehat{I}_{2,\varepsilon,\vartheta}(\tau)$ from the proposed fast GMIF algorithm for the $(D+1)$-dimensional time series $\left\{\left(\mathbf{x}_{t,\vartheta}, x^{(0)}_{t+\tau}\right)\right\}_t$. Obviously, for any τ the mutual information should be zero. However, the estimator $\widehat{I}_{2,\varepsilon,\vartheta}(\tau)$ is a random variable attaining for finite sample size T values varying around zero. For increasing T both, the expected value and the variance of the estimator should approach zero. For the speech signal in Fig. 1 this is revealed in the top three pictures of Fig. 8 for a fixed time segment of duration 64 ms

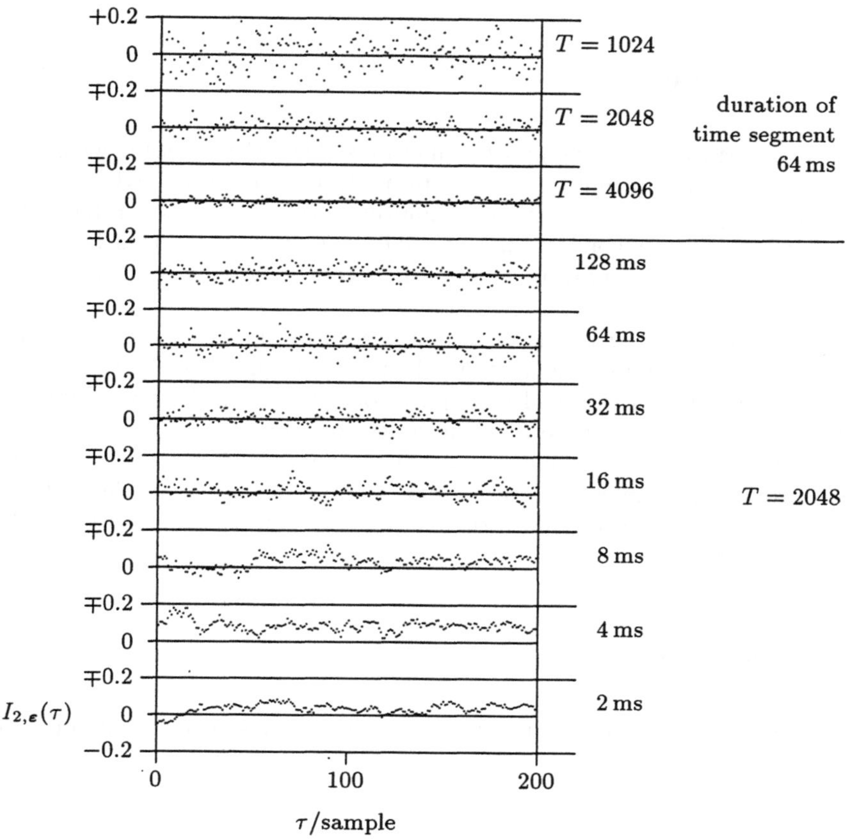

Fig. 8. Mutual information between signal of Fig. 1 and an i.i.d. test signal for different values of the sample size T and fixed time segment (top series of pictures) and vice versa (bottom series of pictures); coarse graining levels $\varepsilon_1 = \varepsilon_0 = 2^{-5}$

and dimension $D = 1$. For $T = 4096$ the variance of the estimator is less than

0.05 bit which is less than 1% of the maximal possible value $-\log_2 \varepsilon_0 = 5$ bit. The variance of the estimator is found to depend only on T but not on the duration of time segment of the speech signal.

If we let, on the other hand, the number of samples T fixed and decrease the duration of the time segment (this is done by interpolation of the speech signal) we get for durations $\lesssim 32$ ms significance for a strictly positive bias of the estimator. The only reason for this can be that on this time scales and level of coarse graining the signal cannot be considered to be stationary. A similar investigation could be performed for dimension $D > 1$. Hence we can detect instationarity from a positive bias of the GMIF estimator. (A more precise investigation of this problem is beyond the scope of this chapter.)

3 Conclusions

We have described in this chapter two methods of signal processing which are of general relevance in multivariate data analysis. With the GMIF method very effectively statistical relations in multivariate time series can be investigated which goes beyond traditional linear analysis techniques based on correlation or Fourier analysis. Our method can be equally well applied to stochastic and chaotic data.

Our investigations were essentially motivated by the existence of fast and robust algorithms. This is a necessity in practical signal analysis, often it essentially determines the field of applications. We have the hope that the reader gets its own experience with the codes given in the appendices. Indeed much faster codes are possible. Time must show in how far the new techniques provide new knowledge on reality. Some applications in this volume are rather encouraging to us (see the chapters by Boker et al.; Herzel et al.; and Hoyer et al.).

Acknowledgment

This work was supported by the Deutsche Forschungsgemeinschaft project Po442/2-1.

Appendix A: Miscellanea for Ranking

A.1 A Code for Ranking

The following Pascal code yields the series of rank numbers $\{R_t\}_{t=1}^{T}$. We prefer absolute ranks (integers) having computational advantages.

```
TYPE T_series     = ARRAY[1..5000] OF REAL;
     T_index_rank = ARRAY[1..5000] OF INTEGER; (* .. max length time series *)

PROCEDURE QuicksortIndex(CONST series : T_series;
                         CONST l,r    : INTEGER;
                         VAR   index  : T_index_rank);
  VAR i, j, d : INTEGER;
      m       : REAL;
  BEGIN
    i := l;
    j := r;
    IF j > i THEN BEGIN
      m := series[ index[(l + r) shr 1] ];
      REPEAT
        WHILE series[ index[i] ] < m DO  i := i+1;
        WHILE series[ index[j] ] > m DO  j := j-1;
        IF i <= j THEN BEGIN
           d := index[i];         (* exchange indices *)
           index[i] := index[j];
           index[j] := d;
           i := i+1;
           j := j-1;
        END;
      UNTIL i > j;
      QuicksortIndex(series, l, j, index);
      QuicksortIndex(series, i, r, index);
    END; (* IF j > i THEN .. *)
  END; (* QuicksortIndex *)

PROCEDURE Ranking(CONST series : T_series;
                  CONST length : INTEGER;
                  VAR   rank   : T_index_rank);
  VAR index : T_index_rank;
      t     : INTEGER;
  BEGIN
    FOR t := 1 TO length DO index[t] := t;
    QuicksortIndex(series, 1, length, index);
    FOR t := 1 TO length DO rank[index[t]] := t;
  END; (* Ranking *)
```

A.2 The problem of Tied Ranks

Suppose we have a time series $\{x_t\}_{t=1}^T$ attaining values in $\{0, 1, \dots, K-1\}$. Let $l(k)$ denote the number of points of the series which are equal to $k = 0, 1, 2, \dots, K-1$, $l(k) = \#\{x_t = k\}$. Then ranking according to (2) yields a series of rank numbers R_t attaining values in $\left\{l(0), l(0)+l(1), \dots, \sum_{k=0}^{K-1} l(k)\right\}$, with the distribution given by $l(k)/T$. In general, this is not the desired uniform distribution in $\{1, 2, \dots, T\}$, which follows only if $l(k) = 1$ or 0 for each k. In order to get a uniform distribution, we could proceed in the following way: Let $x_{t_1}, x_{t_2}, \dots x_{t_{l(k)}}$ denote all points in the series $\{x_t\}$ which are equal to k. Distinguish between these values by setting

$$x_{t_u} \longrightarrow k + a_u \equiv y_{t_u} \quad \text{for } u = 1, \dots, l(k) \ ,$$

where $0 \le a_1 < a_2 < \dots < a_{l(k)} < 1$. Thus we get from the original series $\{x_t\}_{t=1}^T$ another series $\{y_t\}_{t=1}^T$, where all values y_t are different among one another. Summarizing, via

$$\{x_t\}_{t=1}^T \longrightarrow \{y_t\}_{t=1}^T \longrightarrow \{R_t\}_{t=1}^T$$

the series of rank numbers is uniformly distributed in $\{1, 2, \ldots, T\}$. Indeed, any other transformation

$$x_{t_u} \longrightarrow k + \pi(a_u) \equiv y_{t_u},$$

where π is a permutation of $a_1, a_2, \ldots, a_{l(k)}$, leads to a different series of rank numbers. Exactly $\prod_{k=0}^{K-1} l(k)!$ such sequences are possible. In the above Pascal code any permutation π is used which need not to be specified for our purposes.

Appendix B: Miscellanea for the GMI

B.1 A Theorem

Let $\mathcal{P} = \{p_m\}_{m=1}^M$, $\mathcal{Q} = \{q_n\}_{n=1}^N$, and $\mathcal{S} = \{s_{mn}\}_{m,n=1}^{M,N}$ denote complete probability distributions of the discrete random variables ξ, η, and of the random vector (ξ, η), respectively. Suppose that η is uniformly distributed, *i.e., $q_n = \varepsilon = N^{-1}$ for all $n = 1, 2, \ldots, N$. Then*

$$I_2(\mathcal{S}) \equiv H_2(\mathcal{Q}) + H_2(\mathcal{P}) - H_2(\mathcal{S}) \ , \tag{12}$$

where H_2 is defined by (7), satisfies

$$0 \leq I_2(\mathcal{S}) \leq \min\{H_2(\mathcal{P}),\, H_2(\mathcal{Q})\} \ . \tag{13}$$

We have

$$\begin{aligned} I_2(\mathcal{S}) = 0 &\iff \xi \text{ and } \eta \text{ are independent,} \\ I_2(\mathcal{S}) = H_2(\mathcal{Q}) &\iff \eta \text{ is a function of } \xi, \\ I_2(\mathcal{S}) = H_2(\mathcal{P}) &\iff \xi \text{ is a function of } \eta. \end{aligned}$$

Proof: Under the assumptions of the theorem we get

$$\begin{aligned} I_2(\mathcal{S}) &= -\log\,\varepsilon - \log \sum_m p_m^2 + \log \sum_{m,n} s_{mn}^2 \\ &= \log \frac{\sum_{m,n} s_{mn}^2}{\varepsilon \sum_m p_m^2} \\ &= \log \left(1 + \frac{\sum_{m,n} (s_{mn} - p_m \varepsilon)^2}{\varepsilon \sum_m p_m^2} \right) \ . \end{aligned}$$

The argument of the logarithm cannot be less than 1, and $I_2(\mathcal{S}) = 0$ iff $s_{mn} = p_m \varepsilon$ for all $m = 1, 2, \ldots, M$ and $n = 1, 2, \ldots, N$, which means that ξ and η are independent. Moreover, we have

$$\sum_m p_m^2 = \sum_m \left(\sum_n s_{mn} \right)^2 \geq \sum_{mn} s_{mn}^2 \ ,$$

and hence

$$H_2(\mathcal{S}) - H_2(\mathcal{P}) = \sum_m p_m^2 - \sum_{mn} s_{mn}^2 \geq 0 \ .$$

The equality on the right hand side holds iff, for every $m \in \{1, 2, \ldots, M\}$, $s_{mn} = p_m$ for exactly one $n(m) \in \{1, 2, \ldots, N\}$ and $s_{mn} = 0$ else. This means that η is a function of ξ. From (3) we conclude that $I_2(\mathcal{S}) = H_2(\mathcal{Q})$ in this case, but in general

$$I_2(\mathcal{S}) \leq H_2(\mathcal{Q}) = \log N \tag{14}$$

holds. In the same way we get

$$H_2(\mathcal{S}) - H_2(\mathcal{Q}) = \sum_n q_n^2 - \sum_{mn} s_{mn}^2 \geq 0 \ .$$

Hence also $I_2(\mathcal{S}) \leq H_2(\mathcal{P})$, where the equality holds iff ξ is a function of η. Together with the inequality (14) the relation on the right hand side of (13) follows.

B.2 GMI in the Continuous Case

Let $(\boldsymbol{\xi}, \eta) \equiv (\xi_D, \ldots, \xi_d, \ldots, \xi_1, \eta)$ be a $(D+1)$-dimensional random vector where each ξ_d, η, $\boldsymbol{\xi}$, and $(\boldsymbol{\xi}, \eta)$ are described by absolutely continuous measures with densities $p_d(x_d)$, $q(y)$, $p(\mathbf{x})$, and $s(\mathbf{x}, y)$, respectively. We set $\mathbf{x} \equiv (x_D, \ldots, x_d, \ldots, x_1)$ and define the GMI

$$I_2(\boldsymbol{\xi}, \eta) \equiv \log \int\limits_{\mathbb{R}^{D+1}} \frac{s^2(\mathbf{x}, y)}{p_1(x_D) \ldots p_D(x_1) q(y)} \mathrm{d}\mathbf{x}\mathrm{d}y$$

$$- \log \int\limits_{\mathbb{R}^D} \frac{p^2(\mathbf{x})}{p_1(x_D) \ldots p_D(x_1)} \mathrm{d}\mathbf{x} \ ,$$

provided the integrals exist.[6] It has the following properties:

1. $I_2(\boldsymbol{\xi}, \eta) \geq 0$,
2. $I_2(\boldsymbol{\xi}, \eta) = 0 \iff \boldsymbol{\xi}$ and η are independent,
3. For any monotone functions $u_d : \xi_d \to u_d(\xi_d)$, $d = 1, \ldots, D$, and $v : \eta \to v(\eta)$ the GMI is invariant,

$$I_2(\boldsymbol{\xi}, \eta) = I_2\big(u(\boldsymbol{\xi}), v(\eta)\big) \ ,$$

where $u(\boldsymbol{\xi}) \equiv \big(u_1(\xi_1), \ldots, u_D(\xi_D)\big)$.

[6] The definition supposes measures which are absolutely continuous with respect to the Lebesgue measure such that the probability densities exist. This is a natural assumption for real world (noisy) processes. However, for nonlinear dynamical systems generating strange attractors this assumption is invalid – they produce singular natural measures. However, the GMI can be redefined here in terms of Radon-Nikodym derivates (Pompe 1997).

A proof is easily given by first transforming the integrals according to $x_d \to F_d(x_d)$, $y \to G(y)$, where F_d and G are the distribution functions of ξ_d and η, respectively. The discrete GMI of Appendix 3 is obtained by considering a discrete approximation of the transformed integrals on the $(D+1)$- and D-dimensional unit cube, where at least the coarse graining level ε_0 of the y-component must be uniform. For $D = 1$ we get

$$I_2(\xi,\eta) = \log\left(1 + \varphi^2(\xi,\eta)\right) \quad ,$$

with the quadratic contingency

$$\varphi^2(\xi,\eta) \equiv \int\limits_{\mathbb{R}^2} \frac{s^2(x,y)}{p(x)q(y)} \mathrm{d}x\mathrm{d}y - 1 \quad .$$

There are relations for the contingency (e.g., Rényi 1970) which carry to the GMI. For instance, for the correlation coefficient

$$\varrho(\xi,\eta) \equiv \frac{\int\int (\xi - \overline{\xi})(\eta - \overline{\eta}) s(x,y) \mathrm{d}x\mathrm{d}y}{\sqrt{\int (\xi - \overline{\xi})^2 p(x) \mathrm{d}x} \sqrt{\int (\eta - \overline{\eta})^2 q(y) \mathrm{d}y}} \quad ,$$

where $\overline{\xi}$ denotes the expected value of ξ, we get

$$I_2(\xi,\eta) \geq \log\left[1 + \varrho^2\left(u(\xi), v(\eta)\right)\right]$$

for any Borel-measureable functions u and v.

B.3 Fast Code for GMIF Estimation

The following code is for the estimation of the GMIF in the case of (3+1)-dimensional time series. It uses the code of Appendix 3. The case of higher or lower dimensions could be easily derived from it. The following table provides the relation between the notations in Sect. 2 and that in the code:

$x_t^{(0)}$	:	`series0[t]`	
$x_{t-\vartheta_d}^{(d)}$	:	`series`d`[t]`	$d = 1, 2, 3$
ϵ_d	:	`Eps`d	$d = 1, 2, 3$
T	:	`length`	
range of τ	:	`tau_min..tau_max`	
$I_{2,\epsilon,\vartheta}(\tau)$	:	`gmif[tau]`	

```
TYPE T_count = ARRAY[-200..200] OF LONGINT;  (* LONGINT = 4 Byte INTEGER *)
                                             (* INTEGER = 2 Byte INTEGER *)
     T_gmif  = ARRAY[-200..200] OF REAL;     (* min .. max time lag tau  *)

PROCEDURE CountPairs_Cross3(CONST rank0, index1, rank2, rank3 : T_index_rank;
                            CONST Eps0, Eps1, Eps2, Eps3      : INTEGER;
                            CONST length                      : INTEGER;
                            CONST tau_min, tau_max            : INTEGER;
                            VAR   C123                        : LONGINT;
                            VAR   C0123                       : T_count);
```

```
  VAR tau          : INTEGER;
      t1, t2       : LONGINT;
      i, j         : LONGINT;
  BEGIN
    C123 := 0;
    FOR tau := tau_min TO tau_max DO C0123[tau] := 0;
    FOR i := 1 TO length - 1 DO BEGIN
      t1 := index1[i];
      IF ( (t1 <= length - tau_max) AND (t1 > -tau_min) ) THEN BEGIN
        j  := i + 1;
        WHILE ( (j < i + Eps1) AND (j <= length) ) DO BEGIN
          t2 := index1[j];
          IF ( (t2 <= length - tau_max) AND (t2 > -tau_min) ) THEN
            IF ( ABS( rank2[t1] - rank2[t2] ) < Eps2 ) THEN
              IF ( ABS( rank3[t1] - rank3[t2] ) < Eps3 ) THEN BEGIN
                Inc(C123);
                FOR tau := tau_min TO tau_max DO
                  IF ( ABS( rank0[t1 + tau] - rank0[t2 + tau] ) < Eps0 )
                    THEN Inc(C0123[tau]);
              END;
          Inc(j);
        END; (* WHILE *)
      END; (* IF *)
    END; (* FOR i *)
  END; (* CountPairs_Cross3 *)

PROCEDURE GMIF3(CONST series0, series1, series2, series3 : T_series;
                CONST Eps0, Eps1, Eps2, Eps3             : INTEGER;
                CONST length, tau_min, tau_max           : INTEGER;
                VAR   gmif                               : T_gmif);
  VAR rank0, index1, rank2, rank3 : T_index_rank;
      C0123                       : T_count;
      C123                        : LONGINT;
      t, tau                      : INTEGER;
      C0                          : REAL;
  BEGIN
    Ranking(series0, length, rank0);
    FOR t := 1 TO length DO index1[t] := t;
    QuicksortIndex(series1, 1, length, index1); (* rank1 is not needed *)
    Ranking(series2, length, rank2);
    Ranking(series3, length, rank3);
    CountPairs_Cross3(rank0, index1, rank2, rank3,
                      Eps0, Eps1, Eps2, Eps3, length, tau_min, tau_max,
                      C123, C0123);
    C0 := 2*Eps0/length - SQR(Eps0/length);
      (* C0 with error correction *)
    FOR tau := tau_min TO tau_max DO
      gmif[tau] := Ln( C0123[tau] / (C0 * C123) )  / Ln(2);
      (* GMIF in units of bit *)
  END; (* GMIF3 *)
```

References

Aczél J., Daróczy Z. (1975): *On Measures of Information and Their Characterizations.* Academic Press, New York

Bandt Ch., Pompe B. (1993): The entropy profile – a function describing statistical dependences. J. Stat. Phys. **70**, 967-983

Brock W. A. (1991): Causality, chaos, explanation and prediction in economics and finance. In Casti J. L., Karlqvist A. (1991): *Beyond Belief: Randomness, Prediction and Explanation in Science.* CRC Press, Boca Raton, Florida. pp. 230-279

Campbell L. L. (1966): Definition of entropy by means of a coding problem. Wahrscheinlichkeitstheorie verw. Geb. **6**, 113-118

Cox D. R., Hinkley D. V. (1994): *Theoretical Statistics.* Chapman & Hall, London

Eckmann J.-P., Ruelle D. (1985): Ergodic theory of chaos and strange attractors. Rev. Mod. Phys. **57**, 617-656

Fraser A. M., Swinney H. L. (1986): Using mutual information to find independent coordinates for strange attractors. Phys. Rev. A **33**, 1134-1140

Grassberger P., Procaccia I. (1983): Estimation of the Kolmogorov entropy from a chaotic signal. Phys. Rev. A **28**, 2591-2593

Grassberger P., Schreiber Th., Schaffrath C. (1991): Nonlinear time sequence analysis. Int. J. Bifurcation and Chaos **1**, 521-547

Green M., Savit R. (1991): Dependent variables in broad band continuous time series. Physica D **50**, 521-544

Lindgren B. W. (1993): *Statistical Theory.* Chapman & Hall, New York

Pompe B. (1993): Measuring statistical dependencies in a time series. J. Stat. Phys. **73**, 587-610

Pompe B. (1995): A tool to measure dependencies in data sequences. In Ciarlini P., Cox M. G., Pavese F., Richter D. (1996): *Advanced Mathematical Tools in Metrology II.* Proc. Int. Euroconf., 27-30 September 1995, Lady Margret Hall, Oxford. World Scientific, Singapore. pp. 81-99

Pompe B., Heilfort M. (1994): Miogram for short-time signal analysis with an application to speech. In Gruber R., Tomassini M. (1994): *Proc. 6. Joint EPS-APS Int. Conf. on Physics Computing.* European Physical Society, Boite Postale 69, CH-1213 Petit-Lancy 2, Geneva, Switzerland. pp. 85-88

Pompe B. (1997): *Die Messung von Informationsflüssen mit einer verallgemeinerten Transinformation.* Habilitationsschrift, Greifswald (available via www: http://ap01.physik.uni-greifswald.de/~ pompe)

Rabiner L. R., Schafer R. W. (1978): *Digital Processing of Speech Signals.* Prentice-Hall, New Jersey

Rényi A. (1970): *Probability Theory.* North-Holland, Amsterdam

Roberts R. A., Mullis C. T. (1987): *Digital Signal Processing.* Addison-Wesley, Reading, MA

Savit R., Green M. (1991): Time series and dependent variables. Physica D **50**, 95-116

Scheinkman J. A., LeBaron B. (1989): Nonlinear dynamics and stock returns. J. Business **62**, 311-337

Silverman B. W. (1986): *Density Estimation for Statistics and Data Analysis.* Chapman & Hall, London

Takens F. (1983): Invariants related to dimension and entropy. In Atas do 13. Col. brasiliero de Matematicas, Rio de Janeiro

Wolf A., Swift L. B., Swinney H. L., and Vastano J. A. (1985): Determining Lyapunov exponents from a time series. Physica D **16**, 285-317

Analysing Synchronization Phenomena from Bivariate Data by Means of the Hilbert Transform

Michael Rosenblum[1,2] and Jürgen Kurths[1]

[1] Max-Plank-Arbeitsgruppe "Nichtlineare Dynamik" an der Universität Potsdam, Am Neuen Palais 19, PF 601553, D-14415 Potsdam, Germany

[2] A. von Humboldt Fellow. Permanent address: Mechanical Engineering Research Institute, Russian Academy of Sciences, 101830 Moscow, Russia

Abstract. We use the analytic signal approach based on the Hilbert transform to compute the phase difference between two non-stationary signals and find out epochs of phase locking.

1 Introduction

Bivariate data are often encountered in the study of physiological systems. The usual problem in the analysis of these data is whether two signals are dependent or not. As the data are practically always non-stationary, the application of traditional techniques such as cross-spectrum and cross-correlation analysis (Panter 1965) or nonlinear characteristics like generalized mutual information (Pompe 1993) has its limitations.

Another common problem occurs when the signals remind of periodic functions with slowly varying parameters. The natural approach here is to consider two time series as an output of two coupled oscillators, and to quantify their interaction by measuring the phase difference between these signals. As examples we can mention studies of coordinated movements (Fuchs et al. 1996; Tass et al. 1995) and cardiorespiratory interaction (Schieck 1994). Nevertheless, this procedure seems to be not trivial for non-sinusoidal signals (see discussion in Fuchs et al. 1996), and different *ad hoc* methods are used for phase calculation.

In the present work we would like to attract the attention to the *analytic signal approach* based on the Hilbert transform. This technique, widely used in the signal processing (Panter 1965; Rabiner and Gold 1975; Smith and Mersereau 1992), allows one to obtain unambiguously the phase difference for arbitrary signals. Presence of a certain relationship between phases is an indicator of some dependency between components of bivariate data. Thus, this method addresses both problems outlined above. As the Hilbert transform does not require stationarity of the data, variations of that dependency with time can be easily studied.

We relate the discussed method to the phenomenon of *phase synchronization of chaotic systems* recently demonstrated by Rosenblum et al. (1996).

Examples of application of the presented technique to the study of posture control data, visually guided forearm tracking, and interaction of cardiac and respiratory systems are given by Rosenblum et al. (this volume), Tass et al. (1996), and Hoyer et al. (this volume).

2 Instantaneous Phase of a Signal

A consistent way to define the phase of an *arbitrary* signal is known in signal processing as analytic signal concept (Panter 1965; Rabiner and Gold 1975; Smith and Mersereau 1992). This general approach, based on the Hilbert transform and originally introduced by Gabor (1946), unambiguously gives the *instantaneous phase and amplitude* for a signal $s(t)$ via construction of the *analytic signal* $\zeta(t)$, which is a complex function of time defined as

$$\zeta(t) = s(t) + j\tilde{s}(t) = A(t)e^{j\phi(t)} , \tag{1}$$

where the function $\tilde{s}(t)$ is the Hilbert transform of $s(t)$

$$\tilde{s}(t) = \pi^{-1}\text{P.V.} \int_{-\infty}^{\infty} \frac{s(\tau)}{t-\tau} d\tau \tag{2}$$

and P.V. means that the integral is taken in the sense of the Cauchy principal value. The instantaneous amplitude $A(t)$ and the instantaneous phase $\phi(t)$ of the signal $s(t)$ are thus uniquely defined from (1).

As one can see from (2), the Hilbert transform $\tilde{s}(t)$ of $s(t)$ can be considered as the convolution of the functions $s(t)$ and $1/\pi t$. Due to the properties of convolution, the Fourier transform $\tilde{S}(\omega)$ of $\tilde{s}(t)$ is the product of the Fourier transforms of $s(t)$ and $1/\pi t$. For physically relevant frequencies $\omega > 0$, $\tilde{S}(\omega) = -jS(\omega)$. This means that the Hilbert transform can be realized by an ideal filter whose amplitude response is unity, and phase response is a constant $\pi/2$ lag at all frequencies (Panter 1965).

A harmonic oscillation $s(t) = A\cos\omega t$ is often represented in the complex notation as $A\cos\omega t + jA\sin\omega t$. It means that the real oscillation is complemented by the imaginary part which is delayed in phase by $\pi/2$, that is related to $s(t)$ by the Hilbert transform. The analytic signal is the direct and natural extension of this technique, as the Hilbert transform performs the $-\pi/2$ phase shift for every frequency component of an arbitrary signal.

An important advantage of the analytic signal approach is that the phase can be easily obtained from experimentally measured scalar time series. Numerically, this can be done via convolution of the experimental data with a pre-computed characteristic of the filter (Hilbert transformer) (Rabiner and Gold 1975; Smith and Mersereau 1992; Little and Shure 1992). Although Hilbert transform requires computation on the infinite time scale, i.e. Hilbert transformer is an infinite impulse response filter, the acceptable precision of about 1% can be obtained with the 256-point filter characteristic. The sampling rate must be chosen in order to have at least 20 points per average

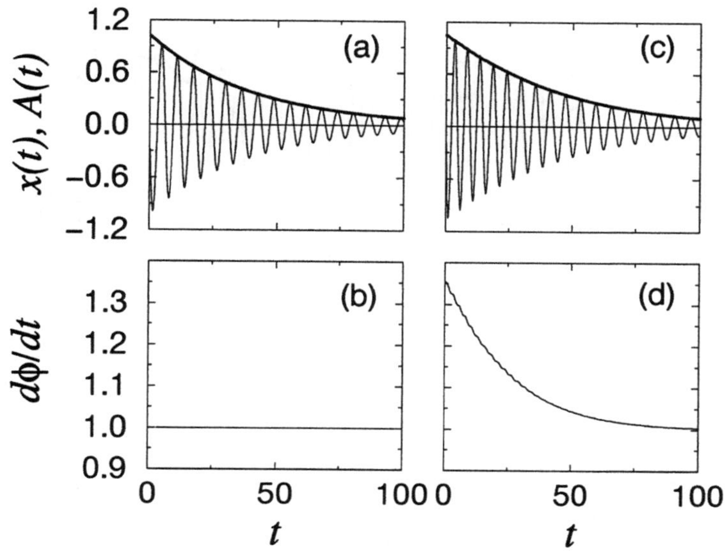

Fig. 1. Free vibrations $x(t)$ of the linear (a) and nonlinear (Duffing) (c) oscillators. The instantaneous amplitudes $A(t)$ calculated via Hilbert transform are shown by thick lines. Corresponding instantaneous frequencies $d\phi/dt$ are shown in (b) and (d)

period of oscillation. In the process of computation of the convolution $L/2$ points are lost at the both ends of the time series, where L is the length of the transformer.

We illustrate the properties of the Hilbert transform by the following examples.

Example 1. Harmonic oscillator. The Hilbert transform of the harmonic oscillation $x(t) = A\cos\omega t + \phi_0$ equals $\tilde{x}(t) = A\sin\omega t + \phi_0$; respectively the phase $\phi(t) = \omega t + \phi_0$. It means, that the *phase portrait* of the harmonic oscillator in coordinates $(x, \tilde{x})$ is a circle for any ω. Note, however, that the often used coordinates $(x, \dot{x})$ and delay coordinates $(x(t), x(t-\tau))$ generally produce an ellipse; more important, the phase obtained from such plots demonstrate oscillations that are the artifact of calculation (compare with the discussion in Fuchs et al. 1996).

Example 2. Damped oscillators. Let us take as the measured signals free oscillations of linear

$$\ddot{x} + 0.05\dot{x} + x = 0 \tag{3}$$

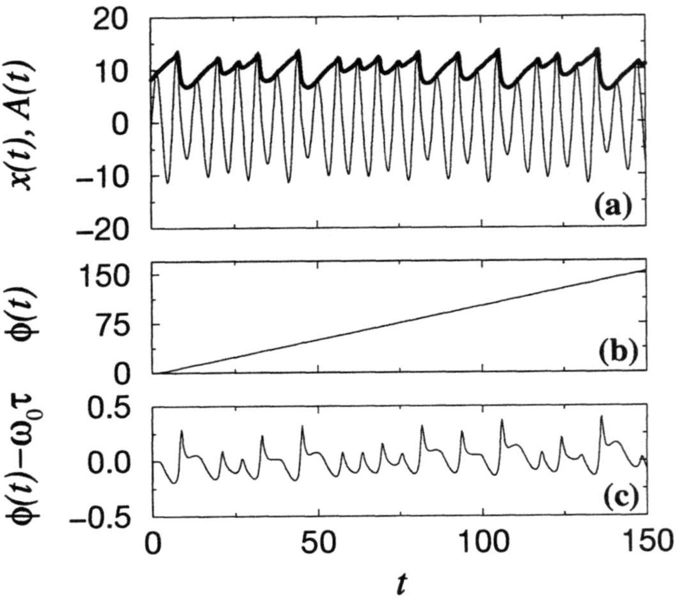

Fig. 2. Solution of the Rössler system $x(t)$ and its instantaneous amplitude $A(t)$ (thick line) (a). Instantaneous phase ϕ grows practically linear (b), nevertheless small irregular fluctuations are seen (c)

and Duffing

$$\ddot{x} + 0.05\dot{x} + x + x^3 = 0 \tag{4}$$

oscillators, and calculate from $x(t)$ instantaneous amplitudes $A(t)$ and frequencies $d\phi/dt$ (Fig. 1). The amplitudes, shown as thick lines, are really envelopes of decaying processes. The frequency of the linear oscillator is constant, while frequency of the Duffing oscillator is amplitude-dependent, as expected. Note, that although only about 20 periods of oscillations have been used, the nonlinear properties of the system can be easily seen from the time series, because frequency and amplitude are estimated in every point of the signal. This method is used in mechanical engineering for identification of elastic and damping properties of a vibrating system (Feldman 1985; Feldman and Rosenblum 1988; Feldman 1994).

Example 3. Rössler oscillator. Let us choose as an observable the x coordinate of the Rössler system

$$\begin{aligned} \dot{x} &= -y - z \, , \\ \dot{y} &= x + 0.15y \, , \\ \dot{z} &= 0.2 + z(x - 10) \, . \end{aligned} \tag{5}$$

Instantaneous amplitude and phase are shown in Fig. 2. The phase ϕ grows practically linear, nevertheless small irregular fluctuations of that growth are seen. This agrees with the known fact that oscillations of the system are chaotic, but the power spectrum of $x(t)$ contains a very sharp peak (Crutchfield et al. 1980).

3 Phase Synchronization of Chaotic Systems

Synchronization of periodic self-oscillatory systems is defined as a phase entrainment

$$|n\phi(t) - m\psi(t)| < \text{const}\,, \tag{6}$$

where n and m are integer numbers. In the presence of noise the phase difference is unbounded and performs a random-walk-like motion. However, if the noise is small, the frequencies are nearly locked, i.e. the relation between them is fulfilled in average:

$$n\langle\frac{d\phi}{dt}\rangle = m\langle\frac{d\psi}{dt}\rangle\,. \tag{7}$$

Phase synchronization of chaotic oscillators (Rosenblum et al. 1996; Pikovsky et al. 1997; Parlitz et al. 1996) is a direct generalization of this classical phenomenon. In the synchronous regime, the phases of interacting chaotic systems become locked, while the amplitudes vary chaotically, and are practically uncorrelated. A weaker type of synchronization has been also demonstrated, where the frequencies are entrained, while the phase difference exhibits a random-walk-type motion. Mutual phase synchronization of two nonidentical chaotic systems has been considered in Rosenblum et al. 1996a. It has been shown, that phase synchronization manifests itself in the Lyapunov spectrum of the coupled system: when the phase locking occurs, one of two zero Lyapunov exponents becomes negative.

The central problem in the study of phase synchronization is to introduce the notion of phase for chaotic oscillating system. There exist no unambiguous and strict definition. Nevertheless, often we can find a projection of the attractor on some plane (x, y) such that the plot reminds us of the smeared limit cycle, i.e. the trajectory rotates around the origin, or any other point that can be chosen as the origin. It means that we can choose the Poincaré section in a proper way. With the help of the Poincaré map we can define a phase, attributing 2π increase to each intersection of the trajectory with the secant surface. If the above mentioned projection is found, we can also introduce the phase as the angle between the projection of the phase point on the plane and a given direction on the plane, i.e. $\varphi = \arctan(y/x)$.

Another possibility is to calculate the instantaneous phase by taking some coordinate of the oscillating system as an observable. Although the analytic signal approach provides the unique determination of the phase of *a signal*, we cannot avoid ambiguity defining the phase for a *dynamical system*, as the result depends on the choice of the observable. Here we face the same problem as in the choice of the appropriate projection mentioned above. However, one can often find an "oscillatory" observable that provides the Hilbert phase ϕ_H in good agreement with our intuition. For example, the z-coordinate is a natural choice for the well-known Lorenz system. The detailed discussion of different definitions of the phase of the system can be found in Pikovsky et al. (1997). For the experimental studies, the phase calculated from the Hilbert transform is mostly convenient.

It is noteworthy that the phenomenon of phase synchronization is observed even when completely different systems, such as the Rössler oscillator and the Mackey–Glass differential-delay system, or the Rössler and the hyperchaotic Rössler oscillators, interact. Phase synchronization even occurs if the systems are qualitatively different, i.e. one is chaotic and another one periodic. Another important feature is that the phase synchronization is observed already for extremely weak coupling, and in some cases can have no threshold, contrary to other types of synchronization of chaotic systems.

4 Calculating Relative Phase

The relative phase, or phase difference of two signals $s_1(t)$ and $s_2(t)$ can be obtained via the Hilbert transform as

$$\varphi_1(t) - \varphi_2(t) = \arctan \frac{\tilde{s}_1(t)s_2(t) - s_1(t)\tilde{s}_2(t)}{s_1(t)s_2(t) + \tilde{s}_1(t)\tilde{s}_2(t)} \,. \tag{8}$$

Let us consider two examples.

Example 1. Two coupled Rössler oscillators. The equations of the coupled system are

$$\begin{aligned} \dot{x}_{1,2} &= -\omega_{1,2}y_{1,2} - z_{1,2} + \varepsilon(x_{2,1} - x_{1,2}), \\ \dot{y}_{1,2} &= \omega_{1,2}x_{1,2} + 0.15y_{1,2}, \\ \dot{z}_{1,2} &= 0.2 + z_{1,2}(x_{1,2} - 10), \end{aligned} \tag{9}$$

where parameters $\omega_1 = 0.89$ and $\omega_2 = 0.85$ define the average frequency of oscillations. They have been chosen in order to work within the frequency region without large periodic windows. To generate the signals with slowly varying parameters, we modulate coupling coefficient $\varepsilon = 0.03 + 0.02\sin(0.01t)$, and calculate the relative phase between $x_1(t)$ and $x_2(t)$. The results are shown in Fig 3. Due to modulation of the coupling, oscillators synchronize and desynchronize repeatedly. From these bivariate data we can easily distinguish time intervals, where the phase difference is constant, i.e. phases are locked. Respectively, we can conclude that within these intervals there is a resonant interaction between the systems, and they are synchronized.

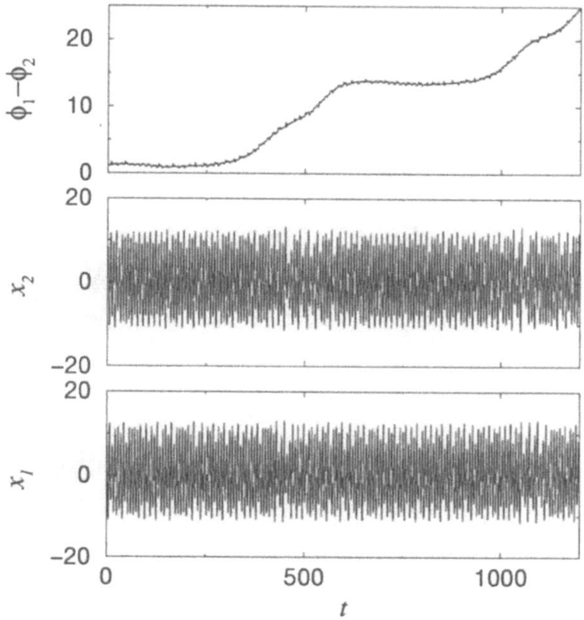

Fig. 3. Phase difference between two coupled Rössler oscillators. The coupling coefficient changes slowly with time. The periods of synchronous motion can be clearly seen

Example 2. Coupled Rössler and van der Pol oscillators. Similar results are found if two completely different systems, namely periodic van der Pol oscillator and chaotic Rössler systems, are coupled:

$$\begin{aligned}
\dot{x} &= -\omega_1 y - z + \varepsilon(u - x), \\
\dot{y} &= \omega_1 x + 0.15y, \\
\dot{z} &= 0.2 + z(x - 10),
\end{aligned} \tag{10}$$

$$\ddot{u} - \mu(1 - u^2)\dot{u} + \omega_2 u = \varepsilon(x - u) ,$$

where $\omega_1 = 0.85$, $\omega_2 = 0.85$, and $\varepsilon = 0.02 + 0.02\sin(0.01t)$. The results are presented in Fig 4.

5 Conclusions

We have described a consistent method of calculation of the phase difference between two time series. We have shown that this method can be effectively

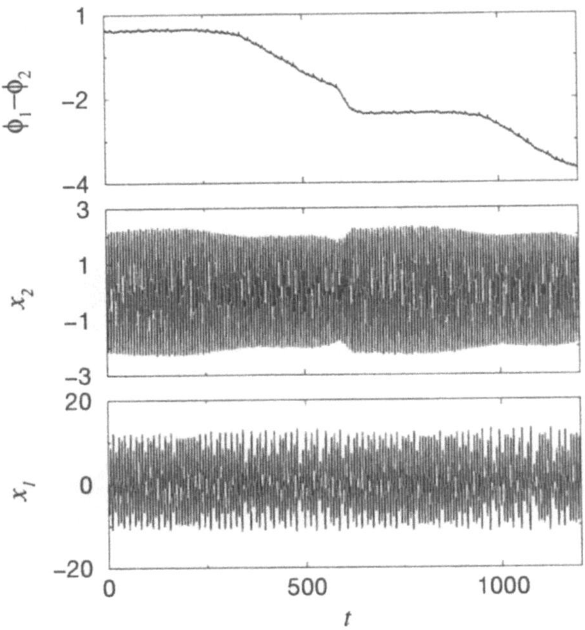

Fig. 4. Phase difference between coupled Rössler and van der Pol oscillators. The coupling coefficient changes slowly with time. The periods of synchronous motion can be clearly seen

used to reveal time-varying weak interaction between self-oscillating systems, which can be either chaotic or periodic.

Let us stress that if the phase difference between components of bivariate data is bounded, it does not necessary mean that the signals are generated by two synchronized oscillatory systems. For example, these signals can be the input and output of some phase-shifting (nonlinear) filter. Nevertheless, the technique can be formally applied; both the assumption on the underlying model and the interpretation of the result depends on the particular problem. This is similar to the usage of the coherence function and phase of the cross-spectra: although the model underlying cross-spectrum calculation is an one input – one output linear system, the technique can be applied to arbitrary bivariate data.

6 Acknowledgments

M.R. is gratefull to Alexander von Humboldt Foundation for financial support.

References

Crutchfield J., Farmer D., Packard N., Shaw R., Jones G., Donelly R.J. (1980): Power spectral analysis of a dynamical system. Phys. Lett. A **76**, 1–4

Feldman M.S. (1985): Investigation of the natural vibrations of machine elements using the Hilbert transform. Sov. Machine Sci. **2**, 44–47

Feldman M.S. (1994): Nonlinear system vibration analysis using Hilbert transform I. Free vibration analysis method "FREEVIB". Mech. Systems and Signal Processing **8**, 119–127

Feldman M.S., Rosenblum M.G. (1988): Computer program for determination of nonlinear elastic and damping properties of a vibrating system. In *Proceedings of the Workshop "Software in Machine Building CAD Systems"*, 89. Izhevsk, 1988, (In Russian)

Fuchs A., Jirsa V., Haken H., Kelso J.S. (1996): Extending the HKB model of coordinated movement to oscillators with different eigenfrequencies. Biol. Cybern. **74**, 21–30

Gabor D. (1946): Theory of communication. J. IEE London **93**, 429–457

Little J.N., Shure L. (1992): *Signal Processing Toolbox for Use with MATLAB. User's Guide.* Mathworks, Natick, MA

Panter P. (1965): *Modulation, Noise, and Spectral Analysis.* McGraw–Hill, New York

Parlitz U., Junge L., Lauterborn W., Kocarev L. (1996): Experimental observation of phase synchronization. Phys. Rev. E. **54**, 2115-2118

Pikovsky A.S., Rosenblum M.G., Kurths J. (1996): Synchronization in a population of globally coupled chaotic oscillators. Europhys. Lett. **34**, 165–170

Pikovsky A.S., Rosenblum M.G., Osipov G.V., Kurths J. (1997): Phase synchronization of chaotic oscillators by external driving. Physica D, **104**, 219–238

Pompe B. (1993): Measuring statistical dependencies in a time series. J. Stat. Phys. **73**, 587–610

Rabiner R, Gold B. (1975): *Theory and Application of Digital Signal Processing.* Prentice–Hall, Englewood Cliffs, NJ

Rosenblum M.G., Pikovsky A.S., Kurths J. (1996): Phase synchronization of chaotic oscillators. Phys. Rev. Lett. **76**, 1804–1807

Schieck M. (1994): Quantifizierung und Modellierung der respiratorischen Sinusarrhythmie. Technical Report 2899, Forschungszentrum Juelich, Juelich

Smith M.J., Mersereau R.M. (1992): *Introduction to Digital Signal Processing. A Computer Laboratory Textbook.* Wiley, New York

Tass P., Wunderlin A., Schanz M. (1995): A theoretical model of sinusoidal forearm tracking with delayed visual feedback. J. Biol. Phys. **21**, 83–112

Tass P., Kurths J., Rosenblum M.G., Guasti G., Hefter H. (1996): Delay induced transitions in visually guided movement. Phys. Rev. E **54**, 2224–2227

Analyzing Spatio-Temporal Patterns of Complex Systems

R. Friedrich, V.K. Jirsa, H. Haken[1], and C. Uhl[2]

[1] Institute of Theoretical Physics and Synergetics, University of Stuttgart, Pfaffenwaldring 57 / IV, D-70550 Stuttgart, Germany
[2] Max-Planck-Insitute of Cognitive Neuroscience Inselstr. 22-26, D-04103 Leipzig

1 Introduction

Rapid progress in the field of noninvasive imaging methods in medicine will provide huge amounts of data sets in the nearest future. Especially imaging methods with high time resolutions like multivariate measurements of the electroencephalogram (EEG) or the magnetoencephalogram (MEG) will allow for a detailed documentation of spatio-temporal processes in biological systems. Therefore, it is of extreme importance to develop methods which allow for a characterization and classification of spatio-temporal processes with special emphasis on medical applications.

It is common belief, supported by experimental and theoretical results, that biological signals contain significant ingredients of nonlinear behaviour. Great insights into the behaviour of complex systems becomes possible, when these systems are close to bifurcation points, where the system's behaviour undergoes a qualitative change [1], [2]. A sound theoretical basis for an understanding of the observed phenomena has been gained and a unified mathematical description has been developed. The ingredients of this description are the so-called order parameters which are related to the cooperative behaviour of the system and which in turn determine its temporal evolution. In spatially extended systems the order parameters are related to coherent spatial structures.

This result, which has proved to be of great importance for physical and chemical systems, yields a strategy to deal with complex biological patterns [2]. This strategy consists of two parts. The first part is related to the kinds of experiments singled out for an examination. Here, it is desirable to drive a complex biological system to a situation where a qualitative change occurs inducing cooperative behaviour by the process of selforganization. The second part relies on an analysis of the data based on the mathematical framework of order parameters, order parameter dynamics and spatial modes, since these are the central quantities underlying the process of selforganization. The outcome of such an approach yields a macroscopic description of the system's behaviour.

The purpose of the present article is to outline this approach to the analysis of complex biological patterns. In Sect. 2 we discuss the mathematical structures underlying the process of self-organization and discuss the notions of order parameters, order parameter dynamics and spatial modes. In Sect. 3 a method is presented which allows to extract all these relevant quantities directly from experimental data. Applications of our method to the analysis of brain electric signals are discussed in Sect. 4. Here, we consider an analysis of MEG patterns derived during a coordination experiment and we discuss results for EEG patterns derived during petit-mal epilepsies.

2 Self-organization in Pattern Forming Systems

Various physical, chemical as well as biological systems are able to undergo spontaneous transitions to spatially ordered states which may additionally undergo temporal evolution. As a necessary condition for the occurence of such transitions these systems must be open systems far from equilibrium since the state of a closed system will approach the state of maximal entropy according to the second law of thermodynamics. The mathematical structures describing the formation of spatio-temporal states are well known and we shall give a brief outline in the following [1], [2].

2.1 State Vectors and Order Parameters

Let us start with the discussion of the notion of spatio-temporal states. We consider a system which is characterized by several quantities q_i, $i = 1, .., n$, forming the state vector $\mathbf{q}$. Since we consider a spatially distributed system the state vector depends on the spatial coodinates $\mathbf{r}$ as well as on the time t:

$$\mathbf{q}(\mathbf{r}, t) = [q_1(\mathbf{r}, t), ...q_n(\mathbf{r}, t)] \quad . \tag{1}$$

A spatio-temporal process is characterized by the n values of the state vector $\mathbf{q}(\mathbf{r}, t)$ at each point $\mathbf{r}$ in space and time t. This characterization apparently needs a huge amount of data.

The temporal evolution of a physical system can be described on the basis of evolution equations which are the mathematical formulations of the basic laws of nature. An evolution equation relates the temporal derivative of the state vector to a nonlinear function of the state vector as well as its spatial derivatives:

$$\frac{\partial}{\partial t}\mathbf{q}(\mathbf{r}, t) = N[\mathbf{q}(\mathbf{r}, t), \nabla, \sigma] \quad . \tag{2}$$

An evolution equation allows one to calculate the state vector $\mathbf{q}(\mathbf{r}, t)$ at a time t from the knowledge of an initial condition, i.e. the state vector at a previous time t_0.

So far we have not taken into account the fact that we deal with systems exhibiting the emergence of cooperative behaviour. After a certain relaxation time the subsystems at different locations $\mathbf{r}$ do not evolve independently if the whole system behaves in a collective way and the state vector takes the form:

$$\mathbf{q}(\mathbf{r},t) = \mathbf{q}[\mathbf{r},\mathbf{u}(t)] \ . \tag{3}$$

Instead of specifying the state vector $\mathbf{q}(\mathbf{r},t)$ at each space time point the pattern-forming process is characterized by a *finite set of variables* $u_i(t)$, $i = 1,..,n_u$. These variables are called *order parameters.* Characterizing a system by order parameters apparently needs less amount of data.

As a consequence of the representation (3) a dynamical system for the order parameters $u_i(t)$ exists. This dynamical system takes the form of a finite set of ordinary differential equations:

$$\dot{u}_i(t) = h_i[u_j(t)] \ . \tag{4}$$

Thus, instead of investigating (2), a system of reduced dimensionality has to be examined. Frequently, the number of order parameters is quite small, so that a complex system can be characterized by the dynamical behaviour of few degrees of freedom. The emergence of order parameters in a complex system is a signature of selforganization. The corresponding reduction of the dynamical degrees of freedom is a result of the cooperative behaviour.

2.2 Mode Decompositions and Order Parameters

It is a well-known fact that a pattern at a certain time t can be represented as a linear superposition of a set of basic patterns, so-called *modes*, which we denote as $\mathbf{v}_i(\mathbf{r})$:

$$\mathbf{q}(\mathbf{r},t) = \sum_i \xi_i(t)\mathbf{v}_i(\mathbf{r}) \ . \tag{5}$$

The amplitudes $\xi_i(t)$ describe the contribution of the mode $\mathbf{v}_i(\mathbf{r})$ to the total pattern. Since the pattern evolves in time, the mode amplitudes are time dependent. The set of modes used for a mode decomposition is not unique.

Now let us turn to patterns of a selforganizing system. One may imagine that the cooperative behaviour of the system leads to well-organized spatial structures. The order parameters have to be connected with these spatial structures. For instance, the order parameters $u_i(t)$ may be amplitudes of certain spatial patterns $\mathbf{v}_i^u(\mathbf{r})$. The state vector takes the following form:

$$\mathbf{q}(\mathbf{r},t) = \sum_i^n u_i(t)\mathbf{v}_i^u(\mathbf{r}) + \sum_i s_i[u_j(t)]\mathbf{v}_i^s(\mathbf{r}) \ . \tag{6}$$

The order parameters obey the following evolution equation:

$$\dot{u}_i(t) = h_i[\mathbf{u}(t)] \ . \tag{7}$$

Here, we have to take into account that a finite number of order parameters exists. However, a mode decomposition of a pattern generally involves infinitely many modes. The remaining modes, denoted as $\mathbf{v}_i^s(\mathbf{r})$, can not exhibit dynamical behaviour of their own. Therefore, their amplitudes can have no explicit time dependence and only depend implicitly on time via the variables $u_i(t)$. The amplitudes of the modes $\mathbf{v}_i^s(\mathbf{r})$ are *enslaved* by the order parameters: The values of the amplitudes of the modes $\mathbf{v}_i^s(\mathbf{r})$ are functions of the order parameters:

$$s_j = s_j[u_i(t)] \ . \tag{8}$$

The representation (6), the dynamical system (7), together with the relation (8) yields a closed description of a wide class of pattern-forming systems. Let us briefly discuss, under which conditions such a behaviour can be expected to occur.

The existence of a state vector according to (6), (7), (8) can be rigorously proven for dissipative physical, chemical and biological systems close to instabilities. In these cases a complex system is driven away from equilibrium until its state becomes unstable. The instability is induced by the spatial modes $v_j(\mathbf{r})$ which get positive growth rates. All other modes remain damped. In such a situation the amplitudes of the unstable modes turn out to be the order parameters. For further details we refer the reader to the two monographs of H. Haken [1],[2].

An extension of the idea of "slaving" has been formulated for certain class of systems. For these systems, the representation (6), (7), (8) remains valid. However, the restriction that the system is close to an instability can be relaxed. The essential property, which allows for such a representation, is a separation of time scales of the modes. The linear growth rates of strongly damped modes and those of the modes corresponding to the order parameters have to be separated by a wide gap. In that case the existence of a so-called inertial manifold, defined by a relation according to (8) can be proven [4].

Summarizing we note that time scale separation is a basic mechanism which leads to the emergence of selforganized behaviour in complex systems. Especially in biological systems it is the mechanism that allows for the spontaneous emergence of functional behaviour. In these cases a representation of the state vector according to (6), (7), (8) applies inducing the existence of order parameters. They turn out to be the central quantities.

Furthermore, the existence of order parameters infers a classification of the behaviour of a complex system in terms of their dynamics. Various classes, like first and second nonequilibrium phase transitions and transitions to oscillatory behaviour by Hopf bifurcation are well-known and have been investigated in detail [1], [2]. Further extensions are the cases where the dynamical behaviour is generated by several order parameters behaving, e.g. like a set of coupled nonlinear oscillators. Each oscillator is related to a time dependent pattern interacting nonlinearly with the patterns corresponding to the other oscillators. Our analysis of the MEG patterns of the coordination experiment of Kelso et al. [14] presented below will reveal such a type of behaviour. Another significant behaviour of order parameter dynamics is the emergence of chaotic behaviour. Chaotic time dependence may arise in systems with three and more variables. Since the order parameters are amplitudes of certain spatial modes which superimpose to the state vector a complex behaviour in time arises although the spatial pattern at each time instant shows coherence. Nowadays a lot of physical systems are known which show chaos of this type, i.e. chaotic behaviour due to the interaction of spatial modes. Below we shall show that such a situation arises in brain waves generated during a stage of petit-mal epileptic seizures.

Finally, let us mention that all attempts to isolate low dimensional chaotic behaviour in brain signals by measuring metric properties like fractal dimensions, Ljapunov exponents etc. are implicitly based on the assumption of the existence of order parameters.

3 Analysis of Spatio-Temporal Structures

For a complex system whose evolution equation (2) is well-known it is straightforward to determine the macroscopic description (6), (7), (8). However, such a macroscopic description may also exist for a complex system whose evolution equation (2) is yet unknown, a situation which is typical for biological systems. Here, one is faced with the inverse problem, i.e. with the task to extract the macroscopic description from the experimental data set.

In this section we would like to discuss methods for the analysis of spatio-temporal signals $q(\mathbf{r},t)$. We assume a discrete spatial resolution of the signal, given e.g. by the position of the SQUIDs or electrodes of the MEG or EEG measurements. The signal $q(\mathbf{r},t)$ can be represented by a vector-function $\mathbf{q}(t)$ consisting of components $q_i(t)$, which correspond to a measurement at one spatial point:

$$q(\mathbf{r},t) \quad \Rightarrow \quad \mathbf{q}(t) \ . \tag{9}$$

One aim of an analysis of spatio-temporal data is to represent it by a set of spatial modes and corresponding amplitudes:

$$\mathbf{q}(t) = \sum_i \xi_i(t)\mathbf{v}_i \ , \tag{10}$$

and thereby to gain information about the underlying system. Since there are several possible choices of the spatial modes $\mathbf{v}_i$, one has to require additional features which one would like to incorporate into this representation.

One possibility is to require a representation of maximal convergence with respect to the L_2 norm, which is given by the so-called Karhunen–Loève (KL) expansion [3]. Another possibility, which focuses on the identification of the underlying dynamics of the system, is based on the representation of a state vector of a selforganizing system in terms of order parameters.

3.1 Karhunen–Loève Expansion

This expansion, which is also known as 'principal component analysis' (PCA) or 'empirical orthogonal function' (EOF) decomposition, can be derived by minimizing the least-square error function R_n,

$$R_n = \langle (\mathbf{q}(t) - \sum_{i=1}^{n} \eta_i(t)\mathbf{w}_i)^2 \rangle \ , \tag{11}$$

by a suitable choice of modes $\mathbf{w}_i$, $i = 1, .., n$. The brackets $\langle \ldots \rangle$ denote time averaging:

$$\langle f(t) \rangle = \frac{1}{t_1 - t_0} \int_{t_0}^{t_1} f(t) dt \ . \tag{12}$$

Variation of (11) with respect to the spatial modes $\mathbf{w}_i$ leads to an eigenvalue problem,

$$C\mathbf{w}_i = \lambda_i \mathbf{w}_i \ , \tag{13}$$

of the correlation-matrix C,

$$C_{ij} = \langle q_i(t) q_j(t) \rangle \ . \tag{14}$$

Since the matrix C is symmetric, the modes $\mathbf{w}_i$ are orthogonal and the eigenvalues λ_i measure the mean contribution of the term $\eta_i(t)\mathbf{w}_i$ to the signal $\mathbf{q}(t)$:

$$\langle \eta_i \eta_j \rangle = \delta_{ij} \lambda_i \ . \tag{15}$$

Ordering the eigenvalues λ_i in descending order,

$$\lambda_1 > \lambda_2 > \ldots \ , \tag{16}$$

the representation of the signal $\mathbf{q}(t)$,

$$\mathbf{q}(t) = \sum_i \eta_i(t) \mathbf{w}_i \ , \tag{17}$$

is the best converging mode-expansion with respect to the L_2 norm. The amplitudes $\eta_i(t)$ are given by projecting the modes $\mathbf{w}_i$ onto the signal $\mathbf{q}(t)$:

$$\eta_i(t) = \mathbf{w}_i \cdot \mathbf{q}(t) \ . \tag{18}$$

Clearly the disadvantage of this expansion lies in a lack of information of the underlying dynamics of the system, i.e. there is no information about an evolution equation of the amplitudes of the form

$$\dot{\eta}_i = f_i[\{\eta_j\}] \ . \tag{19}$$

If one recalls the representation (6), (7), (8) it is evident that the KL modes are not the modes which one would like to determine as collective structures of a system. However, the KL expansion is an useful tool to extract relevant modes out of the signal. Quite often it already yields a qualitative understanding of the underlying system.

3.2 Spatio-Temporal Analysis

As already mentioned above our approach aims at a solution of the inverse problem of structure formation. The goal of our analysis is to detect the order-parameters $u_i(t)$, the order parameter dynamics as well as the corresponding spatial modes $\mathbf{v}_i^{(u)}$ and $\mathbf{v}_i^{(s)}$ of the 'enslaved' modes.

We first introduce a biorthogonal set of modes $\mathbf{v}_i^{(u)}$, $\mathbf{v}_i^{(s)}$, $\mathbf{v}_i^{(u)+}$, $\mathbf{v}_i^{(s)+}$, which we want to extract from the data set:

$$\mathbf{v}_i^{(\alpha)}\mathbf{v}_j^{(\beta)+} = \delta_{ij}\delta_{\alpha\beta} \ . \tag{20}$$

The modes correspond to the mode decomposition of the state vector (3).

In a second step we decide how many order parameters are involved in the dynamical evolution and specify the order parameter dynamics (7). Then we introduce the potential V, which represents a least-square-error function with respect to the orderparameter equations (7) and the enslaved amplitudes (8):

$$V = \sum_i \frac{\langle(\dot{\mathbf{q}}\mathbf{v}_i^{(u)+} - f_i[\{\mathbf{q}\mathbf{v}_j^{(u)+}\}])^2\rangle}{\langle(\dot{\mathbf{q}}\mathbf{v}_i^{(u)+})^2\rangle} + \sum_i \frac{\langle(\mathbf{q}\mathbf{v}_i^{(s)+} - g_i[\{\mathbf{q}\mathbf{v}_j^{(u)+}\}])^2\rangle}{\langle(\mathbf{q}\mathbf{v}_i^{(s)+})^2\rangle} \ . \tag{21}$$

Assuming a polynomial form for the functions f_i and g_i, which may correspond to assumed normalforms of instabilities, the corresponding coefficients as well as the modes $\mathbf{v}_i^{(s)+}$ can be eliminated as a function of the modes $\mathbf{v}_i^{(u)+}$:

$$\{f_i, g_i, \mathbf{v}_i^{(s)+}\} = F[\{\mathbf{v}_i^{(u)+}\}] \ . \tag{22}$$

For the exact calculations we refer the reader to [5]. The reason for this possible elimination lies in the bilinear occurrence of these parameters in the potential V. Inserting these equations back into the potential V, we obtain a nonlinear low-dimensional potential

$$V = V[\{\mathbf{v}_i^{(u)+}\}] \ , \tag{23}$$

which depends on the modes $\{\mathbf{v}_i^{(u)+}\}$ only. The global minimum of the potential can be found by a gradient dynamics,

$$\dot{\mathbf{v}}_i^{(u)+} = -\frac{\partial V[\{\mathbf{v}_j^{(u)+}\}]}{\partial \mathbf{v}_i^{(u)+}} \ , \tag{24}$$

starting with different values for the vector $\mathbf{v}_i^{(u)+}$. This obtained minimum then represents the best choice for the spatial modes $\mathbf{v}_i^{(u)+}$, $\mathbf{v}_i^{(s)+}$ and the coefficients of the functions f_i and g_i with respect to the spatio-temporal dynamics of the signal.

To obtain the spatial modes $\mathbf{v}_i^{(u)}$, $\mathbf{v}_i^{(s)}$ from the modes $\mathbf{v}_i^{(u)+}$, $\mathbf{v}_i^{(s)+}$ on has to solve the equation (20). This can be achieved uniquely, if the number of modes $\mathbf{v}_i^{(u)+}$, $\mathbf{v}_i^{(s)+}$ is equal to the dimension of the vectorspace in which the trajectory $\mathbf{q}(t)$ is moving. If there are less modes, we can use this freedom and maximize the contribution of the modes $\mathbf{v}_i^{(u)}$ and $\mathbf{v}_i^{(s)}$ to the signal by variation of the potential W,

$$W = \langle(\mathbf{q}(t) - \sum_i u_i(t)\mathbf{v}_i^{(u)} - \sum_i s_i(t)\mathbf{v}_i^{(s)})^2\rangle \ . \tag{25}$$

If we summarize the modes $\mathbf{v}_i^{(u)}$, $\mathbf{v}_i^{(s)}$ to $\mathbf{v}_i$ and the amplitudes $u_i(t)$ and $s_i(t)$ to $\xi_i(t)$, the minimum of the potential W is given by

$$\mathbf{w}_i = \sum_j C_{ij}^{(\xi)^{-1}} \langle \xi_j \mathbf{q} \rangle \ . \tag{26}$$

The correlation matrix $C^{(\xi)}$ is thereby defined as

$$C_{ij}^{(\xi)} = \langle \xi_i \xi_j \rangle \ . \tag{27}$$

By this procedure we now have achieved a complete description of the spatio-temporal signal $\mathbf{q}(t)$ in terms of order parameters and 'enslaved' amplitudes and corresponding spatial modes (6).

The only assumption for our approach is the number of order parameters underlying the signal. This assumption can be verified self-consistently by the minima of the occurring potentials. Another technical point we would like to mention concerns the functions f_i and g_i: If there are hints about these functions, e.g. by the KL expansion or due to symmetry arguments, it is advisable to restrict the functions f_i and g_i to these observed classes, since this can simplify the search for the best parameters. Here, a classification of order parameter equations using normal form approaches helps.

4 Applications to Brain Electromagnetic Signals

The elementary unit of the nervous system is the neuron which is divided into three basic components [11]: the dendrites, the cell body and the axon. The dendrites act as the receptive side of the neuron. Synapses on the dendrites convert the inputs from other neurons by initiating electric currents along the dendrites which are spatially integrated at the cell body. There are mainly two kinds of synapses [9],[10]: An excitatory synapse causes a current to flow into the dendrite at the synapse, along the dendritic cable to the cell body. An inhibitory synapse causes a current to flow in the opposite direction. These currents are linearly summed up. If the resulting current at the cell body exceeds a certain threshold, it is converted into a pulse train along the axon. Pools of neurons are structured in cortical columns and tend to synchronize their activity. These coulumns can be regarded as the quasi-microscopic entities leading to spatially coherent behavior. The EEG measures macroscopic quantities which mainly correspond to summed dendritic potentials [11] and the MEG measures macroscopic quantities mainly corresponding to summed dendritic currents [7]. The fact that significant signals can be detected in EEG and MEG measurements shows that coherent behaviour of neuronal activity exists within the central nervous system.

There has been various attempts in modelling the dynamical phenomena in the brain observed at different spatial scales. Microscopic theories are devoted to the behaviour of single neurons. Mesoscopic theories aim at describing the spatio-temporal development of the overall mean level of electromagnetic activity in neuronal populations. The first models of this type were published by Beurle [12] in 1956 and Griffith [13] in 1963. Both used partial differential equations to describe the propagation of a field that describes the overall excitation of the neural network. Wilson and Cowan [8] presented a two-variable description of the electric neuronal activity of a neural population, being composed of an excitatory subpopulation and an inhibitory subpopulation. Nunez presented a two-variable model for the excitatory and inhibitory synaptic activities of spatially localized aggregates [7]. From Nunez equations Jirsa and Haken derived a one-variable field theoretical description [17] in form of a generalized, nonlinear wave equation. For the case of the brain-motor-behaviour experiment by Kelso et al. [14] this field equation can be reduced to a phenomenological model by Jirsa et al. [16] which describes the spatio-temporal dynamics of the order parameters.

All these models are nonlinear and take into account the spatial connection of neurons in the brain. These two properties are necessary ingredients for the emergence of spatio-temporal structures and the developped mathematical models allow for selforganized behaviour leading to a representation of the state vector in terms of order parameters, enslaved modes and the corresponding spatial structures according to (6), (7), (8). This should be especially the case in a situation where a phase transition takes place, since then order parameters arise inevitably. These arguments justify an investigation of EEG and MEG patterns along the lines discussed above.

4.1 Spatio-Temporal Dynamics in the Human MEG

In the following we shall consider the MEG experiment by Kelso et al. [14] where phase transition phenomena are observed in the motor behaviour as well as in the brain signals. A subject was exposed to a periodic succession of acoustic stimuli and asked to press a button in a syncopated motion. The stimulus frequency at the beginning was set to 1 Hz and was increased by 0.25 Hz after 10 stimulus repetitions. Around the frequency of 1.75 Hz the subject switched spontaneously to a synchronized motion. During this experiment the spatially resolved magnetic field data were recorded over the left parieto-temporal cortex, mainly covering the motor and auditory areas, by an array of 37 SQUID detectors.

Detailed analyses conducted by Fuchs et al. [15] revealed that in the pretransition region the registered brain signals oscillate mainly with the stimulus frequency. Stimulus and motor response frequency are locked 1:1. By increasing the frequency a switch in behavior occurs to the posttransition region, where the brain signals mainly oscillate with twice the stimulus frequency. Near the transition point, predicted features of nonequilibrium phase transitions like critical slowing down and fluctuation enhancement [1] are observed in both the behavioral data and the brain signals.

In order to obtain information about the spatio-temporal behavior of the brain signals from the *entire* SQUID array a Karhunen–Loève expansion (KL expansion) was applied [3]. The first KL mode obtained from a KL expansion at each frequency represents about 60% of the entire spatio-temporal signal. It is observed that in the pretransition region the spatial structure of the first KL mode remains constant and mainly oscillates with the stimulus frequency. In the posttransition region a different spatial mode is observed mainly oscillating with twice the stimulus frequency. This mode turns out to be a spatial superposition of the two first KL modes obtained from a KL expansion of the entire time series over all plateaus.

In [16] the experimentally observed phenomena were interpreted as follows: The observed temporal and spatial behavior corresponds to a competition of two order parameters $u_1(t)$ and $u_2(t)$ with the corresponding spatial components $\mathbf{v}_1^{(u)}$ and $\mathbf{v}_2^{(u)}$, where the first dominates in the pretransition region oscillating with the stimulus frequency and the second in the posttransition region oscillating with twice the stimulus frequency. The temporal dynamics underlying the transition has been mathematically modelled as an interaction of two order parameters, where the spatial base of the order parameters has been assumed to be the first two KL modes of the KL expansion over all plateaus. This model reads:

$$\dot{u}_i = f_i(\{u_j\}, y) \qquad \text{with} \qquad i = 1, .., 4 \tag{28}$$

where

$$f_1(\{u_j\}, y) = u_3 \;, \tag{29}$$

$$f_2(\{u_j\}, y) = u_4 \;, \tag{30}$$

$$f_3(\{u_j\}, y) = -\omega_1^2\, u_1 + g_1(u_1, u_2)\, u_3 + \epsilon_1\, y^2\, u_1 + \epsilon_2\, y\, u_2^2 \;, \tag{31}$$

$$f_4(\{u_j\}, y) = -\omega_2^2\, u_2 + g_2(u_1, u_2)\, u_4 + \epsilon_3\, y^2\, u_2 + \epsilon_4\, y\, u_1^2 \;, \tag{32}$$

$$\ddot{y} + \Omega^2\, y = 0 \,, \tag{33}$$

u_1 denotes the time dependent coefficient of the order parameter mode dominating in the pretransition region and u_2 the one dominating in the posttransition region. Since u_1 and u_2 perform oscillatory behavior the first time derivatives u_3 and u_4 are introduced. The functions $g_i(u_1, u_2)$ represent a nonlinear expression which leads to a saturation of the amplitudes of u_1 and u_2 and provides a cross coupling between them. The temporal dynamics of the external auditory stimulus represented by y is given in (33). Here the stimulus frequency serves as the control parameter. The two coupled oscillators (28) are parametrically excited by the external stimulus.

Based on the dynamical system for the order parameters the corresponding spatial modes can be determined according to the procedure discussed above. It turns out that the reconstructed signals closely fit the ones obtained from experiments. For details we refer the reader to [16].

4.2 Petit-Mal Epilepsies

Epileptic seizures are usually divided into two classes. In the case of *partial seizures* the epileptic activity is localized at one or several epileptogenic foci. *Generalized seizures* are characterized by global seizure activity involving both cerebral hemispheres. Petit-mal epilepsy belongs to the class of generalized seizures. It is usually related with an absence which lasts a few seconds. It shows up in EEG signals by a pronounced spike-wave behaviour, which in clinical application is the diagnosis of petit-mal epilepsy. Usually there are three spike-wave cycles per second.

The patterns of electrical potentials obtained from multielectrode measurements during seizure consist of two regions of opposite polarity, which undergoes a characteristic evolution in time. For details we refer the reader to [18], where results of a spatio-temporal analysis has been described. Let us briefly summarize the results.

The state vector $\mathbf{q}(t)$ consists of the values of the potentials measured at the various electrodes. For the analysis of these patterns it is assumed in a selfconsistent manner that the dynamics is governed by three order parameters denoted as $u_1(t)$, $u_2(t)$, and $u_3(t)$. These order parameters are the amplitudes of three spatial modes $\mathbf{v}_i^{(u)}$. The dynamics of the amplitudes $u_1(t)$, $u_2(t)$, $u_3(t)$ are supposed to obey the following system of ordinary differential equations:

$$\begin{aligned} \dot{u}_1(t) &= u_2(t) \quad , \\ \dot{u}_2(t) &= u_3(t) \quad , \\ \dot{u}_3(t) &= f(u_i, \mathbf{a}) \quad . \end{aligned} \tag{34}$$

The nonlinear function $f(u_i, \mathbf{a})$ is expanded in a Taylor series in $u_1(t)$, $u_2(t)$, $u_3(t)$:

$$f(u_i, \mathbf{a}) = a_1 + \sum_i a_i u_i + \sum_{ij} a_{ij} u_i u_j + \ldots \quad . \tag{35}$$

We may represent the dynamics entirely in terms of a differential equation of the third order for the amplitude $u_1(t)$:

$$\frac{d^3}{dt^3} u_1(t) = f[u_1(t), \frac{d}{dt} u_1(t), \frac{d^2}{dt^2} u_1(t), \mathbf{a}] \quad . \tag{36}$$

In principle one should take into account the stable modes [5], [6]. Usually, their contributions to the state vector are of higher order. Therefore, they have been neglected yielding the following form of the state vector $\mathbf{q}(t)$:

$$\mathbf{q}(t) = \sum_i u_i(t)\mathbf{v}_i^{(u)} . \tag{37}$$

The coefficients $\mathbf{a}$ of the Taylor expansion (35) as well as the patterns $\mathbf{v}_i^{(u)+}$ have been determined directly from the experimental time series by an application of the method outlined above. They are determined by the absolute minimum of the following two potentials:

$$\begin{aligned} V_1 &= \langle\{\mathbf{v}_1^{(u)+} \cdot \frac{d}{dt}\mathbf{q(t)} - \mathbf{v}_2^{(u)+} \cdot \mathbf{q(t)}\}^2\rangle \\ &+ \quad \langle\{\mathbf{v}_1^{(u)+} \cdot \frac{d^2}{dt^2}\mathbf{q(t)} - \mathbf{v}_3^{(u)+} \cdot \mathbf{q(t)}\}^2\rangle \ , \end{aligned} \tag{38}$$

$$\begin{aligned} V_2 = \langle\{\mathbf{v}_1^{(u)+} &\cdot \frac{d^3}{dt^3}\mathbf{q}(t) \\ &-f[\mathbf{v}_1^{(u)+} \cdot \mathbf{V}(t), \mathbf{v}_1^{(u)+} \cdot \frac{d}{dt}\mathbf{q}(t), \mathbf{v}_1^{(u)+} \cdot \frac{d^2}{dt^2}\mathbf{q}(t), \mathbf{a}]\}^2\rangle \ . \end{aligned} \tag{39}$$

The absolute minimum of the two potentials with respect to the modes $\mathbf{v}_i^{(u)+}$ and the coefficients $\mathbf{a}$ yields an approximation to the spatial modes as well as the temporal dynamics (34), (36).

It has been shown in [18] that the entire spatio-temporal signal can be reconstructed from a numerical integration of the dynamical systems (34) using the parameter values $\mathbf{a}$ determined by the present method. With these solutions the complete time signal $\mathbf{q}(t)$ can be calculated according to (37). The obtained reconstruction is in close accordance with the experimental data.

An investigation of the differential equations (34) shows that the dynamical behaviour underlying peti-mal epileptic seizures is related to dynamical systems discussed by Shilnikov [19].

5 Summary

Analyzing the temporal as well as the spatial aspects of biological signals seems to become important for the understanding of self-organized spatio-temporal behaviour of complex systems. In the present contribution we have discussed a method which could be denoted as the *inverse problem* of the theory of selforganization. This problem consists in extracting order parameters, spatial modes and their dynamics from experimental data. An application of this method to electromagnetic brain signals has already provided exciting results. Our method is based on a well-defined hypothesis with respect to the nature of the signal under consideration: This signal has to stem from a self-organizing process determined by the dynamics of order parameters. In this respect our approach deviates from purely data-driven methods for analyzing spatio-temporal signals.

References

[1] H. Haken: *Synergetics. An Introduction.* Springer-Verlag, Berlin, Heidelberg, 1983

[2] H. Haken: *Advanced Synergetics.* Springer-Verlag, Berlin, Heidelberg, 1987

[3] H. Haken: *Information and Selforganization.* Springer-Verlag, Berlin, Heidelberg, 1988

[4] S. Yigitbasi: Theorie inertialer Mannigfaltigkeiten und ihre Anwendung auf den Laser, Dissertation, Stuttgart 1995

[5] C. Uhl, R. Friedrich, H. Haken: Reconstruction of spatio-temporal signals of complex systems, Z. Phys. B **92**, 211–219 (1993)

[6] C. Uhl, R. Friedrich, H. Haken: Analysis of spatiotemporal signals of complex systems, Phys. Rev. E **51**, 3890 (1995)

[7] Nunez P.L.: *Neocortical Dynamics and Human EEG Rhythms.* Oxford University Press, Oxford, 1995

[8] Wilson H.R., Cowan J.D.: Excitatory and inhibitory interactions in localized populations of model neurons. Biophys. J. **12**, pp. 1–24 (1972);
Wilson H.R., Cowan J.D.: A mathematical theory of the functional dynamics of cortical and thalamic nervous tissue. Kybernetik **13**, 55–80 (1973)

[9] Braitenberg V., Schüz A.: *Anatomy of the Cortex. Statistics and Geometry.* Springer, Berlin, Heidelberg, 1991

[10] Abeles M.: *Corticonics.* Cambridge University Press, Cambridge, 1991

[11] Freeman W.J.: Tutorial on neurobiology: from single neurons to brain chaos, Intl. J. of Bifurcation and Chaos, **2** 451–482 (1992)

[12] Beurle R.L.: Properties of a mass of cells capable of regenerating pulses. Philos. Trans. Soc. London, Ser. A **240**, 55-94

[13] Griffith J.S.: A field theory of neural nets I: Derivation of field equations. Bull. Math. Biophys. **25**, 111–120 (1963)
Griffith J.S.: A field theory of neural nets II: Properties of field equations. Bull. Math. Biophys. **27**, 187–195 (1965)

[14] Kelso J.A.S., Bressler S.L., Buchanan S., DeGuzman G.C., Ding M., Fuchs A., Holroyd T.: A phase transition in human brain and behavior. Phys. Lett. A **169**, 134–144 (1992)

[15] Fuchs A., Kelso J.A.S., Haken H.: Phase transitions in the human brain: spatial mode dynamics. Intl. J. of Bifurcation and Chaos **2**, 917–939 (1992)

[16] V.K. Jirsa, R. Friedrich, H. Haken, J.A.S. Kelso: A theoretical model of phase transitions in the human brain. Biol. Cybern. **71**, 27–35 (1994);
V.K. Jirsa, R. Friedrich, H. Haken: Reconstruction of the spatio-temporal dynamics of a human magnetoencephalogram. Physica D **89**, 100–122 (1995)

[17] V.K. Jirsa, H. Haken, A derivation of a macroscopic field theory of the brain from the quasi-microscopic neural dynamics. To be published

[18] R. Friedrich, C. Uhl: Spatio-temporal analysis of human electroencephalograms: Petit-mal epilepsy. Physica D **98**, 171–182 (1996)

[19] L.P. Shil'nikov: A case of the existence of a countable number of periodic motions. Sov. Math. Dokl. **6**, 163–166 (1965); Math. USSR Sbornik **10**, 91 (1970)

II Cardio-Repiratory System

Are R-R-Intervals Data Appropriate to Study the Dynamics of Heart?

Alexei Potapov[1,2]

[1] Max-Planck-Arbeitsgruppe "Nichtlineare Dynamik" an der Universität Potsdam,
[2] Permanent address: Keldysh Institute for Applied Mathematics, Russian Academy of Sciences, Miusskaya pl. 4, 125047, Moscow, Russia.
e-mail: `copal@spp.keldysh.ru`

Abstract. We consider the problem of recovering the implied dynamical system, which describes the dynamics of heart, from a time series of RR intervals. The main conclusion is that on small time scales such recovery fails, and on large time scales the correlation integral behaves like that for noisy system. Consequently, it seems that recovery of underlying dynamical system and measuring its parameters (dimension, Lyapunov exponents etc.) from these data is hardly possible and more adequate is application of statistical techniques.

1 Introduction

This work has been initiated by presentations and discussions at the Dresden workshop "Nonlinear Techniques in Physiological Time-Series Analysis". Two participants, Prof. J. Zebrowski and Prof. J. Skinner, have proposed several data sets of RR intervals measured for a number of ill and more healthy, without dangerous disease, persons. The detailed description of the data can be found in this volume. In this text the data sets will be referred to by their names, dp$x$_y, $x = 1, 2, 3$, $y = a, b$ for J. Zebrowski data and pzm, $z = 1, ..., 20$ for J. Skinner data. The main goal was to propose a technique which would allow to distinguish between these two groups only by processing the data.

In this work several methods of analysis of low-dimensional chaos were applied for this purpose, like it was done e.g. in (Babloyantz & Destexhe 1988, Lefebvre et al. 1993, Chaos, V.5, 1995). Unfortunately, it did not enable us to propose some useful classification. Instead, the results lead to the conclusion that RR intervals measured with millisecond precision are not an appropriate object for applying the nonlinear dynamics methods to study the underlying dynamical system, provided it exists (statistical approaches, e.g. entropy based, may prove to be more efficient). This is partially due to the insufficient precision, and partially, probably, because such intervals may be a bad observable $h(\mathbf{x})$ for characterizing the state of the human heart $\mathbf{x}$: too complex function of the state or a function degenerate in some sense. The example of such a bad choice of observable may be the energy for slightly perturbed hamiltonian systems. Due to perturbation, energy does not conserve, but it changes only slightly while the state of the system may change considerably. With a human heart this situation may occur, e.g., if there is a very strong mechanism which stabilize the heart rate variability with respect to some sort of perturbations. Then essential change of human state may cause only small variations of heart rate.

2 Reconstruction of Attractor from a Series of Time Intervals

When one applies the nonlinear dynamics methods, the basic assumption is that there exist a dynamical system $\dot{\mathbf{X}} = \mathbf{F}(\mathbf{X})$, such that the observed time series $x_i = x(t_i)$ is a function of a system's state: $x_i = x(t_i) = h(\mathbf{X})$ (Takens 1981, Eckmann & Ruelle 1985, Sauer et al. 1991). Then, using Takens theorem, one can construct the delay vectors $\mathbf{z}(t) = \{x(t), x(t+\tau), \ldots, x(t+(m-1)\tau)\} = \Lambda(\mathbf{X}(t))$, and for proper m and τ on the system attractor $\mathbf{z}$ will be one-to-one function of the system state $\mathbf{X}$. That is, knowing $\mathbf{z}$ one (in principle) can find $\mathbf{X}$. By processing the set of vectors $\mathbf{z}$ it is possible to determine the characteristics of dynamical system which are invariant under change of variables: dimensions, entropies, Lyapunov exponents. Sometimes it is possible even to reconstruct the equations of motion in the $\mathbf{z}$ representation: $\mathbf{z}(t+\tau) = \mathbf{f}(\mathbf{z}(t))$ (Eckmann & Ruelle 1985, Abarbanel et al. 1993).

RR-intervals data from this viewpoint can be considered as a special choice of observable, resembling the Poincaré section technique or neuron firing (Sauer 1995). The basic assumption for validity of this approach is that the duration of the interval is a function *only* of the system state at its beginning: $\Delta t = \Phi(\mathbf{X}(t))$. Then it is possible to introduce a new dynamical system

$$\mathbf{X}_{n+1} = \mathbf{G}(\mathbf{X}_n), \tag{1}$$

where $\mathbf{X}_n$ and $\mathbf{X}_{n+1}$ are the system state vectors at the beginning and the end of the RR interval. Under such assumptions for the new dynamical system (1) Δt becomes usual observable, and therefore the data processing should provide information about attractor of (1).

It can be shown, that for model examples this approach do work. Let us consider the Lorenz model

$$\dot{X} = 10(Y - X), \qquad \dot{Y} = (28 - Z)X - Y, \qquad \dot{Z} = XY - \frac{8}{3}Z,$$

and successive maxima of the variable Z. Such maxima form the Poincaré section of the Lorenz attractor by the surface $3XY - 8Z = 0$, and this section gives rise to the dynamical system of type (1) with attractor dimension $\nu \cong 1.06$. Along with the series of Z maxima we can consider the series of time intervals between them. Figure 1 shows 2D embedding of attractor for both series, and the behaviour of correlation integral for them. It can be seen that

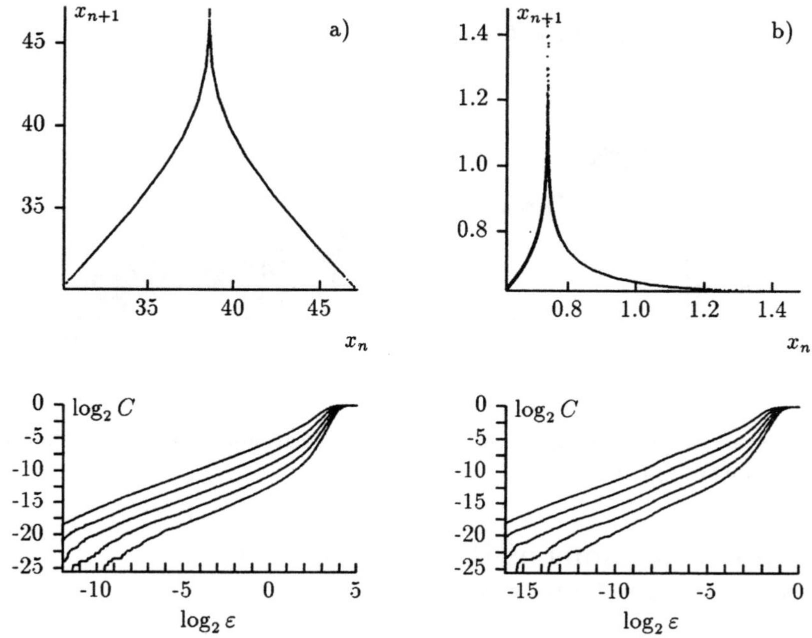

Fig. 1. 2-D embedding of time series (return map) for Poincaré section of Lorenz attractor (upper panel) and correlation integral for the same series in embeddings with $m = 4, 6, 8, 10, 12$. (a) Series of succesive maxima of Z, $x_i = Z(t_i)$ (b) series of time intervals between these maxima, $x_i = t_i$.

the plots are similar, but one for the series of time intervals demonstrates a bit worse linearity. The dimension estimates from both plots are close to 1.06. In this case it proves that the reconstruction from time intervals is more sensitive to the errors of numerical method and special precautions should be made to diminish their influence. Note that for experimental time series there are no such problems, but high precision of measurement may also be needed.

Consequently, the reconstruction of attractor from a series of time intervals is in principle possible, but practically may be rather difficult and require very high precision data.

Below we shall denote the series of RR intervals by x_i, $i = 1, \ldots, N$. In the calculations below they are given in seconds, i.e. most of x_i are close to 1, but usually less than 1.

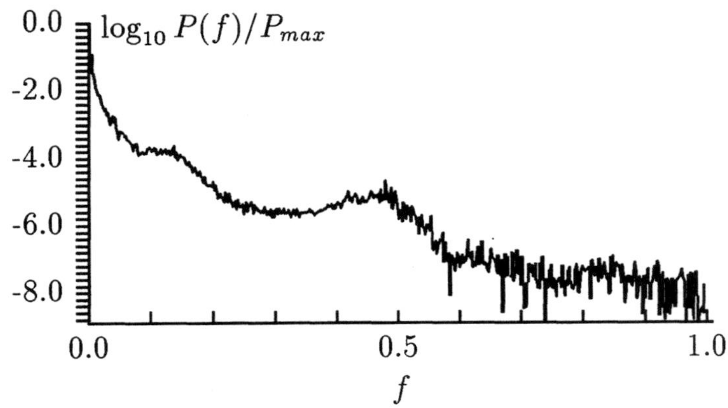

Fig. 2. Example of power spectrum with a peak corresponding to short-time variations. Data set `dp1_b`.

3 Approximations of Equation of Motion for Short-Time Behaviour

The problem of recovering the dynamical system from a series of RR-intervals and measuring its characterictics has been considered e.g. in (Lefebvre et al. 1993). But the results allow neither to reject the dynamical character of data, nor definitely confirm it. One of the reasons for it might be the nonstationarity of data. Here we tried to avoid this problem by recovering the dynamics from short parts of time series, when it could remain stationary.

Usually the variations of x_i have several characteristic time scales. It is generally believed that short-time variations are related with breathing rhythm, while long-time variations – with some other processes. In the power spectrum of x_i series often there are two different spectral peaks, corresponding to those process, which are separated by a through (cf. Fig. 2).

The original idea of using this separation of scales was to try to represent the data as being generated by a low-dimensional dynamical system with coefficients slowly varying in time. We hope that short time dynamics may be rather simple and can be described by a low-dimensional mapping, for example,

$$x_n = a_0 + a_1 x_{n-1} + a_2 x_{n-2} + a_3 x_{n-1}^2 + a_4 x_{n-1} x_{n-2} + a_5 x_{n-2}^2$$
$$+ a_6 x_{n-1}^3 + a_7 x_{n-1}^2 x_{n-2} + a_8 x_{n-1} x_{n-2}^2 + a_9 x_{n-2}^3 . \quad (2)$$

Here the embedding dimension $m = 2$. Since the number of coefficients a_i for this approximation is rather small (here $n_a = 10$), the set of a_i can be determined from a short part of the time series of the length $L_a \cong 3n_a$. As

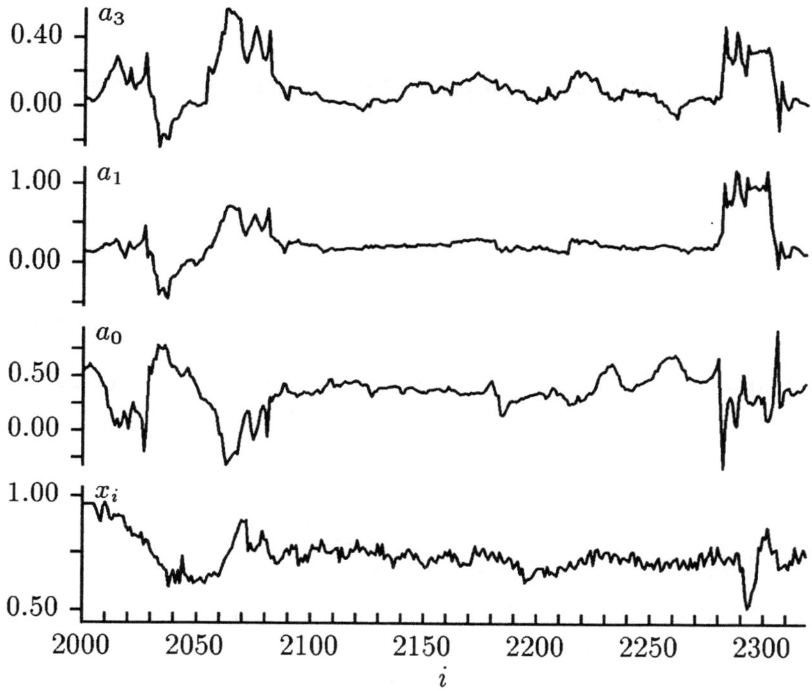

Fig. 3. The behaviour of coefficients a_i. Data set dp1_b. This figure correspond to approximation (2). For higher-dimensional embeddings a_i behave similarly, in particular almost alwais $|a_i| < 1$.

experiments show, for shorter L_a spurious oscillations may arise due to bad conditioning of the least square fit matrix, for this reason also it has been necessary to use special regularization methods. When the coefficients a_i are determined, the L_a window is moved one point to the right and the next set of coefficients is determined. By such a way, instead of a single series x_i we obtain n_a seria for $a_j(t)$.

We used various approximations of type (2) with embedding dimensions m from 2 to 5, power of the polynomial from 2 to 5 and tried to take into account all or only some of the nonlinear terms, to keep the number of unknown coefficients not too large. But surprisingly all the results looked similar.

The typical behaviour of a_j is shown in Fig. 3. It can be seen that there are intervals of almost constant a values (small variations may be caused by bad matrix conditioning) separated by short parts of abrupt a change. As also can be seen, often such a instabilities coincide with abrupt changes in x_i.

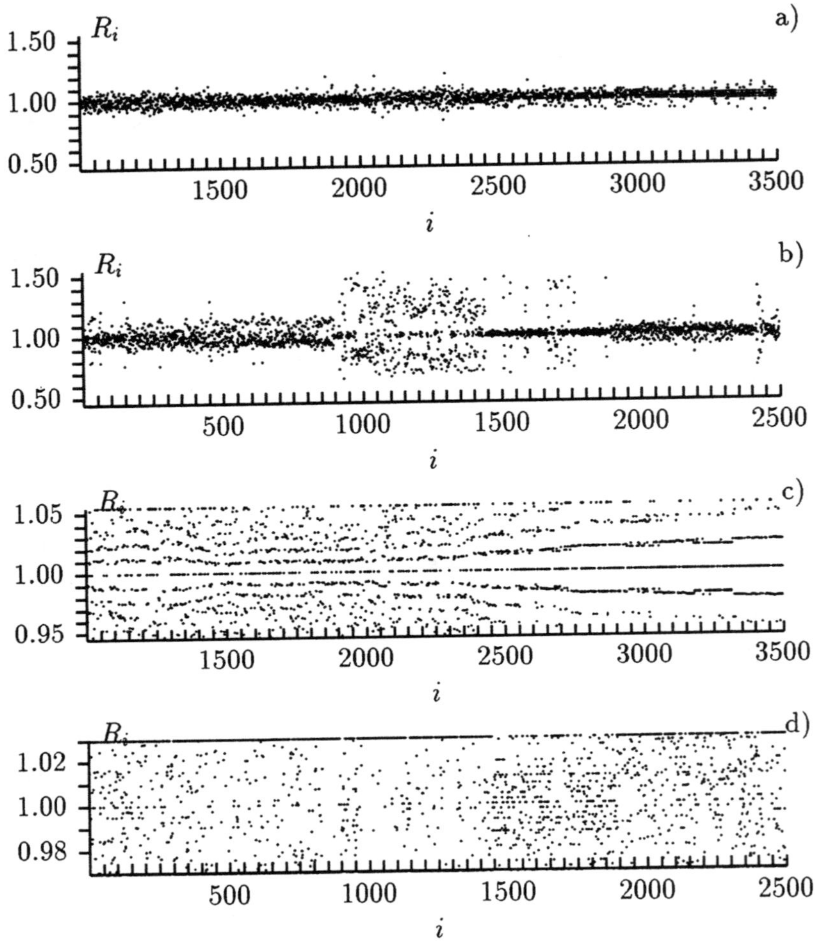

Fig. 4. Variations of $R_i = x_i/x_{i-1}$. Data sets dp1_b (a,c) and p15m (b,d)

Such changes in x_i may be related with the well-known extrasistola phenomenon. The proportion of such events varies from person to person and can be visible on the plots of the function $R_i = x_i/x_{i-1}$. The example (part of the data array dp1_b) is shown in Fig. 4a, and Fig. 4b shows part of p15m series plot. Figures 4c and 4d shows the enlarged central parts for the same data as in Figs. 4a and 4b. Here the new effect appears – points become situated along the lines. Most probably this is due to finite data precision, because these lines repeat the variations of local average of x. This effect does not allow the detailed study of the structure of oscillations near to 1.

Therefore, the first problem of this local approximation technique comes from nonstationarity – rare and abrupt changes. The second one arises due to the fact that almost always the obtained set of coefficients a_i correspond to very simple dynamics – fast convergence to a fixed point $x_{i+1} = x_i$, often even $\sum_{k>0} |a_k| < 1$ (remind, that usually $x < 1$, and for essentially nonlinear behaviour a_k for nonlinear terms must be greater than 1). This may mean, that the short-time behaviour arise not due to low-dimensional dynamics. From the viewpoint of local approximations the situation is most probably described by simple linear or nonlinear autoregression model with external noise

$$x_n = \sum a_k P_k(x_{n-1}, \ldots, x_{n-m}) + \xi_n$$

without any internal dynamics except convergence to the fixed point, which corresponds to regular heart beat without variability. It proves that qualitatively this result does not depend on the specific choice of m and the order of polynomial (2), unless n_a becomes too large for approximation to be local in time, when such approach becomes less justified. That is, low-dimensional short-time approximations leave chaos to external perturbations.

4 Correlation Integral: Comparison with Noise

The next step of the data analysis was the dimension calculations. The results on the estimates of correlation dimension for RR-intervals data were reported earlier, e.g. (Lefebvre et al. 1993). The plots of correlation integral presented there almost do not have linear part and the reported dimension estimate was at least 4. The plots of correlation integral for the data analyzed in this paper look similar to the previously mentioned results. The examples for the data series `dp1_b` and `p15m` are shown in Fig. 5. At large scales the slope of $\log C(\varepsilon)$ vs $\log \varepsilon$ plot gradually increase, and on small scales the effects of finite series size and data precision become important. As a rule, there is no linear part, and only sometimes there are traces of linearity.

As it was mentioned above, the results of approximation of short-time equations of motion enabled the interpretation, that the data were generated by a simple dynamical system perturbed by noise. Therefore, it seems reasonable to find out, whether the results of dimension calculations can be interpreted in a similar way, that is, can purely noisy data give similar plots of correlation integral.

Usually for the calculation of correlation integral one uses norms L_1 or L_2. In the L_2 norm the distance between two vectors is calculated as

$$\rho_{ij} = \sqrt{\sum_{k=0}^{m-1} |x_{i+k} - x_{j+k}|^2} = \sqrt{\sum_{k=0}^{m-1} \delta^2_{i+k,j+k}}.$$

Let us suppose that δ_{ij} and $\delta_{i+1,j+1}$ are independent random numbers. Their distribution density $p(\delta)$ may be calculated from a time series, and it appears, that for almost all time seria analyzed this distribution was close to

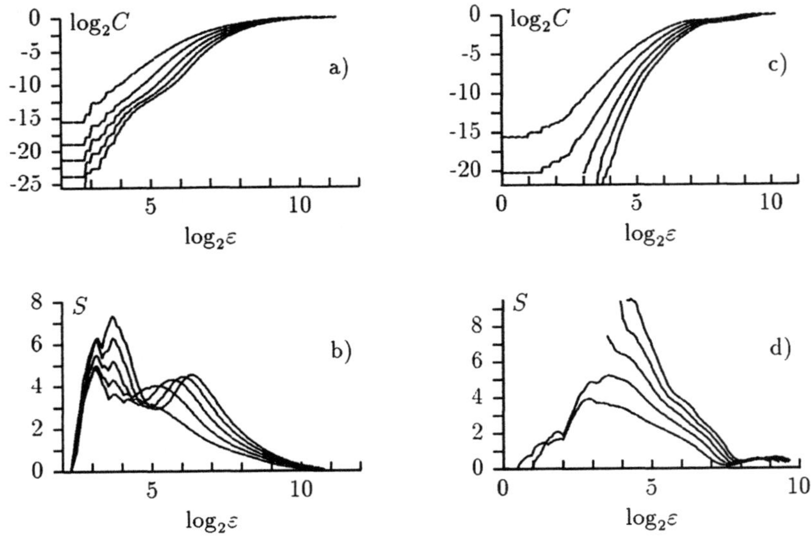

Fig. 5. Plots of correlation integral and its slope for data sets dp1_b (a,c) and p15m (b,d). The embedding dimension $m = 4, 6, 8, 10, 12$.

"semi-Gaussian" (Gaussian for $\delta \geq 0$, and $p = 0$ otherwise) in the domain, where it is concentrated, for small δ values, until p decreases by 2–3 orders of magnitude. The plots of $\log(p(\delta)/p_{\max})$ vs δ^2 for a number of data sets are shown in Fig. 6, and in most cases the dependence is close to linear. So we assume that δ distribution density is $p(\delta) = A_1 \exp(-\delta^2/2\sigma^2)$ for $\delta \geq 0$ and $p = 0$ otherwise.

Then in m-dimensional embedding ρ will have the so-called χ-distribution with the density $p_m(\rho) = A\rho^{m-1} \exp(-\rho^2/2\sigma^2)$, and in case of even $m = 2k$ the correlation integral $C_m(\varepsilon) = \int_0^\varepsilon p_m(\rho)d\rho$ can be calculated analytically. The slope (dimension estimate) also can be obtained,

$$S = \frac{d \log C_m}{d \log \varepsilon} = \frac{\varepsilon p_m(\varepsilon)}{C_m(\varepsilon)} = mS_0(\varepsilon, m)\,,$$

where the function $S_0(\varepsilon, m)$ weakly depends on m , and therefore we come to the conclusion, that the ratio S/m for noise should be almost independent from m, in contrast with chaotic systems, when it (at least on small scales) tends to zero. Similar approach has been proposed in (Schreiber 1993) to determine noise level in chaotic systems.

The plots of analytically calculated $\log C(\varepsilon)$ and corresponding S/m are shown in Fig. 7.

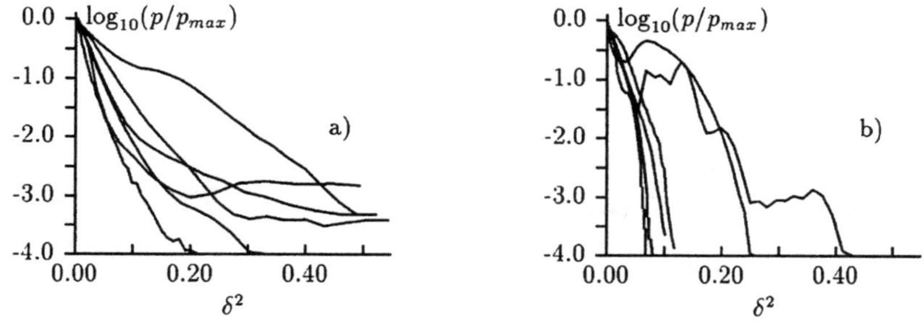

Fig. 6. Plots of probability distributions for $\delta_{ij} = |x_i - x_j|$ for all sets $\mathbf{dp}x\_y$ (a) and $\mathbf{p}z\mathbf{m}$, $z = 1, 5, 7, 11, 15, 17$ (b).

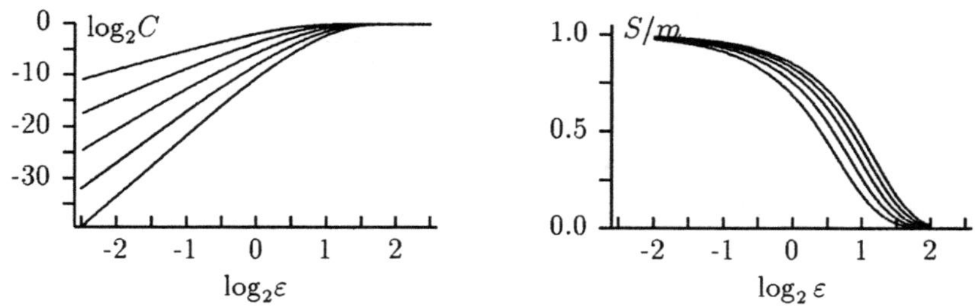

Fig. 7. Plots of correlation integral and its slope divided by m for pure noise. $\sigma = 1$, $m = 4, 6, 8, 10, 12$.

The plots of S/m, (m=4, 6, 8, 10, 12) for a number of data sets are presented in Fig. 8. It is interesting, that sometimes the plots resemble a combination of two similar noisy patterns, as in Figs. 8b and 8c. Probably, they correspond to two different processes – small clutter about mean value and large amplitude outbursts. But most important is another feature – when m increases these plots tend to some "limiting" curve, and therefore it is hard to distinguish, whether the underlying system is chaotic or stochastic.

5 Conclusions

The main conclusion for this work is that by applying several methods of nonlinear dynamics it seems to be impossible to tell whether time seria of RR intervals are chaotic or stochastic. This may be due to several reasons.
1. Accuracy of data is not high enough. Fig. 4 definitely shows that most small scale oscillations occur near to the threshold of data discretization (about 1ms). The detected noise may be partially caused by discretization errors.

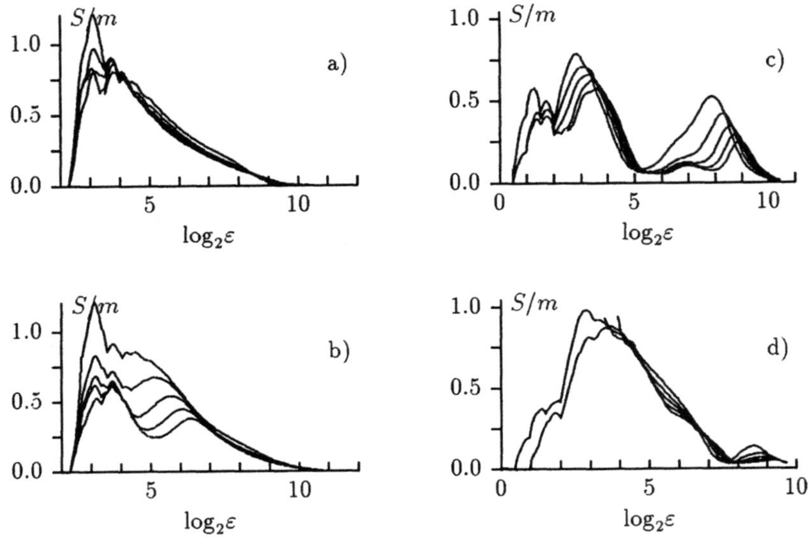

Fig. 8. Plots of S/m for a) dp1_a, b) dp1_b, c) p5m, d) p15m. The embedding dimension $m = 4, 6, 8, 10, 12$.

But if, for example, the function **G** in the RR-dynamical system (1) is too complex and resemble that for random number generators, then the dynamics of heart in the RR-intervals approach can be properly reconstructed only from those small-scale oscillation, which are beyond the measurement precision (Malinetskii et al. 1993). In other words, RR intervals may be inadequate observable for reconstruction of heart attractor, provided the latter exists.
2. Some conditions of Takens theorem may be violated in this approach, which also prevents proper reconstruction of dynamics.
3. Nonstationarity of the data (see e.g. Isliker & Kurths 1993) also can cause similar effects.But since the reconstruction of dynamical system for short-time variations proved to be unsuccessful, then probably other localized measures (Mayer-Kress 1994) may also be affected by the same factors.

Nonetheless, statistical approaches may prove to be more helpful, e.g. (Pompe 1993, Ivanov et al. 1996). Also for cardiograms themselves (not RR-intervals) the application of dynamical techniques works, as reported e.g. in (Babloyantz & Destexhe 1988, Chaos, V.5, 1995, Anishchenko et al. 1992).

Acknowledgements

This work has been done during author's one month visit to Potsdam. I would like to thank Prof. J. Kurths for hospitality and helpful discussions and to acknowledge support from Ministerium für Wissenschaft, Forchung und Kultur des Landes Brandenburg. I would like to thank Prof. J. Skinner and Prof. J. Zebrovski for giving me the numerical data, and I am grateful to M. Rosenblum, P. Saparin and M. Zaks for cooperation and useful comments.

References

Abarbanel H.D.I., Brown R., Sidorowich J.J., Tsimring L.S. (1993): The analysis of observed chaotic data in physical systems. Rev. Mod. Phys. **65** 1331

Anishchenko V.S., Postnov D.E., Saparin P.I., Safonova M.A. (1992): Diagnostics of self-oscillating systems by methods of nonlinear dynamics. Applied Nonlinear Dynamics **1**, 10

Babloyantz A., Destexhe A. (1988): Is the normal heart a periodic oscillator?. Biol. Cybernetics **58**, 203

Chaos, **5**, (1995): 1–215 present some recent results in processing of physiological time series

Eckmann J.P., Ruelle D. (1985): Ergodic theory of chaos and strange attractors, Rev. Mod. Phys. **57** 617

Isliker H., Kurths J. (1993): A test for stationarity: finding parts in time series apt for correlation dimension estimates. Int. J. Bifurc. Chaos **3** 1573

Ivanov P.Ch., Rosenblum M.G., Peng C.-K., Mietus J., Havlin S., Stanley H.E., Goldberger A.L. (1996): Scaling behaviour of heartbeat intervals obtained by wavelet-based time series analysis. Nature **383** 323

Lefebvre J.H., Goodings D.A., Kamath M.V., Fallen E.L. (1993): Predictability of normal heart rhythms and deterministic chaos. Chaos **3** 267

Mayer-Kress G.(1994): Localized measures for non-stationary time-series of physiological data. Integrative Psychological and Behavioral Science **29** 203

Malinetskii G.G., Potapov A.B., Rakhmanov A.I. (1993): Limitations of delay reconstruction for chaotic dynamical systems. Phys. Rev. **E 48** 904

Packard N.H., Crutchfield J.P., Farmer J.D., Shaw R.S. (1980): Geometry from a time series. Phys. Rev. Lett. **45** 712

Pompe B. (1993): Measuring statistical dependencies in a time series. J. Stat. Phys. **73** 587

Sauer T. (1995): Interspike interval embedding of chaotic signals. Chaos **5**, 127-132

Sauer T., Yorke J.A., Casdagli M. (1991): Embedology. J. Stat. Phys. **65** 579

Schreiber T. (1993): Determination of the noise level of chaotic time series. Phys. Rev. E **48** R13

Takens F. (1981): Detecting strange attractors in turbulence, In Dynamical systems and turbulence. Lecture Notes in Mathematics Vol. 898. Springer, Berlin, Heidelberg, p. 336

New Nonlinear Algorithms for Analysis of Heart Rate Variability: Low-Dimensional Chaos Predicts Lethal Arrhythmias

Low-Dimensional Chaos in Heartbeats

James E. Skinner[1], Jan J. Żebrowski[2], and Zbigniew J. Kowalik[3]

[1] Totts Gap Institute, Bangor, USA
[2] Institute of Physics, Warsaw University of Technology, Warsaw, Poland
[3] Institute of Experimental Audiology, University of Münster, Münster, Germany

Abstract. Reduced autonomic control of heartbeat intervals occurs with advanced heart disease and is an independent risk factor for mortality in cardiac patients. Such loss of control is manifested in the heartbeat intervals as a reduction in the total variability, including contributions made by oscillatory reflexes. Animal studies suggest that although the loss of autonomic control may arise following acute coronary artery obstruction, myocardial infarction, or other cardiological events, it may also arise periodically from psychological stress and other transient nonstationary influences mediated by the nervous system. Recent clinical studies of high-risk patients suggest that the deterministic measures of heartbeat dynamics may be more sensitive and specific predictors of risk of death than the more usual stochastic ones, such as the mean, standard deviation or power spectrum. In the present study, several new algorithms based in deterministic chaos theory were applied to a data set made from 20 high-risk patients, each of whom had documented nonsustained ventricular tachycardia (VT) and 10 of whom manifested lethal ventricular fibrillation (VF) within 24 hr. Only the algorithms which measured the time-dependent dimensional complexity (D2i, PD2i) in the data were able to discriminate those patients that later manifested VF. The algorithm which treated the problem of data nonstationarity (PD2i) had the highest sensitivity (100%) and specificity (100%) ($P < 0.001$, binomial test). Those algorithms which detected order in the data (stochastic-surrogates, determinism, largest Lyapunov exponent, entropy) clearly showed all data to contain low-dimensional chaos, but the order itself did not discriminate between VF and VT risk. It is concluded that among the nonlinear measures of heart rate variability, the ones that quantify the time-dependent complexity, as opposed to detecting the order, are best able to predict clinical risk of sudden cardiac death.

1 Introduction

Sudden cardiac death is predominantly due to ventricular fibrillation (VF) and it accounts for over 500 000 yearly fatalities in the United States alone (Rapaport 1988). A low ventricular ejection fraction or a high degree of premature ventricular complexes observed in a 24 hr electrocardiogram are noninvasive indicators of risk (MPRG 1983). Although their sensitivity is statistically significant, their predictive power for a given individual (i.e., specificity) is not very good, nor do they suggest when the lethal event might occur (Pratt et al. 1987). Based upon recent insight into the involvement of the autonomic nervous system and higher cortical centers in animal models of sudden cardiac death (Skinner et al. 1975a; Gillis et al. 1976; Verrier and Lown 1981; Skinner et al. 1981; Parker et al. 1990), the relationship of the neurocardiac activities to cardiac vulnerability to VF is being closely examined (Billman et al. 1982; Hull et al. 1990). In patients with a myocardial infarction, the standard deviation of spontaneously varying interbeat intervals and the sensitivity of interbeat intervals to forced changes in blood pressure have both been shown to be prospective predictors of mortality (Wolf 1972; Myers et al. 1986; Martin et al. 1987; Kleiger et al. 1987; La Rovere et al. 1994; Bigger et al. 1988; Rich et al. 1988; Bigger et al. 1989; Kleiger et al. 1990). Power-spectral analysis of oscillations in the heartbeat intervals suggests an increase in sympathetic as well as a decrease in parasympathetic oscillatory activities in the vulnerable individuals (Bigger et al. 1989; Lombardi et al. 1987; Myers et al. 1986; Malliani et al. 1991).

It has been proposed that fluctuations in heart rate manifest deterministic chaos (Mayer-Kress et al. 1988; Babloyantz and Destexhe 1988; Skinner et al. 1991). The use of stochastic analytic predictors, such as the mean, standard deviation, or power spectrum, may therefore be inappropriate for describing the dynamics of the heartbeats. In the conscious pig, the chaotic dimension of the heartbeats is immediately reduced following acute coronary artery occlusion; the degree of the reduction discriminates which occlusions will later result in VF and which will not (Skinner et al. 1991). Stochastic measures used on the same small data sets have no predictive power (Skinner et al. 1991; Skinner 1994a).

Recent clinical studies (Skinner et al. 1993; Vybiral and Skinner 1993) show results similar to the animal experiments. In blinded retrospective studies, it was shown that a temporal reduction in the heartbeat dimension occurs in those patients who later manifest fatal VF, but it does not occur in those who survive (sensitivity = 100%, specificity = 83%, $P < 0.01$). The reduction in dimension in the humans was approximately the same as that in the pigs who manifested VF following coronary artery occlusion, but in the humans there were no signs of acute ischemic injury.

Transient alterations in heartbeat dimension are produced in conscious pigs by neurocardiac perturbations, such as psychological stress and intracerebral propranolol (Skinner et al. 1991; Skinner 1994b). Psychological stress reduces heartbeat dimension and increases vulnerability to fatal VF, whereas intracerebral beta-blockade increases dimension and decreases vulnerability to VF. These data strongly suggest that neurocardiac mechanisms can produce nonstationary alterations in heartbeat data relevant to the assessment of cardiac risk.

The purpose of the present study is to compare the predictive abilities of the new algorithms based in deterministic chaos theory with those based in stochastic models, when each is used on the same data set from high-risk patients. Of the new nonlinear algorithms for assessing heart rate variability, some require data stationarity and some do not. Some quantify the order and some only detect order. The results will show that time- dependent quantification of order in the heartbeats, by an algorithm that addresses the problem of data nonstationarity in a new way, clearly discriminates among the high-risk patients the ones that will manifest VF within 24 hr from those that will manifest only VT within the next three years.

2 Methods

2.1 Subjects

The P1–P20 subjects were patients of the Cardiology Section of Baylor College of Medicine and are the same as those used in a previous clinical study (Skinner et al. 1993). It is simple to observe physiological differences between high-risk cardiac patients and normal subjects; one does not need sophistocated heart rate variability studies to document these differences. The appropriate data set in which to examine predictors of outcome is one in which the controls are documented by independent clinical criteria to be at the same high risk as the experimental subjects, but, in the long-term, have a negative outcome. For the P1–P10 subjects, 10 Holter tapes were randomly selected from the archives that were recorded from patients who manifested VF on the day of the recording (VF patients). These were compared to 10 tapes (P11–P20 subjects) chosen at random from patients who, like the VF subjects, had documented nonsustained VT, but with no VF observed for at least three additional years (VT patients).

All P1–P20 subjects had coronary artery disease accompanied by premature ventricular complexes plus episodes of nonsustained VT. Medications varied widely and were used in multiple combinations. The following drugs and numerical sequences refer to subjects in the VF and VT groups, respectively, taking that drug: calcium channel blockers 3, 10; nitrates 4, 5; digitalis 8, 7; diuretics 7, 2; beta-receptor blockers 2, 4; vasodilators - 1, 8; angiotensin converting enzyme inhibitors 1, 2; class-1 antiarrhythmics 1, 5. Both groups had comparable cardiac dysfunctions and medical histories as will be presented in the results section.

Two additional subjects, one manifesting VF in the Holter record and the other manifesting only nonsustained VT, had 24 hr analyses performed on their heartbeat intervals after digitization of the electrocardiograms at 512 Hz. These subjects were used to document the validity of the use of 12 min samples of data from the P1–P20 subjects.

Twelve additional high-risk subjects were from the National Institue of Cardiology, Warsaw. These subjects had electrocardiograms digitized at 128

Hz, and because of the high descretization noise in the heartbeat intervals (see the results section), these data were not evaluated by all algorithms. All data were viewed and reviewed by a qualified cardiologist.

2.2 Data Acquisition

The analog electrocardiographic tapes from the Holter-monitors of the P1–P20 subjects were digitized by a MassComp computer (12 bits, 512 Hz) and consecutive interbeat intervals (RR) were constructed using a three- point convexity operator set to the width of the R wave; the output of the operator is maximum when points 1 and 3 lie near the baseline and point 2 is in the peak of the R wave); threshold detection (60% maximum dv/dt) of the convexity operator output provides stable R wave detection that is within 1 ms of that determined by an experienced cardiologist in 95% of the cases (Negoescu et al. 1993).

In each of the P1–P20 subjects, the number of data points between each identified R wave were counted and were not converted to msecs, because this would increase the absolute value of the descretization noise (see the results section). Each tape had three epochs selected for analysis: A, immediately after the beginning of the recording; C, immediately before VF or at matched comparable times in the control tapes; B, midway between the A and C records. Each A, B, C epoch was around 12 min in duration. Both linked composites of the A, B and C records and the individual records themselves were analyzed. For patients P1–P4 and P11–P14, all A, B, C subepochs were sampled from arrhythmia-free parts of the tapes.

For the independent VF and VT subjects, the entire 24 hr record was digitized at 512 Hz and the RR intervals made as above. For the 12 patients from Warsaw, the RR intervals were made with a scanner (Del Mar Avionics) from 24 hr electrocardiograms digitized at 128 Hz.

2.3 Analytic Algorithms: Stochastic Measures

Power Spectrum (Fourier Analysis) of Stationary Data This measure presumes data stationarity. The use of a Hamming window (i.e., to enhance middle frequencies to compensate for edge effects) is desirable for short epochs of data, but for the present epochs of around 1 000 heartbeats each it was considered to be unnecessary. The discrete Fourier transform was preferred over the fast method, as the data lengths varied somewhat in size. Conversion of interval data from the beat domain to the time domain was achieved by linear interpolation between fixed resampled RR measures (4 resamples for the shortest RR interval determined the resampling rate). Integration of the power in the 0.05 to 0.15 Hz range represents the power band in the Traube–Hering–Meyer oscillations (THM) shown to be of mixed sympathetic and parasympathetic origin and produced by a mechanism regulating beat-to-beat blood pressure (see (Negoescu et al. 1993)). Integration of the power in the 0.25 to 0.40 Hz range represents the power band in the respiration-associated oscillations (RSA) that are shown to be of parasympathetic origin and produced by a mechanism regulating the respiratory sinus arrhythmia. The peak-power ratio (THM/RSA) has been associated with risk of VF (Myers et al. 1986; Lombardi et al. 1987; Malliani et al. 1991).

Mean Standard Deviations (MSD) of Successive 5-min Subepochs of Stationary Data The serial standard deviations of 5-min subepochs of stationary RR intervals have been shown to have a mean that is smaller in moderately high-risk patients who experience VF than in those who do not (Wolf 1972; Rich et al. 1988; Bigger et al. 1989; Kleiger et al. 1990). To achieve data stationarity those intervals altered by an arrhythmia were removed.

2.4 Analytic Algorithms: Deterministic Measures

Correlation Dimension (D2) of Stationary Data This is the classical approach for calculating the correlation dimension developed by Grassberger and Procaccia (Grassberger and Procaccia 1983) and commonly used in many published studies. It requires data stationarity, embedding dimensions of at least $2 \times D2+1$, and Tau selection appropriated to the first zero crossing of the autocorrelation function (i.e., first minimum approaching the zero crossing may also be used). For RR interval data, Tau = 1 is generally determined. Any noise in the data will produce spurious results. Any data oversampling or undersampling (i.e., digitization rate) must also be carefully considered to eliminate near-neighbor effects or stroboscopic effects due to poor Nyquist samplings. As data filtering can also lead to spuriously low calculations, the resulting D2 was compared to that of a randomized-phase surrogate (Theiler et al. 1992), to be sure that the data were not of a stochastic origin (e.g., high-dimensional noise). Another surrogate was also used, one in which the sample and surrogate probability distributions were the same (Schreiber, this volume). The parameters used were: linearity criterion, LC = 0.3 (± 15% of second derivitive of correlation integral); convergence criterion, CC = 0.40 (maximum standard-deviation of slope mean for embedding dimensions m = 9 to 12); Tau = 1.

Pointwise Correlation Dimension (D2i) of stationary data This algorithm is the same as D2, with the exception that instead of all possible vector difference lengths being in the rank-ordered D2 set, which is used for making the correlation integral, only those vector difference lengths relative to a fixed reference vector are made (Mandelbrot 1982; Farmer et al. 1983). This makes the dimensional estimate time-dependent, since the position of the reference vectors is temporal, instead of global, like the D2 algorithm. Mayer-Kress et al. (1988) used D2i to evaluate three types of physiological time series (EEG, RR's of EKG, EMG), and they suggested that it is less sensitive to data nonstationarities than the D2 algorithm. D2i, however, still presumes data stationarity in the underlying model (Farmer et al. 1983). It is just more robust when stationarity is violated, because of the dominance effect of the reference vector (Mayer-Kress et al. 1988). The parameters used were: linearity criterion, LC = 0.3 (± 15% of second derivitive of correlation integral); convergence criterion, CC = 0.40 (maximum standard deviation of slope mean for m = 9 to 12); Tau = 1.

Point Correlation Dimension (PD2i) of Either Stationary or Nonstationary Subepochs This algorithm developed in Skinner's laboratory (Skinner et al. 1991; Skinner et al. 1994) addresses the problem of data nonstationarity. By restricting the scaling to the small log $-r$ region of the correlation integral of the conventional D2i, only vector difference lengths similar to those that the reference vector is in will contribute to the slope; i.e., the small, log $-r$, vector-difference lengths will only occur in abundance if the comparison vector is in a subepoch that is stationary with respect to the one the reference vector is in. The result (as will be illustrated in Figs. 1 and 2 below) is that the slope is insensitive to nonstationary changes in the data; this insensitivity occurs because the comparison vectors in nonstationary subepochs all make larger, log-r, vector-difference lengths, and hence do not significantly affect the slope in the small log-r scaling-region. The PD2i algorithm is different from what is called the "local dimension" of an attractor (Judd and Mees 1991), because the small log-r values are not made by just the near neighbors. The PD2i applied to RR data from high-risk subjects, both animal (Skinner et al. 1991; Skinner 1994a) and human (Skinner et al. 1993; Vybiral and Skinner 1993), shows low-dimensional excursions only in those subjects that later (i.e., within 24 hr) manifest VF; thus PD2i can assess risk of death among high-risk cardiac patients. The parameters used were: linearity criterion, LC = 0.3 (± 15% of second derivitive of correlation integral); convergence criterion, CC = 0.40 (maximum standard-deviation of slope mean for m = 9 to 12); Tau = 1; Plot Length, PL = 0.15 (restricted length of scaling region of correlation integral is from r = 0 to 15% of the total number of r values).

Determinism (Det) of Stationary Subepochs This is a relatively new algorithm (Kaplan and Glass 1992) that has been used on biological data for determining whether or not a given series is deterministic or is stochastic. It measures the local direction of prediction vectors as a function of their position in phase space. For random data none of the vectors will be oriented in a predictable direction with a time increment, whereas those that are deterministic will show some (i.e., high) predictability. The parameters used were m = 10, n = 4, passes = 10 to 15.

Dynamic Determinism (DynDet) of Both Stationary and Nonstationary Subepochs This algorithm (Mühlnickel et al. 1994), like PD2i, presumes that nonstationarities will arise in the data stream. It is basically the same algorithm as Det in that it looks at the mean orientation of vectors as a function of their positions in phase space. DynDet, however, shows higher levels of determinism than the static version (Det), that is, when used on known chaotic data. It is also more sensitive when comparing control and experimental data (i.e., it produces larger F values in statistical tests); note that this same larger F effect is also seen when comparing D2 (static) and PD2i (dynamic) values used on the same data. The parameters used in the present study were $m = 10$, $n = 4$, passes = 10 to 15.

Approximate Entropy (ApEn) of Both Stationary and Nonstationary Subepochs This is the algorithm developed by Pincus (Pincus 1991) to circumvent the need to distinguish between stochastic and deterministic data series; i.e., it makes no presumption as to whether the serial data show random variation around a local mean or are completely determined in their variation. The relative contributions of high-dimensional (e.g., noise) and low-dimensional data (e.g., sinewave) in nonstationary data, however, will determine the algorithmic output. The algorithm has been shown to be quite sensitive to RR interval changes between some control and experimental groups (Pincus and Viscarello 1992; Pincus and Goldberger 1994). Pincus suggested that a defined single value of r be used in the ApEn calculation for comparisons between subjects (i.e., 20% of the standard deviation of the dynamic range of the data). Storella and associates (Storella et al. 1995) suggested the use of a range of r that spans this single value, so that the peak or a peak-related value can be used in the comparison; when applied to RR data of humans data, however, they found that ApEn showed a greater sensitivity to anesthesia (36% reduction) than peak ApEn (12% reduction), although both were statistically significant; PD2i showed a reduction similar to ApEn (33% reduction). The parameters used in the present study were $m = 2$, $r =$ 1 to 10, Tau = 1.

Pattern Entropy (PatEnt) for Stationary and Nonstationary Data This algorithm was developed by Żebrowski and associates to analyze 24 hr heartbeat data from cardiac patients (Żebrowski et al. 1994). This measure is 3-dimensional, as opposed the more conventional 1-dimensional entropies. Because it uses joint-probabilities, it has higher values when more order exists in the data. It has been shown to have a statistically significant sensitivity in discriminating RR intervals recorded from the high-risk VF patient (electroconverted) and the RR's observed in the same patient after a 1-year survival. The running window version has less sensitivity to data nonstationarities. Thus far PatEnt has only been used on 24 hr RR's made from data digitized at 128 Hz (Del Mar Avionics scanner). The parameters used in the present study were: window length = 400; Tau = 2; bin width = 15 ms.

Largest Lyapunov Exponent for Stationary and Nonstationary Data (DynLyap or "Chaoticity") This algorithm was recently developed by Kowalik and Elbert (1994) to assess divergence in a data series. Since the seminal work of Poincare (Poincare 1892) on mathematical divergence and chaos, the largest Lyapunov exponent (Lyapunov 1892) having a positive value has been the standard by which a data series is judged to be chaotic. To set closer boundaries than positive and negative infinity for random to periodic data, respectively, the new algorithm has been modified so that periodic data converges to a positive value around 6 and random data converges to a positive value of 1. Chaotic data will produce values between these two extremes. To treat the problem of data nonstationarity, the DynLyap ("Chaoticity") algorithm uses a running window that passes through the data series and calculates the largest Lyapunov exponent in the window. A similar strategy is often used for assessing the power spectrum in data that may contain nonstationarities. Parameters used were: window length = 512; m = 10; Tau = 1; step = 1.

To confirm that this new algorithm provides results similar to the conventional methods, an implementation of the modified Wolff algorithm using a running window (Kowalik et al. this volume), was also employed.

2.5 Statistics

All data were analyzed in a coded, blinded fashion to preclude experimenter bias. For data that had normal distributions, random sampling and heterogeneity of variance, the Student t-test was used. For nonparametric comparisons the binomial probability test was employed (Winer 1962). Ten percent of the data were re-analyzed as a control for operator error.

3 Results

3.1 Assessment of Noise Levels in the Data

For a heartbeat interval of 500 ms (i.e., a short one, at the lower range normally encountered in patients), which is digitized at 512 samples per second, the uncertainty in calculating the RR interval is 2 samples out of 256; exactally where the R wave was located within the samples on each end of the 256 is unknown. This error, called descretization noise, can range between +1 or −1 samples out of the 256. For a heartbeat interval of 1000 ms this error reduces to half (i.e., 2/512, which is the same as ± 1/512). Thus the range of values of the descretization noise in percent, as RRs range between 500 and 1000 ms, is expressed as: ± 2.0 to ± 1.0 / digitization rate. If the digitization rate is only 128 Hz, then the descretization noise is increased four-fold.

If the data (number of samples between R waves in integers) are multiplied by a factor to convert samples to ms, such that 1 sample = 1 integer = 1 ms, (e.g., for data digitized at 512 Hz, the conversion factor is, 1000 ms / 512 samples = 1.953), then the value of the discretization noise is also multiplied by this same amount. The signal-to-noise ratio is not altered by the conversion, but the absolute value of the noise is.

The discretization noise is additive with those of other sources of Gaussian-distributed error as the variances squared (i.e., square-root of the sum of the squared values). The uncertainty of the algorithm for detecting the RR intervals, which employs a convexity operator to increase accuracy, has been found to be ± 1 ms (Negoescu et al. 1993); if the data are expresses as 1 ms = 1.953 samples (i.e., digitized at 512 Hz), then this is an equivalent RR algorithmic noise of ± 0.512 samples.

The wow-and-flutter uncertainty (w/f) of the Holter tape-recorder is a large source of noise that can increase if the batteries for the Holter monitor are not fully charged. The w/f was measured in the P1–P20 tapes by observing the timing signal (a constant frequency) placed on track 4. Using a calibrated Tektronix oscilloscope with a voltage-triggered sweep, it was found that over an interval equivalent to 500 ms on the tape (i.e., the playback speed was 46 times that of the record speed), the triggered signal showed a w/f variation that had a standard deviation of ± 0.23%; the variation was doubled at a longer interval equivalent to 1000 ms. The ± 0.23% to ± 0.46% values are inclusive for all P1–P20 tapes. Note that w/f error increases with interval length, whereas that of the descretization noise decreases. Thus the absolute value of the total error (in integers) for each of the P1 - P20 data files, digitized at 512 Hz, not converted from samples to ms (i.e., 0.512 samples = 1 ms), and having RRs ranging from 500 ms to 1000 ms, is: descretization noise = ± 2 integers to ± 1 integer; RR algorithmic noise = ± 1 ms × 1 integer/1.953 ms = 0.51 integer; w/f noise = ± 0.23% × 256 integers = 0.59 integers, to 0.46% × 512 integers = 2.36 integers; total (var. sq.) = ± 2.15 to 2.61 integers.

The ECG data processed by the Del Mar Avionics scanner uses 128 Hz for the digitization rate and multiplies the counted samples in each RR interval by a conversion factor of 7.81, so that 1 integer = 1 ms. The minimum descretization noise, without the 7.81 conversion, is four-fold larger (i.e., ± 2 out of 128 instead of ± 2 out of 512 for a 500 ms RR interval). With other noise sources remaining the same, the total noise calculated is dominated by the four-fold greater descretization error and is large enough to produce spurious results in some of the algorithms: descretization noise = ± 8 integers to ± 4.0 integers; RR algorithmic noise = ± 1 ms x 1 integer/1.953 ms = 0.51 integer; w/f noise = ± 0.23% x 256 integers = 0.59 integers to 0.46% x 512 integers = 2.36 integers; total (var. sq.) = ± 8.04 to ± 6.15 integers.

Adding ± 8 integers of noise to the data, which is above the ± 5 integer limit for the PD2i algorithm, will produce spurious results for both the PD2i and D2i algorithms (in (Skinner et al. 1994); also see Fig. 1 below). This amount of noise presumably would contaminate the outputs of the other nonlinear algorithms, as all of them are very sensitive to noise. Thus the total noise of the P1 - P20 data, which have not been converted to 1 sample = 1 ms = 1 integer, is well below the ± 5 integer noise-tolerance level, but that of the data produced by the Del Mar Avionics scanner is too high; reducing the absolute noise level in the data, by dividing by 7.81, would still leave too much noise contamination for a meaningful result.

3.2 Time-Dependent Dimensional Algorithms (D2i, PD2i)

Figure 1 shows the performance of the two time-dependent algorithms on computer-generated and electronic sine-generated data that have been concatenated to form an obviously nonstationary data epoch. The Henon data were generated by a map-function (iterative) and have a natural Tau = 1 (Tau determined by first minimum or zero-crossing of the autocorrelation function). The Lorenz data were generated by partial differential equations (continuous) and were made by adjusting dv/dt to 0.01 so that the first minimum of the autocorrelation function resulted in a Tau between 1 to 2. The 6 kHz sine-wave data were generated by an electronic generator (with 60 Hz ripple superimposed) so that the digitized values of each cycle would not be exactally the same, and again Tau = 1 was indicated. The random data were generated by a white noise generator, and these have a natural Tau = 1. Each subepoch had 1 200 data points.

As seen in the upper left of Fig. 1, both the PD2i and D2i algorithms accurately track the dimension of the periodic sine-wave data in time. For the aperiodic Lorenz and Henon data there is considerable point-to-point variation. The subepoch mean and standard deviation of the D2i values can be seen to be larger than those of the PD2i. Low-dimensional estimates for the random noise are accepted by the D2i algorithm that are rejected by the PD2i, using the same linearity and convergence criteria. Compared to the classical D2 values, calculated with a large data length (e.g., > 1 000 000 data points), the subepoch means of PD2i in these present nonstationary data (1 200 data points) differ by less than 4%.

The reason for the greater PD2i accuracy is illustrated in Fig. 2. By restricting the scaling length of the small $\log -r$ region (connected arrows), those points that still fall within the linearity criterion (i.e., those points to the right of the connected arrows) are not included in the slope calculation. Note that the PD2i and D2i algorithms employed the same calculation parameters (linearity criterion, convergence criterion, minimum scaling length, Tau, etc.), with the exception that the D2i algorithm does not have the scaling-length restriction of PD2i.

As illustrated in the upper left and upper right of Fig. 1, the time- dependent PD2i values are statistically significantly different at all points in time from those of the randomized-phase surrogate (SUR). Although both

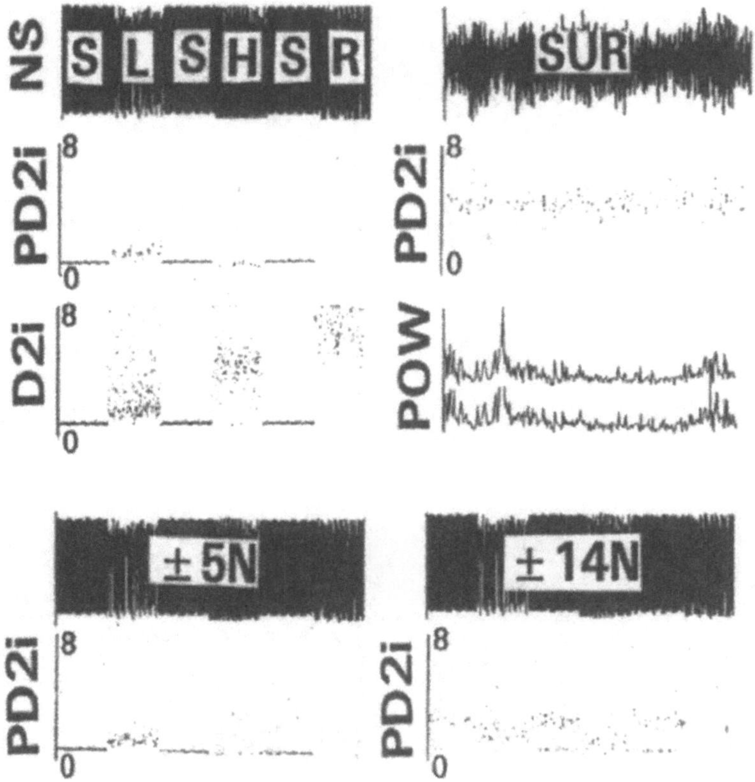

Fig. 1. Performance of time-dependent dimensional algorithms (D2i, PD2i) on nonstationary data (NS) made by concatenating 1 200 data points of subepochs made by known generators (S = sine, electronic; L = Lorenz; H = Henon; R = random). Upper left: individual PD2i's and D2i's with algorithms run on the same data (NS) using the same parameters for linearity, convergence and Tau. Upper right: randomized phase surrogate (SUR) made from the NS data; only the PD2i's are shown and these are statistically significantly different from those at the immediate left; POW shows that the power spectrum for the NS and SUR are identical. Lower left: the NS data have had ± 5 integers of random noise (± 5 N) added point-by-point to them; the PD2i's are seen to be essentially the same as those at the upper left. Lower right: the NS data have had ± 14 integers of random noise added; the PD2i's are seen to be spurious, when compared to those at the upper left.

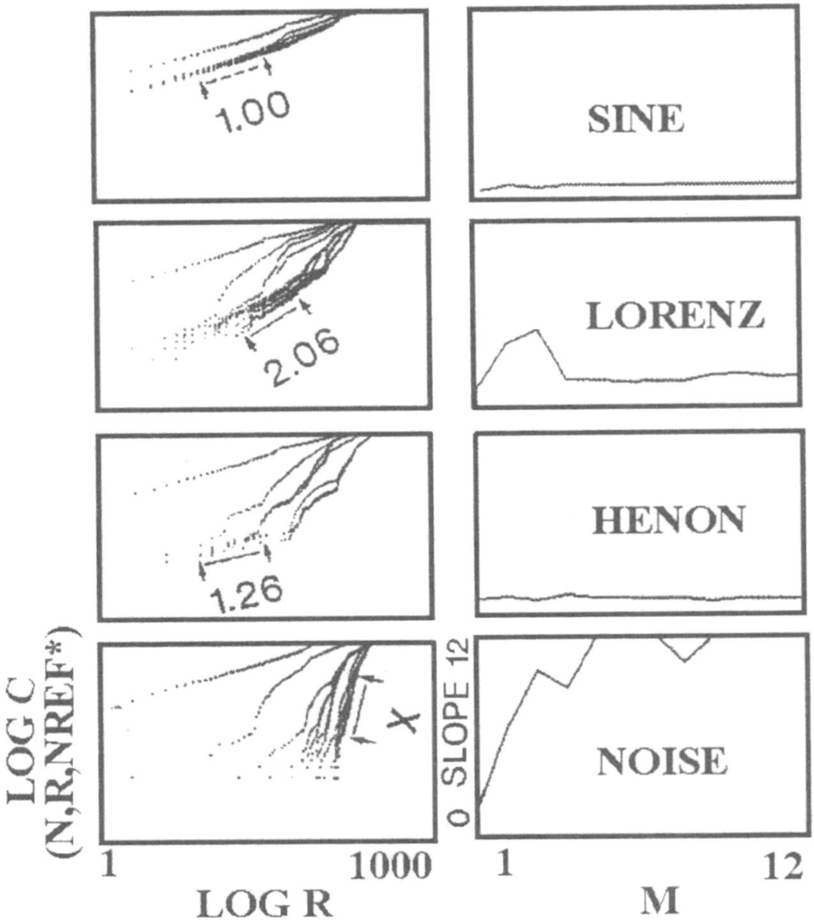

Fig. 2. PD2i correlation integrals (LOG C vs LOG R) and convergence plots (SLOPE vs M) for the NS data seen in the previous figure, when the reference vector is in each of the four types of data. The connected arrows in the left panels show the restricted scaling length observed in the 12th embedding dimension along with the slope value. In the right panels, the "flatness" of the values from the 9 to 12 embedding dimensions (m) show clear convergence of slope vs m, except for the noise.

the nonstationary data and the corresponding surrogate have exactally the same power spectrum (POW), the surrogate consistently has a larger PD2i value. This is also the case when the comparisons are made between each stationary subepoch and its randomized-phase surrogate.

Figure 1 (lower half) illustrates another important point about the nonlinear algorithms - noise tolerance. Noise in the data is anathema to all nonlinear devices. There is, however, a way to deal with this problem when small amounts of noise are involved. The best way to understand this solution is by the "metaphor of the thread": up close a piece of thread looks three dimensional; at arms length it looks only two dimensional; when placed across the room at the limit of visibility it looks only one dimensional; thus dimension is a function of resolution (i.e., the amplitude resolution of the data). The trick is to get the noise in the data below the limit of resolution, without altering the fine-grain structure of the signal.

The trick in the PD2i algorithm is based on the fact that random noise with an amplitude of $\pm$ 5 integers or lower is set to be resolved as a straight horizontal line (i.e., set to have a dimension of zero, instead of a value between 0 and 0.5). The signal-to-noise ratio of the data, however, must be large enough so that the dimension of the signal is not significantly altered if the amplitude is reduced (i.e., reduced to a value such that the dynamic range of the signal remains above $\pm$ 5 to $\pm$ 100 integers); this reduction then places the noise level below the $\pm$ 5 integers of the noise tolerance limit. At the lower left of Fig. 1, it is seen that adding $\pm$ 5 integers to the nonstationary data (NS, seen at the upper left) does not significantly alter the PD2i values (compare upper left to lower left); adding $\pm$ 14 integers (lower right), however, causes severe errors and spurious results.

Because the noise in the P1–P20 data ranges from $\pm$ 2.15 to $\pm$ 2.61 integers, it can be analyzed without converting it from samples (expressed in integers) to 1 ms (i.e., by multiplying the integer value of the samples by 1/0.512 or 1.953). These data can still be analyzed if converted, because the noise level would still be below $\pm$ 5 integers. The RR data based on a 128 Hz digitization of the ECG, however, cannot be analyzed by PD2i or D2i, as the noise content is four-fold larger and will cause the type of spurious results seen in Fig. 1, lower right. Reducing the noise level by multiplying the data by 0.25 would be expected to produce spuriously low dimensional estimates, as the dynamic range of the data would be reduced well below $\pm$ 5 to $\pm$ 100; that is, random data reduced to below $\pm$ 100 integers would result in a spurious reduction of dimension from infinity to approximately 6.0 (Skinner et al. 1994).

The PD2i and D2i algorithmic results applied to the P1–P20 data are presented in Table 1. In column two, a description of the histograms of the PD2i and D2i distributions are indicated; many of them, especially in the VF group (P1 to P10), were not Gaussian in appearence and therefore preclude the usual parametric statistical evaluations. For those that do have a

Table 1. Time-dependent dimension (D2i and PD2i) predicts imminent ventricular fibrillation (VF) among high-risk cardiac patients (P1 to P20), each of whom had previously documented ventricular tachycardia (VT). The data analyzed were three concatenated samples of RR intervals made from three separate 12-min electrocardiograms sampled at the beginning, middle and end of the useful data on a 24 hr Holter tape. All electrocardiograms were digitized at 512 Hz and converted to RR intervals using an algorithm with a convexity operator; the data are expressed in units of digitization samples between successive R waves; the total noise level was below ± 3 integers. Abbreviations: RP-Sur = randomized-phase surrogate of data; Dist = description of distribution of PD2i or D2i values, where G = Gaussian, TP,G = two peaks but otherwise Gaussian, SL = skewed to the left, PinW = peak in white noise (i.e., flat distribution with a narrow peak); Pts = number of data points analyzed; Max/Min = maximum/minimum integers of data; %n = number of PD2i or D2i estimates meeting all criteria; Mean = mean PD2i or D2i averaged over all estimates; SD = standard deviation of the Means; Min = minimum values in distribiution (at least 10 values in an 0.02 dimensional interval of histogram). The global dimensions (D2) of the VF subjects (P1 - P10) were: 3.41, 1.69, 3.13, 1.56, 1.94, 1.18, 1.05, 1.72, 1.35, 3.45; those for the VT group (P11 - P20) were: 3.89, 3.80, 3.43, 4.06, 5.41, 3.92, 3.65, 4.56, 3.13, 2.57 (mean D2 for VF group is 2.05 and that for VT group is 3.84; $P < 0.01$, t- test); the D2 measure has questionable validity, as the data were nonstationary; overlap of VF and VT distributions of the mean PD2i's suggest that it is not a good discriminator of VF/VT outcome (i.e., P1, P3, P10 overlap P11, P20).

				PD2i				D2i				PD2i of RP-Sur			
Data	Dist	Pts	Max/Min	%n	Mean	SD	Min	%n	Mean	SD	Min	%n	Mean	SD	Min
P1	TP,G	3205	463/298	83.6	2.53	0.73	1.2	84.5	2.57	0.85	1.2	57.9	3.53	0.65	1.8
P2	SL	3184	430/333	89.1	2.08	0.63	1.2	89.1	2.08	0.63	1.2	60.6	3.47	0.67	1.6
P3	G	3935	383/329	78.1	2.90	0.86	1.0	80.2	2.98	0.97	1.2	62.1	3.62	0.68	1.0
P4	TP,G	4096	355/248	96.3	1.73	0.56	0.8	96.5	1.73	0.57	0.8	0.2	3.64	0.09	3.6
P5	PinW	3403	836/213	17.3	4.30	3.85	0.6	16.7	4.33	3.84	0.6	10.5	6.68	1.05	4.2
P6	TP,G	3425	417/181	68.1	1.08	0.46	0.6	68.8	1.11	0.61	0.6	33.1	5.45	0.94	2.8
P7	G	2337	627/262	71.1	2.51	1.87	0.6	70.9	2.50	1.84	0.8	25.8	4.59	1.09	2.4
P8	PinW	3359	591/140	34.7	3.21	1.84	0.8	21.4	5.36	2.83	1.8	12.7	6.18	0.96	3.8
P9	TPSL	5286	250/209	96.5	2.05	0.66	1.0	96.5	2.05	0.66	1.2	75.4	2.60	0.73	1.0
P10	SL,W	2415	900/200	17.9	2.28	1.32	0.8	4.6	2.75	2.10	0.6	1.7	5.06	2.01	1.8
P11	G	2681	529/325	74.0	3.30	0.88	1.6	77.5	2.91	0.76	1.6	53.0	3.65	0.71	1.6
P12	G	2341	567/425	86.8	2.97	0.77	1.6	86.7	2.97	0.76	1.6	50.4	3.84	0.80	2.2
P13	TP,G	2691	588/321	79.1	3.04	1.06	1.6	79.4	3.04	1.05	1.6	48.9	3.68	0.83	1.5
P14	G	3279	434/290	82.5	3.20	0.73	1.8	82.6	3.20	0.73	1.8	56.2	3.67	0.74	1.8
P15	GinW	2763	499/262	35.6	5.46	2.30	2.0	35.8	5.47	2.30	2.0	23.6	5.52	1.21	3.2
P16	G	2619	567/300	53.0	4.02	1.03	1.8	53.0	4.02	1.03	1.8	35.1	4.84	0.87	3.2
P17	G	3093	450/308	80.7	2.58	0.70	1.6	80.7	2.58	0.70	1.6	51.6	3.86	0.77	2.0
P18	G	2377	443/366	82.5	2.63	0.71	1.6	82.5	2.63	0.71	1.6	53.8	3.65	0.80	1.6
P19	G,SL	3663	360/278	87.8	2.04	0.63	1.4	87.8	2.04	0.63	1.3	66.1	3.60	0.69	1.6
P20	G	3442	440/290	54.7	4.26	1.67	2.2	54.8	4.26	1.67	2.2	23.4	5.87	0.93	3.2

Gaussian distribution, however, there is a clear reduction of the mean PD2i compared to its randomized-phase surrogate (RP surrogate, $P < 0.001$, t-test).

A criterion was adopted for predicting VF (1.2), a value which is based on the mean PD2i observed during the first minute following complete coronary artery occlusion in conscious pigs (Skinner et al. 1991). Both pigs and humans have the same dimension for RRs recorded from the normal heart during rest (i.e., approximately 3.0) (Skinner et al. 1991; Skinner et al. 1993). Using this prediction criterion, the primary result of the present study is that PD2i's < 1.2, as indicated in column 8 (PD2i/Min), occurred in all subjects who manifested VF within 24 hr (P1 to P10), and PD2i's > 1.2 occurred in all subjects (P11 to P20) who manifested only VT and survived for at least 3 years (sensitivity = 100%, specificity = 100%, $P < 0.001$, binomial probability test).

The D2i algorithm (Min D2i) also shows good discriminability between the VF and VT subjects, with one exception, P8. It was observed in P8 that few periods of data occurred in which the predominantly randomized RR intervals during atrial fibrillation converted to a normal sinus rhythm. For D2i this resulted in inaccurate slopes in the correlation integral, because the linear scaling region was quite contaminated by the noise. These same P8 data, when analyzed by PD2i, resulted in an uncontaminated (or less contaminated) linear scaling region, with a slope and Min PD2i that were less than the 1.2 criterion and predicted VF.

If the absolute level of the noise was increased, by converting the data so that 1 integer = 1 ms (i.e., multiplying each data point by 1.953), the D2i (Min D2i) lost its VF/VT discriminability; 6 of 10 VF subjects had Min D2i's greater than 1.2 and these overlapped with 6 subjects in the VT group ($P > 0.5$). The PD2i (Min PD2i), however, did not lose its discriminability; the PD2i results, after the data conversion, remained essentially the same as those in Table 1, although there was a reduction in the number of acceptable PD2i's (%n) of about 40%.

There was found to be a greater mean PD2i for the VT group compared to the VF group (3.35 compared to 2.47; $P < 0.01$, t-test), and the same held true for the D2i means, although the difference was not as large (3.37 to 2.75, $P < 0.01$, t-test). Because of the overlap of the distributions of the individual mean PD2i's of the two groups, the sensitivity and specificity for mean PD2i, as a measure of risk, is substantially degraded compared to that of Min PD2i, for which there was found to be no overlap.

Figure 3 shows the relationship of each PD2i to the single RR interval that served as the first coordinate for each of the 12 reference vectors (i.e., for embedding dimensions 1 through 12). In some subjects there were obvious nonstationary changes in the mean heartbeat interval during the A, B and C subepochs (e.g., P4, P13, P14, P17). For others nonstationarties were not apparent (e.g., P3, P9, P18). For some subjects the variation within a single

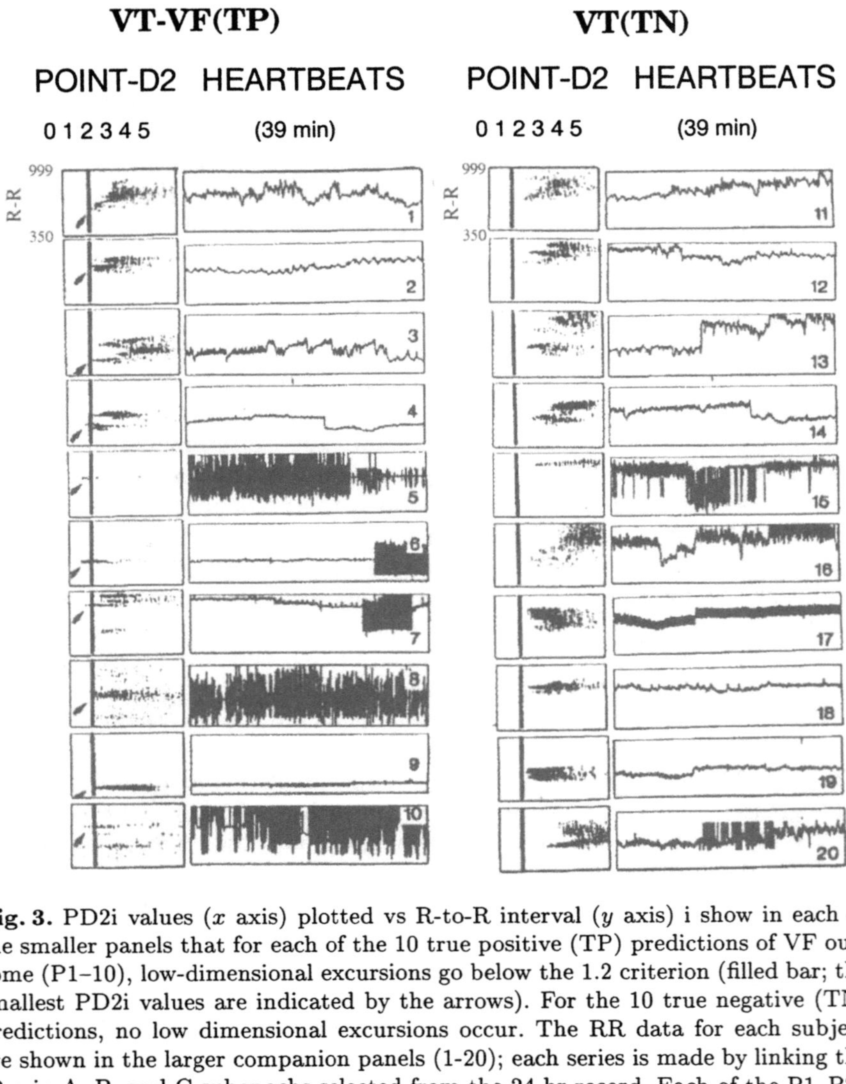

Fig. 3. PD2i values (x axis) plotted vs R-to-R interval (y axis) i show in each of the smaller panels that for each of the 10 true positive (TP) predictions of VF outcome (P1–10), low-dimensional excursions go below the 1.2 criterion (filled bar; the smallest PD2i values are indicated by the arrows). For the 10 true negative (TN) predictions, no low dimensional excursions occur. The RR data for each subject are shown in the larger companion panels (1-20); each series is made by linking the 12-min A, B, and C subepochs selected from the 24 hr record. Each of the P1–P20 subjects had documented nonsustained ventricular tachycardia (VT), which indicated that each was at high-risk of sudden cardiac death. The VT-VF subjects (left columns) manifested VF at the end of the RR data shown, and the VT subjects (right columns) manifested only nonsustained VT during the next 3 years.

A, B, or C subepoch is greater than that between the subepochs (e.g., P1, P3, P11). Each VF subject (P1–P10), in one or more of the A, B and C subepochs, had at least one systematic excursion of PD2i < 1.2. An "excursion" means that temporally the PD2i reduced from a higher to lower value with continuous points in between. The right side of the filled bar is 1.2 dimensions and the left is 1.0 dimensions; the arrows indicate values that extend into or through the filled bar. Low-dimensional excursions below 1.2 did not occur in the VT patients (P11–P20).

3.3 Order in the Data: Surrogates, Lyapunov Exponents, Determinism and Entropies

The randomized-phase surrogate maintains an important statistical property of the original data, the power spectrum. One of the important presumptions in making a parametric statistical test of significance between a given data set and its surrogate is that both are Gaussian distributed around a mean. Table 1, column 2, suggests that this presumption may often be invalid. A surrogate that has a probability distribution the same as that of the data from which it was made may perhaps be a better surrogate for statistical purposes, e.g., the PDE-surrogate (Schreiber, this volume).

Table 2 shows in 3 VF and 3 VT subjects that for 2 of the 14 cases in which the PDE-surrogate could be calculated, the surrogate contains the salient features of the data identified by the PD2i algorithm (i.e., the systematic excursion from a higher dimension to the same PD2i). But in 12 of the 14 cases it does create a "stochasticized" surrogate with different Min PD2i's; in 3 of these 12 different Min PD2i's, a change occurred in the classification of risk (asterisk).

There are other ways than the surrogate method to decide if the data have variations that are ordered (i.e., not random). That is, rather than rejecting the null hypothesis that the data are some type of "filtered" noise (i.e., are stochastic, but with specific statistical properties), it may turn out to be easier to demonstrate that the data are deterministic, and therefore cannot be stochastic. Furthermore, some of these deterministic detectors of order may show VF/VT discriminability.

The traditional indication of low-dimensional chaos being present in deterministic data is that the largest Lyapunov exponent will be a positive number. Kowalik and Elbert (1994) developed a measure similar to the largest Lyapunov exponent, but with the convergence of the values for periodic data being toward a moderate positive number, instead of minus infinity, and that for random noise being toward a value of 1, instead of plus infinity. These new convergence points enable truly chaotic data to be bounded by a relatively small range of numbers that are all positive.

As seen in the upper part of Fig. 4, the Kowalik and Elbert algorithm (which uses a running window) works well on nonstationary data made from concatenations of subepochs of sine, Lorenz, Henon, and random data. The

Table 2. Comparison of the PD2i of some of the P1–P20 data with the PD2i of the equal probability density surrogate (PDE-Sur). The concatenated data files were separated into the individual A, B, C subepochs of RR-intervals to achieve as much stationarity as possible. For each subepoch a series of PDE-surrogates were made. The surrogate file evaluated by PD2i was the one closest to the mean PDE and its number in the series of the 19 generated by the software is indicated as the extension of the PDE file. The electrocardiograms were digitized at 512 Hz and converted to RR intervals using an algorithm with a convexity operator; the data are in integers and expressed in units of digitization-samples between successive R waves.

PD2i					PD2i of RP-Sur						
Data	%n	Mean	SD	Min	Data	%n	Mean	SD	Min	Pts	Max/Min
P1A	87.1	2.65	0.56	1.8	P1A-9	89.0	2.70	0.63	1.8	1079	421/311
P1B	87.1	2.65	0.56	1.8*	P1B-3	74.6	2.43	0.88	1.0*	997	463/314
P1C	88.3	2.10	0.72	0.8	P1C-2	79.0	2.17	0.82	1.0	1119	431/298
P5A	0.3	3.43	0.00	3.4	P5A-17	1.4	4.22	0.96	2.6	1149	836/232
P5B	3.1	4.05	0.73	2.8	P5B-15	3.8	4.34	0.50	3.4	1149	761/213
P5C	38.2	1.04	1.01	0.6	P5C-5	7.3	1.98	0.88	1.0	1100	650/273
P8A	12.9	2.83	0.70	2.0	P8A-12@	2.9	4.36	0.55	3.0	1118	564/159
P8B	7.5	2.65	1.14	1.2*	P8B-10@	1.1	4.25	0.62	3.4*	1119	591/179
P8C	7.9	2.63	0.92	1.4	P8C-12@	1.8	4.68	0.16	4.4	1117	575/140
P12A	77.7	3.45	0.57	2.6	P12A-6	76.1	3.32	0.73	2.0	757	567/472
P12B	86.2	2.84	0.60	1.4*	P12B-15	76.9	2.76	0.81	1.0*	775	500/425
P12C	89.9	2.36	0.60	1.4	P12C-11	93.4	2.42	0.68	1.4	801	534/457
P13A	87.3	2.36	0.58	1.4*	P13A-16	87.6	2.42	0.71	1.2*	1100	395/321
P13B	69.9	3.15	0.74	1.8	P13B-11	63.6	2.77	0.72	1.6	824	526/409
P13C	59.3	3.71	0.56	2.6	P13C-11	62.5	3.65	0.72	2.2	762	588/431
P20A	65.2	3.50	0.68	2.0	P20A-19	62.9	3.65	0.55	2.4	1252	385/281
P20B	24.4	3.44	0.84	1.8	P20B- @	NA	NA	NA	NA	1132	440/280
P20C	53.2	3.68	0.58	1.6	P20C-6	36.1	4.36	0.44	3.0	1053	440/280

@ PDE-surrogate could not be made successfully.
* Changes prediction by Min PD2i (i.e., $PD2i < 1.2$ predicts VF, $PD2i > 1.2$ predicts no VF in these subjects, each of which has documented VT). Upper half are VF subjects (P1, P5, P8), lower half are VT controls (P12, P13, P20).

random and periodic data define the boundaries (i.e., around 1 and 6) and the Lorenz and Henon data show that truly chaotic data have values that lie between these two extremes. Although these two chaotic subepochs have different dimensions (Lorenz = 2.06 dimensions; Henon = 1.46 dimensions) they show equivalent means and ranges of chaoticity as seen in the upper part of Fig. 4.

When the "chaoticity" measure is applied to the RR intervals of the P1–P20 data, it clearly indicates low-dimensional chaos being present in each

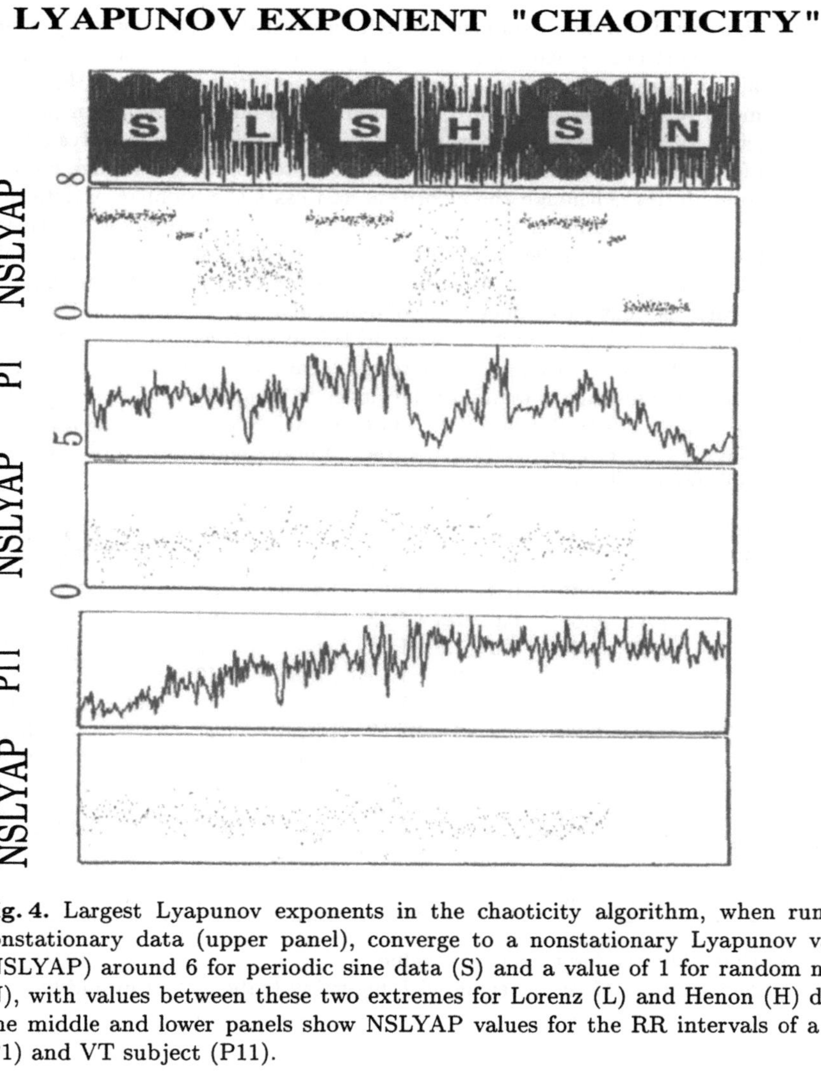

Fig. 4. Largest Lyapunov exponents in the chaoticity algorithm, when run on nonstationary data (upper panel), converge to a nonstationary Lyapunov value (NSLYAP) around 6 for periodic sine data (S) and a value of 1 for random noise (N), with values between these two extremes for Lorenz (L) and Henon (H) data. The middle and lower panels show NSLYAP values for the RR intervals of a VF (P1) and VT subject (P11).

heartbeat series. The middle and lower panels of Fig. 4 show representative results for VF subject P1 and VT subject P11, and they are very similar. Table 3 shows the results for all of the subjects. The degree of chaoticity between the VF and VT groups is not statistically significantly different, and therefore this algorithm does not have the VF/VT discriminability characteristic of the time-dependent dimensional algorithms.

The mean chaoticity for all subjects, however, is statistically significantly larger than that of noise and smaller than that of sine (periodic) data ($P <$

Table 3. Largest Lyapunov exponents in RR interval data calculated by a new "Chaoticity" algorithm by Kowalik and Elbert. The output is an average of the largest Lyapunov exponents within a window that runs through the data. For white noise the mean value converges toward 1.0 coming from minus infinity; for periodic data (e.g., sine waves) it converges toward a high value coming from positive infinity (e.g., 6.3); for chaotic data the time-dependent point-to-point values range between these two boundaries. Although the algorithm addresses the problem of data non-stationarity, it still presumes stationarity within the window. Window length was 280 data points; Tau = 1; embedding dimension = 10; jump-step in data for beginning of each running window was 4. Abbreviations: Data = file of RR intervals of high-risk cardiac patients with documented ventricular tachycardia (P1–P10 manifested ventricular fibrillation within 24 hr; P11–P20 survived for at least 3 yrs); %N = number of positive values out of total; Mean = mean chaoticity or Lyapunov exponent values; SD = standard deviation of Mean; Low = lowest positive value; Hi = highest positive value. SIN, LOR and RAN are data generated by a sine-wave device (i.e., electronically generated), Lorenz equations, and a random number generator (white noise).

Data	%N	Mean	SD	Low	Hi	Data	%N	Mean	SD	Low	Hi
P1	84.5	1.96	0.57	0.4	3.8	P11	81.9	1.72	0.52	0.8	3.0
P2	84.8	1.55	0.45	0.6	3.0	P12	79.0	1.67	0.55	0.6	2.4
P3	87.2	1.81	0.62	0.6	2.4	P13	81.4	1.84	0.77	1.0	2.6
P4	87.9	1.32	0.64	0.6	1.8	P14	85.1	1.46	0.52	0.6	2.2
P5	84.7	1.68	0.45	1.0	2.0	P15@	81.0	1.49	0.71	0.4	3.0
P6	84.3	1.51	1.09	0.2	4.8	P16	81.2	1.70	0.62	0.6	3.0
P7	77.5	1.66	0.96	0.2	4.4	P17	83.8	2.27	0.52	1.2	3.2
P8	84.1	1.88	0.93	0.2	4.2	P18	79.5	1.33	0.44	0.4	2.2
P9	90.5	1.28	0.40	0.6	2.0	P19	85.7	1.23	0.57	0.2	2.4
P10	79.9	3.14	0.92	2.2	3.0	P20	84.3	1.55	0.66	0.2	2.8
SIN	83.3	6.13	0.43	6.8							
LOR	83.6	3.72	0.98	6.8							
RAN	83.9	0.91	0.25	1.8							

*%N less than 100% is due to negative values.
@ P15 Higher running Mean in region with bigeminy (subepoch B in middle).

0.001, *t*-test), but the presumption about homogeneity of variance required for parametric tests may not be valid in these comparisons. The nonoverlap of the distributions between the P1–P20 data and the periodic and random time-series, however, supports the statistical significance indicated in non-parametric comparisons ($P < 0.01$, binomial probability test).

Figure 5 shows that the local Lyapunov exponent in the P1–P20 patients is indeed positive and indicative of low dimensional chaos in the data. This algorithm, like that for chaoticity, uses a running window, but has output values that converge to the same numbers as those for the classical method proposed by Lyapunov (Lyapunov 1892). P6 is the only subject that shows

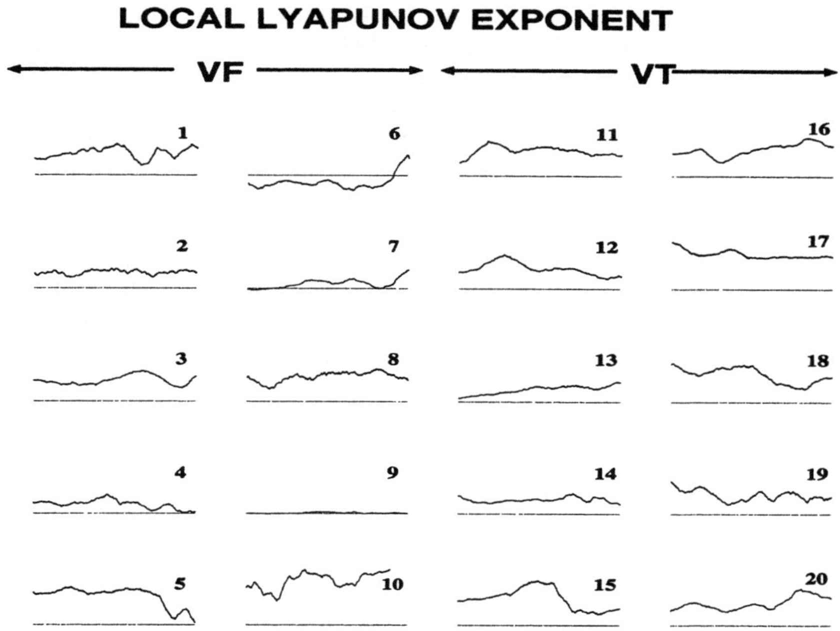

Fig. 5. Local Lyapunov exponent in the P1–P20 patients. This algorithm (Kowalik et al. this Volume) output, unlike the "chaoticity" one, has values that converge to the same numbers as the classical ones (Lyapunov 1892). The x axis is the heartbeat number for the same RR data shown in Fig. 3. The y axis is the value of the largest Lyapunov exponent in the running-window (calibration at lower left is 0.00 to 0.25). The horizontal line corresponds to the 0.00 value, which represents quasiperiodicity in the data series; plus infinity represents random noise; negative value corresponds to periodicity. Note that all subjects show small positive exponents indicitative of low-dimensional chaos; only P6 shows negative values.

negative values indicative of periodicity; a sinusoidal alternate-beat alternans was present throughout the A, B, and C subepochs except toward the end of C, during which the heartbeats desynchronized. In P9 a quasiperiodic 6-beat oscillation occurred throughout the A, B, and C subepochs. All other subjects, in both the VF and VT groups, showed similar profiles of temporal change in the largest Lyapunov exponents that occurred most noticably at points where there were obvious changes in the data stationarity. In all cases the time-related values were small and positive, and therefore indicated low-dimensional chaotic variation in the data.

Another way to show that the data are deterministic is to look at the attractor in phase space. The mean vector-orientation following a time increment is a way to measure the amount of short-term predictability (i.e., determinism) in a given data series. An algorithm that determines the mean predictability was developed by Kaplan and Glass (Kaplan and Glass 1992) as a way to judge whether or not a given data series was stochastic or deterministic. Most finite data series, however, are found to be neither completely deterministic, with a value of 1, nor stochastic with a value of 0; rather they manifest values between the two extremes. This ambiguity about whether or not the data contain order may be enhanced by violations of data stationarity.

Mühlnickel et al. (Mühlnickel et al. 1994) developed a "dynamic" method of measuring the average vector orientations that addresses the problem of data nonstationarity. This measure, "dynamic determinism" (DynDet), shows larger values than the Kaplan/Glass algorithm when applied to finite data made by low-dimensional chaotic generators (e.g., Lorenz data).

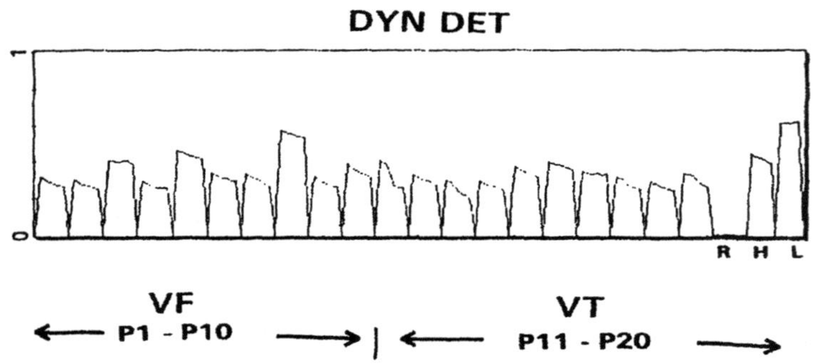

Fig. 6. Determinism, calculated by the dynamic method (DYN DET), shows a value of 0 for random noise (R) and intermediate and high values for Henon (H) and Lorenz (L) data. The values for the VF subjects (P1–P10, serially, left to right) and VT subjects (P11–P20, continuing serially, left to right) are also shown. The top of each of the 23 upward bars shows the consecutive values for 10 through 15 passes.

Figure 6 shows the results of the application of the DynDet algorithm to known chaotic and stochastic data (right) and to the P1–P20 data. These results clearly show that the physiological data are in the same ranges as the low-dimensional chaotic data. Although some of the VF subjects show more determinism (P3, P5, P8) than any of the VT controls, there is no statistically significant difference between the two groups.

Entropy is another way to look at order in the data. A measure called approximate entropy (ApEn) (Pincus 1991) was shown by Pincus and associates (Pincus and Viscarello 1992; Pincus and Goldberger 1994) to distinguish differences in RR data between some types of experimental groups. This measure, however, did not distinguish between the VF and VT subjects in the present study.

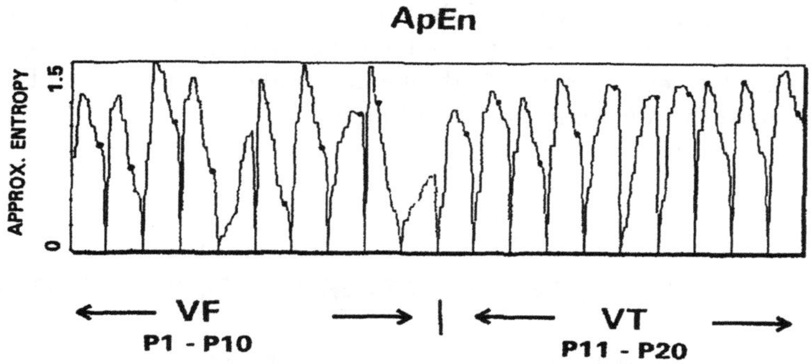

Fig. 7. Approximate entropy (ApEn), expressed as both a range of 10 r values (top of each upward bar; r = 1 to 10 integers) and as that single value where r is 20% of the standard deviation of the dynamic range of the data (dot), shows differences in appearance between the VF subjects (P1–P20; serially, left to right) and VT subjects (P11–P20; continuing serially, left to right), but statistically (t-test) these differences are not significant.

Figure 7 shows the ApEn values for each P1–P20 subject for a range of r values that exhibits the peak ApEn and spans the r-value suggested by Pincus (i.e., 20% of the standard deviation of the data; dot). Neither the peak ApEn nor the 20%-SD ApEn was statistically significantly different between the VF and VT groups ($P > 0.05$, t-test). Clearly, however, there was something different in the appearence; 2 of the VF subjects did not show a clear peak within the fixed 1 to 10 integer r-range (P5, P10), 3 VF subjects showed large peaks that reduced rapidly with increasing r (P2, P7, P9). Only 5 VF subjects showed profiles similar to those of the VT controls (P1, P2, P4, P6, P8).

Pattern entropy (PatEnt) is an algorithm developed by Żebrowski and associates (Żebrowski et al. 1994) for observing higher dimensional entropies (e.g., 2 and 3 dimensional). Because joint-probability is used in the calculations, PatEnt has the property that it will be large for processes that are well ordered and stationary and small for those that are disordered and nonstationary. Previous application of the PatEnt algorithm to the 128 Hz Del Mar Avionics- generated RR-data showed sustained high levels of patterned entropy in subjects at recent risk of VF (Żebrowski et al. 1994). Table 4 shows that this type of result (indicated by "#") occurred in only 3 of the P1–P20 subjects and did not discriminate between the VF and VT outcomes.

Of interest in the P1–P20 data, however, is a change of PatEnt (indicated by "-" in the first column) from a high level (ordered) to a low level (disordered) just before VF. This pattern occurred in only 4 of the 10 VF subjects, but in 0 of the 10 VT subjects. Although promising as a discriminator between groups, this paticular shift of PatEnt did not show high enough specificity for use in individuals ($P > 0.05$, not significant). The opposite change, from a low level in b to a high one in c (indicated by "+"), also did not discriminate the VF from the VT subjects.

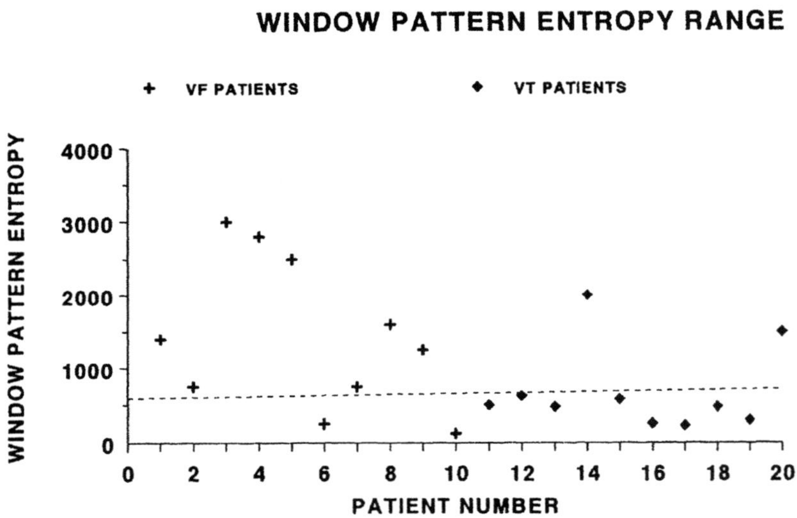

Fig. 8. Window pattern entropy range during the 12-min subepoch just prior to VF (subepoch C) for the P1–P10 patients and at a matched time of day for the P11–P20 subjects. The VF patients (P1–P10, crosses) show a mean windowed PatEnt range that is higher, reflecting more order, than that of the VT subjects (P11–P20, diamonds). The dashed line is an arbitrary separatrix that best divides the two groups. The ordinate values reflect a three dimensional joint probability (see in Żebrowski et al. 1994).

Figure 8 shows a result that was noticed after unblinding the calculations. During the C subepoch, just before VF occurred in the VF subjects

Table 4. Sensitivity and specificity of the patterned entropy algorithm in predicting imminent ventricular fibrillation (VF) in high-risk cardiac patients (P1–P20), each of whom had previously documented ventricular tachycardia (VT). The data analyzed were three separate samples of RR intervals made from three separate 12-min electrocardiograms (a, b, c) sampled at the beginning, middle and end of the useful data on a 24 hr Holter tape. All electrocardiograms were digitized at 512 Hz and converted to RR intervals using an algorithm with a convexity operator; the data are expressed in units of digitization samples between successive R waves; the total noise level was less than ± 3 integers. Abbreviations: Cumulative PE = Cumulative patterned entropy (minimum, maximum, mean) over epoch; Windowed PE = windowed patterned entropy (Min, Max, Mean), where window length = 400 beats; bin width = 15 ms; Tau = 2. P1–P10 showed VF within 24 hr; P11–P20 showed only nonsustained VT during the next three years.

	Cumulative PE			Windowed PE				Cumulative PE			Windowed PE		
VF	Min	Max	Mean	Min	Max	Mean	VT	Min	Max	Mean	Min	Max	Mean
P1a	2857	3341	3065	2286	3964	3262	P11a	178	2585	1093	1089	2935	1793
P1b	879	2920	1264	991	2005	1404	P11b	644	1580	1368	1095	2953	1728
P1c+	951	3195	2145	1799	3216	2374	P11c+	1580	2598	2491	2224	2695	2473
P2a	1237	4486	4112	3829	4776	4316	P12a	2476	4413	3773	2550	4379	3495
P2b	2710	4151	3002	2890	4857	3977	P12b	2473	4172	2892	2177	4147	2656
P2c +	2953	4528	4393	4083	4957	4552	P12c+	2696	4396	4154	3881	4475	4249
P3a	3496	4497	3734	3503	4639	3972	P13a	2925	3895	3285	2492	3905	3218
P3b	1718	3759	2014	1684	2756	2243	P13b	1956	3502	2726	1980	3524	2758
P3c -	1053	2320	1499	914	3942	2452	P13c	1439	2034	1525	1523	1913	1689
P4a#	1128	4941	4267	3188	5158	4453	P14a	1594	3612	3484	3034	4940	3920
P4b#	3884	4990	4313	2446	4965	3571	P14b	3611	4018	3826	3416	4635	4082
P4c#	3868	5164	4422	2376	5198	4309	P14c	2751	3751	3122	2451	4487	3673
P5a	427	3868	466	385	603	465	P15a	2238	3530	2385	1970	2697	2202
P5b	467	694	601	386	866	587	P15b	792	2456	1555	782	3758	2413
P5c+	528	3287	2842	1847	4327	3134	P15c+	2456	3775	3642	3426	3929	3664
P6a	3252	5232	5153	4269	5199	4934	P16a	897	3698	1623	917	2465	1553
P6b	5163	5236	5213	4572	5199	5078	P16b	899	1886	1606	1392	1855	1542
P6c-	2140	5194	2186	1951	2192	2059	P16c	1093	1465	1183	1044	1250	1204
P7a	2140	3650	3291	2804	4220	3515	P17a	1093	1860	1724	1822	2362	2132
P7b	3557	4518	3807	2929	4502	3937	P17b	1812	2506	2462	2360	2559	2466
P7c-	1292	3737	1681	1469	2341	1911	P17c	2288	2448	2328	2210	2375	2288
P8a	1292	2177	1889	1780	2805	2306	P18a#	2303	3976	3837	3555	4885	4312
P8b	1687	1856	1764	1577	2008	1811	P18b#	3713	4473	4239	3526	4605	4221
P8c	1525	2253	1896	1509	3074	2257	P18c#+	3713	5236	5112	4795	5193	5006
P9a	2187	5064	5033	4829	5130	5038	P19a#	3945	4987	4182	1846	4676	3665
P9b	4835	5008	4895	4668	5111	4903	P19b#	3902	4346	4103	3175	4759	4035
P9c-	890	4950	1092	434	1710	1083	P19c#+	4054	5136	5108	4911	5158	5074
P10a	367	1239	497	361	1048	772	P20a	3220	5125	3372	3045	4051	3532
P10b	259	682	432	161	676	349	P20b	2022	3497	2266	1981	3025	2397
P10c	153	261	173	159	257	210	P20c	1621	2200	1862	1483	2911	2159

[+] increase in c epoch compared to b; − = decrease in c compared to b;

[#] sustained high levels;

[−] neither the sensitivity nor the specificity is significant.

or at equivalent times of day in the VT patients, it was noticed that the range of the Max/Min values in Table 4 was greater in the VF subjects (i.e., large excursions to more order) than in the VT ones. The arbitrary separatrix (dashed line) shown in Fig. 8 illustrates this point. Scrutiny of Table 4, however, shows that, in the VT controls, there is still considerable range variation in the A and B subepochs, a finding which suggests that the lack of it in the C subepochs may be adventitious.

Although the noise content of the 128 Hz Del Mar Avionics data precluded PD2i analysis, as shown numerically in the first section above, the PD2i algorithm was still run on the first 16 000 RR segments of each 24 hr record. The results showed, as expected, that neither Min PD2i nor mean PD2i had significant VF/VT discriminability. When the absolute value of the noise was reduced to its lowest level by dividing the data by 7.81 (the conversion factor that changed the original number of samples between R waves to ms), discriminability was still unapparent, as shown in Table 5.

3.4 Clinical Results and 24-Hour RR Observations

Table 6 shows for the P1–P20 subjects that in the VF patients there are low dimensions (i.e., PD2i's < 1.2) during 53 of the 60 halves of each A, B, C subepoch (columns 3 - 5). This result indicates that each 12-min subepoch, which has around 1 000 RR intervals, is a sufficient sample of the 24 hr record from which to judge risk. This finding was confirmed by the PD2i analysis of the 24 hr RR's of the independent VF subject whose Holter tape was digitized at 512 Hz; one or more low-dimensional excursions occurred within each 12-min segment throughout the 24 hr (this subject manifested VF during the 24th hour of the Holter tape). Twenty-four-hour analysis of the VT control showed no low-dimensional excursions below 1.2 during any part of the 24 hr period.

Table 6 also shows that the Min PD2i of the A epoch, which occurred at
Table 6 also shows that the Min PD2i of the A epoch, which occurred at well as the C epoch which occurred only 12 min before VF onset. In P3, P4 and P6 the separation between the time of recording (i.e., prediction of VF) and the actual occurrence of VF was at least 20 hours (Table 5, column 9); these results confirm the similar result observed in the 24 hr record of the independent VF subject. The clinical data presented in Table 6 do not suggest any drug therapies that could have produced the differences in clinical outcome between the VF and VT subjects, but the beta power is too low for any meaningful observation. The important things to note in the clinical evaluations of the VF and VT subjects are that the heart-rate-variability standard deviations, the left ventricular ejection fractions, the coronary artery anatomies, the total 24 hr ectopy rates and the relevant cardiovascular histories are quite similar between the two groups. The following comments were noted in the P1–P20 records: P1, type II aortic aneurysm; P2, heart failure; P3, five days after back surgery; P4, evaluation of ventricular arrhythmias; resuscitated in

Table 5. Point correlation dimension (PD2i) of RR intervals from Warsaw patients. The patterned entropy values for each of these Warsaw subjects has been previously published (Żebrowski et al. 1994).

		Original Data				Noise Reduction × 0.512				Noise Reduction × 0.128			
Data	Diag.	%N	Mean	SD	Min	%N	Mean	SD	Min	%N	Mean	SD	Min
DP1a	VF	27.2	4.82	1.41	2.0	63.8	3.77	1.32	1.6	93.2	1.88	0.87	0.8
DP1b	NOVF	20.5	4.79	1.71	0.8	51.0	4.24	1.48	0.6	90.2	2.46	0.90	0.8
DP2a	NOVF	33.8	4.94	1.79	1.2	66.8	2.99	1.92	0.8	91.2	1.74	1.64	0.8
DP2b	VF	31.2	4.58	1.80	1.2	60.8	3.26	1.82	1.2	91.3	1.71	1.63	0.8
DP3a	VF	29.2	5.04	1.80	1.2	59.0	3.79	1.95	1.0	86.2	1.19	0.41	0.8
DP3b	NOVF	16.9	5.77	1.42	1.8	52.3	4.42	1.08	2.0	95.6	1.59	0.45	1.0
DZI	NOVF	6.7	5.46	1.87	1.2	24.3	5.54	1.42	1.4	57.9	3.96	1.43	1.4
FA	NOVF	01.7	7.43	1.63	4.4	08.9	7.38	0.87	5.0	21.6	6.80	0.80	4.8
KZR	NORM	31.9	4.71	1.30	1.8	69.4	3.19	1.24	1.0	93.4	1.31	0.82	0.8
PCZS	NOVF	6.8	4.77	2.27	0.8	18.4	5.65	2.27	1.6	44.3	5.23	1.95	2.0
RAD	NORM	19.0	5.27	1.37	1.8	49.8	4.46	1.24	1.8	89.8	2.39	0.93	1.2
ROZ	NOVF	11.2	4.45	2.57	0.8	20.5	5.09	2.84	0.6	48.0	4.73	2.29	1.2

The PatEnt values discriminated between RRs made with the Del Mar Avionics Scanner from an electrocardiogram recorded after recent VF (DP3a) and RRs recorded from the same subject 1 year later (DP3b) and between high-risk patients with documented VF (DP1a, DP2b, DP3a) and low-risk (DP1b, DP2a, DP3b, DZI, FA, PCZS, ROZ) or normal (KZR, RAD) subjects. Each epoch analyzed was the first 16 000 RRs in the 24 hr series. The RRs were made by the Del Mar Avionics Scanner from electrocardiograms digitized at 128 Hz. Abbreviations: Diag.= clinical outcome (VF = recent VF; 1yr = 1 year later); Original Data = PD2i calculated on original data in which 1 integer = 1 ms; Noise Reduction × 0.512 = each integer of original data was multiplied by 0.512 before PD2i was calculated; Noise Reduction × 0.128 = each integer of original data was multiplied by 0.128 before PD2i was calculated.

hospital; P5, evaluation of ventricular arrhythmias; P6, heart failure; P7, one day after mitral commisurotomy, resuscitated in hospital; P8, ten days after non-Q infarction; P9, four days after exploratory laparotomy; P10, meningeoma; autopsy consistent with ant. lat. ischemia; old septal infarct; P11, three weeks after MI; P12, post-MI stratification; P13, not classified; P14, not classified; P15, coronary artery bypass graft; P16, NA; P17 coronary artery bypass graft, pre work up; P18, coronary artery bypass graft, pre work up; P19, NA; P20, four weeks post-MI.

Table 6. Clinical data P1–P20 and results from a 24 hr study of a VF and a VT subject (A, B, C 12-min samples from 24 hr Holter; 512 Hz digitization).

		$E(<1.2)$	RR-SD	Holter									History
Pt	Dg	A–B–C	A–B–C	tl	vf	H	EF	PVC	A	S	Meds	Cor An	I A F T D
P1	VF	01-01-11	19 16 16	8	21:00	i	NA	84	65	M	P	NA	+ + − − −
P2	VF	11-11-01	6 18 15	9	21:00	i	28	33	56	M	DHSF	50%LM	+ + + − −
P3	VF	11-11-11	25 14 18	21	07:30	i	50	10	52	F	VN	Min	+ + − − −
P4	VF	01-01-11	7 7 5	24	06:30	i	60	80	61	F	DHe	Min	N N N N N
P5	VF	NA-01-11	134 133 18	18	10:00	o	32	270	73	M	PPrDF	60%RC	− − + − +
P6	VF	11-11-10	4 5 69	24	09:30	i	33	407	63	F	NDVF	Min	− + + − −
P7	VF	11-11-11	9 10 24	4	15:45	i	60	900	42	M	DCoF	Min	+ − − − +
P8	VF	11-11-11	33 37 21	3	14:00	i	35	506	57	M	DiN	NA	+ + − − −
P9	VF	11-01-11	6 18 6	14	11:00	i	NA	105	80	M	NDQ	NA	− + − − −
P10	VF	11-11-11	10 14 NA	9	19:45	i	NA	200	79	M	FD	70%LAD	+ − + + −
P11	VT	00-00-00	25 24 26	23	21:00	i	67	114	58	M	Mp	50%RC	+ − − + −
P12	VT	00-00-00	15 14 12	24	21:00	i	50	188	63	M	DiNHe	99%RC	+ + − + −
P13	VT	00-00-00	14 18 22	24	07:30	o	61	2	57	M	DiN	NA	N N N N N
P14	VT	00-00-00	17 28 14	24	06:30	o	NA	61	NA	M	NfDDiF	NA	N N N N N
P15	VT	00-00-00	32 17 23	23	10:00	i	25	45	53	M	DDi	TVD	+ + + − −
P16	VT	00-00-00	28 44 54	24	09:30	o	87	1	53	F	Di	NA	N N N N N
P17	VT	00-00-00	37 38 38	24	15:45	i	50	2	66	M	QD	100%RC	− + − − +
P18	VT	00-00-00	85 7 17	24	14:00	i	50	29	68	M	QPNfD	100%LAD	− + − − −
P19	VT	00-00-00	7 14 9	23	11:00	i	32	108	63	M	NCNfD	99%LAD	− + − + −
P20	VT	00-00-00	21 24 26	24	19:45	i	14	1152	59	M	NfD	NA	+ + − + −
Nr1	-	00-00-00	86 94 80	24	NA	o	NA	NA	33	M	NA	NA	NA
Nr2	-	00-00-00	86 98 84	24	NA	o	NA	NA	25	M	NA	NA	NA
Nr3	-	00-00-00	71 55 56	24	NA	o	NA	NA	50	M	NA	NA	NA

VF	VF	11-11-11	(1 or more excursions of PD2i < 1.2 in each 12-min segment)
VT	VT	00-00-00	(no excursions of PD2i < 1.2 in any 12-min segment)

Pt = Patient number: for Pts 1-20 (blinded study); 3 normal subjects are included for comparison; VF and VT are subjects that underwent 24 hr PD2i analysis. Dg = clinically diagnosed arrhythmias; VF = nonsustained ventricular tachycardia ending in ventricular fibrillation; VT = nonstained ventricular tachycardia; Nr = normal. $E(<1.2)$ = Excursions of PD2i <1.2; where: 10 = at least one systematic excursion occurred during the first, but not the second, half of the 12 min A,B, or C subepoch; 01 = at least one excursion during the second but not first half; 11 = one excursion during both 6-min periods, and 00 = no low-dimensional excursions. A = 12-min subepoch in the initial part of the available 24 hr data; B = mid epoch, C = final epoch. RR-SD = standard deviation of RR intervals (ms) during a 2 to 3 min stationary period without arrhythmias or artifacts ; A = initial epoch, B = mid epoch, C = final epoch. VF and VT not significantly different ($F = 1.6$, df = 109). Holter Tape; tl = tape length in hours; vf = time of occurrence of VF; H = hospital inpatient (i) or outpatient (o). EF = ejection fraction (in percent); VF and VT not significantly different ($F = 0.48$, df = 31). PVC = premature ventricular complexes per hour, averaged over the full length of the Holter-monitored electrocardiogram (Marquette, Inc.); VF and VT not significantly different ($F = 0.45$, df = 37). A = Age; S = Sex; M = male, F = female.Meds = Medications at time of Holter:
At = Atenolol; Mp = Metroprolol; B = Bumetanide; Mx = Mexilitine; C = Captopril; N = Nitrates; Co = Coumadin; Nf = Nifedipine; D = Digoxin; P = Propranolol; Di = Diltiazem; Pc = Procainamide; E = Enalapril; Pr = Prazosin; F = Furosemide; Q = Quinidine; H = Hydralazine; S = Spironolactone; Hc=hydrochlorothiazide; V = Verapamil. CorAn = Coronary Anatomy; percentage occlusion on coronary angiography; RC = right coronary; LM = left main; LAD = left anterior descending; CX = circumflex; TVD = triple vessel disease; Min = minimal coronary disease. History = medical history; I = myocardial infarction; A = angina; F = heart failure; T = hypertension; D = diabetes mellitus. N or NA = not available or not applicable.

4 Discussion

4.1 Noise and Nonstationarities

Noise input destroys the deterministic feature of a nonlinear system. Any error in the initial conditions will have highly amplified consequences at the output. Noise is also anathema to nonlinear analytic techniques. For example, an almost imperceptible increase in the amount of noise in the data can make the analytic result, say the dimension, jump quickly and nonlinearily to infinity. Therefore the assessment of noise is of paramount importance in the application of nonlinear methods, and the presence of any noise must be addressed if it is unavoidably in the data.

A noise-tolerance level has been preset in the PD2i algorithm so that when the absolute value of the total noise is less than ± 5 integers, then its reconstructed dimension is observed as zero dimensional (Skinner et al. 1994). This strategy, however, dictates that the absolute value of the dynamic range of the signal must be greater than ± 5 to ± 100 integers, else it will produce spuriously low dimensional estimates that will invade the range of biological data (i.e., 0.5 to 6.0) (Skinner et al. 1994). Therefore the RR data based on 128 Hz digitization of the ECG cannot be meaningfully analyzed by the PD2i, as the dynamic range is only + 128 to + 64 integers (i.e., RRs of 1 000 to 500 ms) and the lowest absolute value of the noise is above ± 8 integers. This degree of noise contamination is also likely to cause spurious results when the other nonlinear algorithms are used, because of their extreme sensitivities to noise.

For each of the dimensional algorithms (D2, D2i and PD2i), it is the smallest log-r region that contains the most abundant vector-difference lengths in the correlation integral. As illustrated in Fig. 2, those values above the restricted scaling region, in proportion to the value of r, become increasingly more likely to be contributed by vector-difference lengths in which the reference vector and comparison vector are not in the same species of data. It is this type of contamination that makes the D2i slope calculations different from those of the PD2i. Because there is no restriction in the scaling region, the D2i algorithm will show the type of scaling errors demonstrated in Fig. 1, that is, if the data contain noise or are nonstationary. If the data are uncontaminated (i.e., stationary and without noise), then PD2i and D2i will produce the same values (Skinner et al. 1994).

The PD2i and D2i outputs are approximately the same for the P1–P20 data (Table 1). Therefore the data must be relatively free of noise, a conclusion which is confirmed independently by the noise analyses. Because D2i requires data stationarity for maximum accuracy and PD2i does not, the similar results of the two algorithms suggest that the P1–P20 data must have sufficient stationary for the D2i to work. Because Min PD2i discriminates P8 correctly and Min D2i does not, the sensitivity and specificity of the PD2i algorithm is greater in the discrimination of VF-risk among high-risk patients.

This greater sensitivity, due to the novel method of resolving the problem of data nonstationarity, may become more important in future clinical studies when different types of data are encountered.

The P1–P20 data set represents a "best case" with respect to noise and nonstationarity contamination. If slightly more noise is in the data (e.g., doubled), the greater sensitivity and specificity of PD2i relative to D2i is markedly increased. In the data in which the absolute noise level is increased by a factor of 1.952 (i.e., the P1–P20 data were converted to 1 integer = 1 ms), the sensitivity and specificity of Min PD2i remains the same, but the Min D2i no longer maintains any VF/VT discriminability.

4.2 Detecting Order in the Data Variations

The surrogate method is a way to test whether or not order exists in a data series, that is, as opposed to the data variations being stochastic (i.e., random), but with specific statistical properties. For example, for each P1–P20 data file, the corresponding randomized-phase surrogate is a "stochasticized" version of the original data, but with the statistical property that the power spectrum is the same. When the Min PD2i's of each original data file are compared statistically to those of its randomized-phase surrogate, the result is clear – in every case the comparison leads to a rejection of the null hypothesis that the original data variations are stochastic.

With a different statistical property held constant, however, the null hypothesis may not be rejected. For each PDE-surrogate, which has the same probability distribution as its corresponding data, the Min PD2i's are again found to be different: 7 are decreased, 5 are increased, and only 2 remain unchanged. This alteration, however, causes 3 of the 14 cases to manifest the opposite prediction category. So, again, "stochasticizing" the data alters a critical feature, the order in the variations.

An alternative approach to test for order in the data is to examine them for deterministic chaos or for deterministic predictability (i.e., vector flow). The algorithms selected for the present study, ones which treat the problem of data nonstationarity (chaoticity, local Lyapunov exponent and determinisms), show that both the VF and VT data series have high degrees of order; that is, they have order at least as high as that in data made by known deterministic systems (Table 3 and Figs. 4 and 6). Thus the stochastic-surrogate and determinism approaches both demonstrate that the data variations are indeed deterministic, and therefore they cannot be stochastic.

The outputs of the algorithms which test for determinism and/or nonlinearity (DynDet, chaoticity, local Lyapunov exponents) are themselves unable to discriminate any differences in the order that exists in the heart rate variabilities of the VF and VT patients of the P1–P20 data set (Table 3 and Figs. 4, 5, 6). A lack of discriminability is also found with the entropy measures (ApEn and PatEnt), for they also do not discriminate effectively between the two groups (Table 4 and Figs. 7 and 8). Thus it is not the detection of order in the variations that is the critical feature of the data that leads to the high discriminability of risk, but what the order represents specifically. That is, it is not the detection, but rather the quantification, of the order that is crucial in predicting cardiovascular risk. The new "running-window" algorithms for chaoticity or the Local Lyapunov Exponent (Kowalik et al. this Volume) show promise for future VF/VT discriminability, as they can track data nonstationarities (Figs. 4, 5).

The Windowed pattern entropy algorithm, which also uses a running window, does show risk discriminability in the Del Mar Avionics, 128 Hz, 24 hr, RR data (Żebrowski et al. 1994). Given the high noise content of the latter data, the PatEnt algorithm apparently has a relatively high noise-tolerance level.

The understanding of why PatEnt works in the noisy data, may be recognized in the solution of the Fokker–Planck entropy equations proposed by Borland (this Volume). This formulation enables the partitioning of entropy into a stochastic and a nonstochastic (i.e., deterministic) component. Thus the reason the PatEnt algorithm works in the Del Mar Avionics data may be because it detects the relative shift of noise into the deterministic component. But the comparison of PD2i and PatEnt in the P1–P20 data (Tables 1 and 4) indicates that to achieve high discriminability when the RRs are all high-risk, it is still the algorithmic quantification of the order, not its simple detection, that is required.

4.3 Complexity and Self-organization

The PD2i and D2i results suggest that the time-dependent dimension of the order in the heartbeats is at least one quantification that is associated with increased risk. Mayer-Kress and associates (Mayer-Kress et al. 1988) theorized that a reduction in D2i observed in physiological data might be due to "cooperation" among competing subsystems that control the output (i.e., data) of the overall system. When the dimension is high, each subsystem is independently controlling the output and the complexity of the system has a higher number of degrees of freedom. During "cooperation", some or many of the subsystems synchronize their control of the output and the reconstructed degrees of freedom become reduced, proportionally. They further hypothesize that this "cooperation" is what would be observed during a process of "self-organization" within the system.

This cooperative self-organization is a pivotal concept for the field of cardiovascular research, for it addresses what the quantification of the order in the heartbeat dynamics might mean physiologically. In support of the concept, Skinner et al. (1996) have reported on the dimensional changes of the heartbeat dynamics that occur in a highly simplified neurocardiac system, the isolated rabbit heart, during the accumulating challenge of anoxia/ischemia. In this preparation only the afferent-efferent loops (i.e., the competing subsystems) of the intrinsic cardiac neurons can control its behavior. This control, although simplified, is still representative of the more complex neurocardiac system, as the heartbeats continue to show the same types of inotropic and chronotropic responses to stimulation of the cardiac mechano- and chemoreceptors that are characteristic of the intact condition. In the intact heart the QT and RR-QT subintervals of each heartbeat are negatively correlated, respectively, to the strength of contraction during systole and the rate of contraction during dyastole. Thus these two subintervals of the heartbeat reflect the inotropic and chronotropic outputs of the system(s) that together regulate the heartbeats.

In the simplified neurocardiac preparation, during the control state, the subintervals show a small but independent variation, and the PD2i's of each series are identical, around 1.0. Immediately after the onset of ischemia, an event which stimulates both mechano- and chemo-receptors in the heart, the PD2i's of each subinterval series briefly increase and separate, while the relationship between the jointly plotted subintervals now manifests a linear negative slope. After additional accumulating ischemia, the PD2i's again briefly increase and separate before reducing and returning to an identical value. At this point in time, however, the temporal dynamics of the subintervals shows a right angle turn in their joint plot that is now a linear positive correlation.

Preceding each of the changes in the subinterval joint plot (i.e., in the current cardiac function), there is an increase and separation of the PD2i's of the subintervals that, once the new relationship is established, reduces and reconverges to the same value. The interpretation is that the divergent PD2i's indicate that the system is "reorganizing" and the identical lower values indicate that the newly organized control system is in place – that is, that the afferent-efferent loops controlling the inotropic and chronotropic outputs are again "cooperating" and are again part of the "same" self-organized system.

The implication is that the reduced PD2i and D2i of the heartbeats of the VF patients is evidence that a unique control system has been organized, in which considerable cooperation (i.e., low dimension) exists among the various afferent-efferent loops of the autonomic nervous system. It would seem that it is this specific low-dimensional organization within the nervous system, for whatever reasons, that is the harbinger of VF.

4.4 Dynamical Arrhythmogenesis

Why a specific reduction in the dimension of the system(s) controlling the heartbeats would lead to increased risk of VF is not yet fully understood. Skinner (Skinner 1995) has proposed, and supported with some preliminary data, the hypothesis that the reduction leads to the disappearance of a protective mechanism that, when present, prevents the heartbeats from entering a dangerous part of the dynamical field that results in arrhythmogenesis. This arrhythmogenic subfield is thought to be present in all hearts and forms naturally as a consequence of the nonlinear dynamics of the underlying excitable medium. The protective mechanism may also be dynamical and depend on the dimension of the heartbeat generator being above 1. This theoretical view is based on the important observations of several investigators (Rozenstraukh et al. 1970; Allessie et al. 1976; Winfree 1987; Shibata et al. 1988).

The existance of the arrhythmogenic subfield has been previously demonstrated by Winfree (Winfree 1987). He constructed and examined the dynamics of a mathematical model that employed only the Hodgkin-Huxley equations of excitability. This model predicted the potential occurrence of a re-entrant rotating spiral wave, as occurs in some forms of VT. Shibata and associates (Shibata et al. 1988) showed that indeed these rotating waves could be produced in the cardiac ventricle by injection of a cross-field current. Rozenstraukh (Rozenstraukh et al. 1970) showed that such rotors, as they are also called, could be produced in the cardiac atria by neural stimulation. Allessie et al. (1976) showed that an induced rotor could be split, repeatedly, into multiple rotors by altering the excitability of the medium with ischemia. This latter experiment suggests a mechanism by which VT can degererate into VF during myocardial ischemia.

The dynamical theory proposed by Skinner (Skinner 1995) is more general than the ones which propose myocardial ischemia to be the cause of VF. Besides incorporating the deleterious effects of ischemia, this theory can explain the induction of VF in the normal animal heart by neurogenic means (e.g., brain stimulation and stroke), and the occurrence of sudden cardiac death in the apparently normal heart in humans (Skinner 1995).

In summary, VF risk is predicted by low-dimensional excursions of a quantifier of the deterministic order that exists in the heartbeats. The lowering of dimension may represent a reduction in the complexity of the autonomic controllers of the heartbeat through a mechanism of self-organization that arises as an adaptive response to the net afferent input. The low dimension itself is not the cause of VF, but rather serves as an enabling condition for the occurrence of a dynamical accident. The actual cause of VF is proposed to be the everchanging organization (self-organization) among all the afferent-efferent loops of the nervous system, peripheral and central, that together determine the beat-to-beat variations in the cardiac cycle. This accident is more likely to result in fatal VF under the noncausal potentiating effect of myocardial ischemia.

4.5 Conclusions

The time-dependent quantifiers of order in heartbeat data (D2i and PD2i) show a high degree of sensitivity and specificity in the prediction of VF risk among high-risk cardiac patients that the detectors of order do not possess. The D2 and D2i dimensional algorithms require data stationarity, a requirement which cannot be met by most biological systems over a long period of time, and a requirement which certainly cannot be met by the heartbeat generator during acute ischemia (Skinner et al. 1991). In contrast, the PD2i algorithm reveals the occurrence of nonstationarities in time, and it can operate well even when there are small amounts of noise and transient artifacts in the data (e.g., premature ventricular beats).

Psychological stress and slow wave sleep both reduce the mean PD2i's of the RR intervals in pigs (Skinner et al. 1991). These behavioral states have been associated with increased vulnerability of the ischemic heart to VF (Skinner et al. 1975a; Skinner et al. 1975b). The reduction of heartbeat PD2i's thus appears to be sensitive to the neurocardiac regulatory phenomena that are associated with increased risk in an experimental model of sudden cardiac death. Despite this apparent association, the present data still do not allow a definitive conclusion with respect to causes or mechanisms of VF. New theoretical considerations, however, are being examined that may later illuminate what is thought to be an underlying dynamical disorder.

The clinical significance of this present study is that the single subject, known to be at high risk of sudden death, has specific changes in heart rate variability that can be detected by the algorithms based in deterministic chaos theory. The changes in the heartbeats are detectable hours (and perhaps days) before the occurrence of the lethal arrhythmogenesis itself. The strength of the association between the deterministic measures and lethal arrhythmogenesis may be due to the facts that, (i) the heart, both with (Mayer-Kress et al. 1988; Babloyantz and Destexhe 1988; Skinner et al. 1991) and without (Guevara et al. 1981; Chialvo and Jalife 1987) its innervation, has deterministic chaotic dynamics, and (ii) the new nonlinear algorithms presume the heartbeat variation to be determined, with each RR interval being caused by the physiological generator, unlike the older stochastic measures (e.g., the standard deviation) which presume the beat-to-beat variation to be noise, but with specific statistical properties (Kaplan and Cohen 1990).

The PD2i algorithm appears to have the highest sensitivity and specificity in discriminating VF-risk among the high-risk patients. The reasons for this greater VF/VT discriminability appear to be, (i) the PD2i is a quantifier of the order that exists in the heartbeat dynamics (i.e., its complexity), and (ii) the PD2i addresses the problem of data nonstationarity in a novel way. Thus the choice of the appropriate cardiac measure of heart rate variability determines its sensitivity and specificity in predicting lethal arrhythmogenesis. Choosing the right measure is expected have a significant impact on diagnosis and treatment in clinical cardiology.

Acknowledgement. We would like to thank Thomas Schreiber, for performing the surrogate-data test (Table 2) and to Wanda Popławska for discussions on the application of pattern entropy to cardiac data.
Research Support: National Institutes of Health, NS27745 (JES).

References

Allessie, M.A., Bonke, F.I.M., Schoman, F.J.G. (1976): Circus movement in rabbit atrial muscle as a mechanism of tachycardia II: The role of nonuniform recovery of excitability. Circulation Research **39**, 168–177.

Babloyantz, A., Destexhe, A. (1988): Is the normal heart a periodic oscillator? Biol. Cybernetics **58**, 203–211.

Bigger, J.T., Kleiger, R.E., Fleiss, J.L., Rolnitzky, L.M., Steinman, R.C., Miller, J.P. (1988): Multicenter post-infarction Research Group: Components of heart rate variability measured during healing of acute myocardial infarction. Am. J. Cardiol. **61**, 208–215.

Bigger, J.T., La Rovere, M.T., Steinman, R.C., Fleiss, J.L., Rottman, J.N., Rolnitzky, L.M., Schwartz, P.J. (1989): Comparison of baroreflex sensitivity and heart period variability after myocardial infarction. J. Am. Coll. Cardiol. **14**, 1511–1518.

Billman, G.E., Schwartz, P.J., Stone, H.L. (1982): Baroreceptor reflex control of heart rate: a predictor of sudden cardiac death. Circulation **66**, 874–880.

Chialvo, D.R., Jalife, J. (1987): Non-linear dynamics of cardiac excitation and impulse propagation. Nature **330**, 749–752.

Farmer, J.D., Ott, E., Yorke, J.A. (1983): Dimension of chaotic attractors. Physica D **7**, 153–180.

Gillis, R.A., Corr, P.B., Pace, D.G., Evans, D.E., DiMicco, J., Pearle, D.L. (1976): Role of the nervous system in experimentally induced arrhythmias. Cardiology, **61**, 37–49.

Grassberger, P., Procaccia, I. (1983): Measuring the strangeness of strange attractors. Physica D **9**, 183–208.

Guevara, M.R., Glass, L., Shrier, A. (1981): Phase locking, period-doubling bifurcations, and irregular dynamics in periodically stimulated cardiac cells. Science **214**, 1350–1353.

Hull, S.S., Evans, A.R., Vanoli, E., Adamson, P.B., Stramba-Badiale, M., Albert, D.E., Foreman, R.D., Schwartz, P.J. (1990): Heart rate variability before and after myocardial infarction in conscious dogs at high and low risk of sudden death. J. Am. Coll. Cardiol. **16**, 978–985.

Judd, K., Mees, A.I. (1991): Estimating dimensions with confidence. International Journal of Bifurcation and Chaos **1**, 467-470.

Kaplan, D.T., Cohen, R.J. (1990): Is fibrillation chaos? Circ. Res. **67**, 886–892.

Kaplan, D.T., Glass, L. (1992): Direct test for determinism in a time series. Phys. Rev. Lett. **68**, 427–430.

Kleiger, R.E., Miller, J.P., Bigger, J.T., Moss, A.J., Multicenter Post-infarction Research Group (1987): Decreased heart rate variability and is association with increased mortality after acute myocardial infarction. Am. J. Cardiol. **59**, 256–262.

Kleiger, R.E., Miller, J.P., Krone, R.J., Bigger, J.T. Multicenter Postinfarction Research Group (1990): The independence of cycle length variability and exercise testing on predicting mortality of patients surviving acute myocardial infarction. Am. J. Cardiol. **65**, 408–411.

Kowalik, Z.J., Elbert, T. (1994): Changes of chaoticness in spontaneous EEG/MEG. Integrative Physiological Behavioral Science **29**, 270-282.

La Rovere, M.T., Specchia, G., Mortara, A., Schwartz, P.J. (1988): Baroreflex sensitivity, clinical correlates and cardiovascular mortality among patients with a first myocardial infarction: a prospective study. Circulation **78**, 816–824.

Lombardi, F., Sandorne, G., Pernpruner, S., Sala, R., Garimoldi, M., Cerutti, S., Baselli, G., Pagani, M., Malliani, A. (1987): Heart rate variability as an index of sympathovagal interaction after acute myocardial infarction. Am. J. Cardiol. **60**, 1239–1245.

Lyapunov, A.M. (1947): The general problem of the stability of motion. Commun. 1892, Soc. Math. Kharkov. In Annals Math. Studies 1947, 17, Princeton Univer. Press, DC:NASA.

Malliani, A., Pagani, M., Lombardi, F., Cerutti, S. (1991): Cardiovascular neural regulation explored in the frequency domain. Circulation **84**, 482–492.

Mandelbrot, B.B. (1982): *The Fractal Geometry of Nature.* Freeman, New York.

Martin, G.J., Magid, N.M., Myers, G., Barnett, P.S., Schaad, J.W., Weiss, J.S., Lesch, M., Singer, D.H. (1987): Heart rate variability and sudden death secondary to coronary artery disease during ambulatory electrocardiographic monitoring. Am. J. Cardiol. **60**, 86–89.

Mayer-Kress, G., Yates, F.E., Benton, L., Keidel, M., Tirsch, W., Poppl, S.J., Geist, K. (1988): Dimensional Analysis of nonlinear oscillations in brain, heart, and muscle. Math. Biosci. **90**, 155–182.

MPRG (Multicenter Postinfarction Research Group) (1983): Risk stratification and survival after myocardial infarction. Engl. J. Med. **309**, 31–336.

Mühlnickel, W., Rendtorff, N., Kowalik, Z.J., Rockstroh, B., Miltner, W., Elbert, T. (1994): Testing the determinism of EEG and MEG. Integrative Physiological Behavioral Science **29**, 262–269.

Myers, G.A., Martin, G.J., Magid, N.M., Barnett, P.S., Schaad, J.W., Weiss, J.S., Lesch, M., Singer, D.H. (1986): Power spectral analysis of heart rate variability in sudden cardiac death: comparison to other methods. IEEE Trans. Biomed. Eng. BME **33**, 1149–1153.

Negoescu, R., Skinner, J.E., Wolf, S. (1993): Forebrain Regulation of Cardiac Function Spectral and Dimensional analysis of RR and QT Intervals. Integrative and Physiological Behavioral Science **28**, 331–342.

Parker, G.W., Michael, L.H., Hartley, C.J., Skinner, J.E., Entman, M.L. (1990): Central beta- adrenergic mechanisms may modulate ischemic ventricular fibrillation in pigs. Circ. Res. **66**, 259–279.

Pincus, S.M. (1991): Approximate entropy as a measure of system complexity. Proc. Natl. Acad. Sci. **88**, 2297–2301.

Pincus, S.M., Goldberger, A.L. (1994): Physiological Time-Series Analysis: What Does Regularity Quantify? Am. J. Physiol. **266**, 1643–1656.

Pincus, S.M., Viscarello, R.R. (1992): Approximte Entropy: A Regularity Measure for Fetal Heart Rate Analysis. Obstet. Gynecol. **79**, 249–55.

Poincare, J. (1892): *Les Methodes Nouvelles de la Mechanique Celeste,* Vols 1-3. Gauthier-Villars, Paris. (English transl. (1967): NASA TTF-450–453, Washington).

Pratt, C.M., Theroux, P., Slymen, D. (1987): Spontaneous variability of ventricular arrhythmias in patients at increased risk for sudden death after acute myocardial infarction: consecutive ambulatory electrocardiographic recordings in 88 patients. Am. J. Cardiol. **59**, 278.

Rapaport, E. (1988): Sudden cardiac death. Am. J. Cardiol. **62**, 1–6.

Rich, M.A.W., Saini, J.S., Kleiger, R.E., Carney, R.M., teVelde, A., Freedland, K.E. (1988): Correlation of heart rate variability with clinical and angiographic variables and late mortality after coronary angiography. Am. J. Cardiol. **62**, 714–717.

Rozenstraukh, L., Kholopov, A.V., Yushmanova, A.V. (1970): Vagus Inhibition-Cause of Formation of Closed Pathways of Conduction of Excitation in the Auricles (in Russian). Biofizika **15**, 690–700.

Shibata, N., Chen, P.-S., Dixon, E.G., Wolf, P.D., Danieley, N.D., Smith, W.M., Ideker, R,E. (1988): Influence of shock strength and timing on induction of ventricular arrhythmias in dogs, Am. J. Physiol. **255**, H891–H901.

Skinner, J.E. (1994a): Low-dimensional chaos in biological systems. Biotechnology **12**, 596–600.

Skinner, J.E. (1994b): The role of the central nervous system sudden cardiac death: heartbeat dynamics in conscious pigs during coronary occlusion, psychologic stress and intracerebral propranolol. Integrative Physiological and Behavioral Science **29**, 355–361.

Skinner, J.E. (1995): Neurocardiology: how stress produces fatal cardiac arrhythmias. In Podrid P.J., Kowey, P.R. (1995): *Cardiac Arrhythmia: Mechanisms, Diagnosis and Management* Williams & Wilkins, Baltimore pp. 195–209.

Skinner, J.E., Lie, J.T., Entman, M.L. (1975a): Modification of ventricular fibrillation latency following coronary artery occlusion in the conscious pig: the effects of psychological stress and beta-adrenergic blockade. Circulation, **51**, 656–667.

Skinner, J.E., Mohr, D.N., Kellaway, P. (1975b): Sleep-stage regulation of ventricular arrhythmias in the unanesthetized pig. Circ. Res. **37**, 342–349.

Skinner, J.E., Reed, J.C. (1981): Blockade of a frontocortical-brainstem pathway prevents ventricular fibrillation of the ischemic heart in pigs. Am. J. Physiol. **240**, H156–H163.

Skinner, J.E., Carpeggiani, C., Landisman, C.E., Fulton, K.W. (1991): The correlation dimension of the heartbeat is reduced by myocardial ischemia in conscious pigs. Circ. Res. **68**, 966–976.

Skinner, J.E., Pratt, C.M., Vybiral, T. (1993): A reduction in the correlation dimension of heart beat intervals preceeds imminent ventricular fibrillation in human subjects. Am. Heart J. **125**, 731–743.

Skinner, J.E., Molnar, M., Tomberg, C. (1994): The point correlation dimension: performance with non-stationary surrogate data and noise. Integrative Physiological and Behavioral Science **29**, 217–234.

Skinner, J.E.; Wolf, S.G.; Kresh, J.Y.; Izrailtyn, I., Armour, J.A., Huang, M-He (1996): Application of Chaos Theory to a Model Biological System: Evidence of Self-Organization in the Intrinsic Cardiac Nervous System. Integrative Physiological and Behavioral Science **31**, 122–146.

Storella, R.J., Kandell, R.B., Horrow, J.C., Ackerman, T.S., Polansky, M., Zietz, S. (1995): Nonlinear measures of heart rate variability after fentanyl-based induction of anesthesia. Anesth. Analg. **81**, 1292–1294.

Theiler, J., Eubank, S., Longtin, A., Galdrikian, B., Farmer, J.D. (1992): Testing for nonlinearity in time series: the method of surrogate data. Physica D **58**, 77–94.

Verrier, R.L., Lown, B. (1981): Autonomic nervous system and malignant cardiac arrhythmias. In Weiner, H., Hofer, M.A., Stunkard, A.J. (1981): *Brain, Behavior, and Bodily Disease.* Raven, New York, pp. 273–291.

Vybiral, T., Skinner, J.E. (1993): The point correlation dimension of R-R intervals predicts sudden cardiac death among high risk patients. Computers in Cardiology, 257–260.

Winer, B.J. (1962): *Statistical Principles in Experimental Design* McGraw-Hill, New York.

Winfree, A.T. (1987): *When Time Breaks Down.* University Press, Princeton, NJ.

Wolf, S. (1972): Sudden cardiac death In *The Artery and the Process of Arteriosclerosis: Measurement and Modification* Plenum, New York, pp. 213–275.

Żebrowski, J.J., Popławska, W., Baranowski, R. (1994): Entropy, pattern entropy, and related methods for the analysis of data on the time intervals between heartbeats from 24-h electrocardiograms. Phys. Rev. E **50**, 4187–4205.

Nonlinear Analysis of the Cardiorespiratory Coordination in a Newborn Piglet

Dirk Hoyer[1], Reinhard Bauer[1], Bernd Pompe[2], Milan Paluš[3], Jan J. Żebrowski[4], Michael Rosenblum[5], and Ulrich Zwiener[1]

[1] Institut für Pathophysiologie, Friedrich-Schiller-Universität, D-07740 Jena, Germany
[2] Institut für Physik, Ernst-Moritz-Arndt-Universität Greifswald, Friedrich-Ludwig-Jahn-Straße 16, D-17487 Greifswald, Germany
[3] Institute of Computer Science, Academy of Sciences of the Czech Republic, Pod vodárenskou věží 2, 18207 Prague 8, Czech Republic
[4] Institute of Physics, Warsaw University of Technology, ul. Koszykowa 75, 00-662 Warszawa, Poland
[5] Alexander von Humboldt fellow: Max-Planck-Arbeitsgruppe "Nichtlineare Dynamik", Universität Potsdam, Am Neuen Palais 19, D-14415 Potsdam, Germany; permanent address: Mechanical Engineering Research Institute, Russian Academy of Sciences, 101830 Moscow, Russia

Abstract. We investigate the cardiorespiratory system of a newborn piglet during REM and non-REM sleep as well as general anesthesia, hypoxia, and cholinergic blockade. The coordinated behavior of heart rate fluctuation and respiratory movement reflects essential capabilities of the autonomic coordination. A corresponding multivariate data analysis was done by means of several nonlinear methods: generalized mutual information, redundancy and surrogate data, window pattern entropy, and computation of phase relations. Some of them are applied for the first time in this context.

1 Introduction

The stable operation of organisms is based on the complex coordination between several physiological subsystems. In the present paper we use different nonlinear techniques of multivariate time series analysis to address corresponding interactions within the autonomic nervous system (ANS).

Relevant nonlinear properties of the heart rate dynamics are confirmed which may improve concepts of medical treatment. The diagnostic of sudden cardiac death risk, e. g., could be improved by nonlinear analysis in comparison to the conventional linear heart rate analysis (Kurths et al. 1995). Nonlinearities of the respiratory movement (RM) and the heart rate fluctuation (HRF) in animals were confirmed by surrogate data testing (Hoyer et al. 1996b).

Untill now, most of the nonlinear investigations were performed for a single quantity only, such as HRF. However, essential functions of the ANS are realized by complex coordinations which include particular nonlinear interactions (multi-matched feedback loops including delays, saturations, sensibilization, adaptation). Therefore only a corresponding nonlinear analysis of the cardiorespiratory coordination is adequate to these features of the ANS. Hence we need methods for a *nonlinear and multivariate* data analysis.

Important characteristics of physiological processes are nonstationarities. They arise, e. g., due to switching between different sleep states or slow transitions resulting from several internal and environmental influences. Frequently changing short-term states are also typical for physiological processes. Hence we need methods for a *short-term* data analysis.

In Sect. 2 the cardiorespiratory system is introduced. In Sect. 3 different methods are used for the analysis of the coupled behavior of the RM and the HRF of a newborn piglet. These investigations focus on fundamental features during the perinatal development which are essential for the diagnosis of early brain damages. The investigations of the short-term mutual information focus on slowly varying system properties. This nonlinear analysis is done in comparison with a linear one (Subsect. 3.1, Pompe 1996). The redundancy and surrogate data test the existence of nonlinearities of the cardiorespiratory system (Subsect. 3.2, Paluš 1996). With the window pattern entropy some changes of the system behavior are indicated (Subsect. 3.3, Żebrowski et al. 1994, Żebrowski et al. 1995a). Phase shifts and lockings are important properties of the cardiorespiratory coordination. They are considered by means of the Hilbert transform (Subsect. 3.4, Rosenblum et al. 1996).

2 Materials and Recording Method

2.1 Physiological System

Primarily, the cardiovascular and respiratory system is designated to guarantee an adequate continuous organismic supply with oxygen and nutrients, and an adjusted clearance of metabolic waste products. Therefore, blood flow through the body and gas exchange through the lungs are adjusted to the momentary situation by a combination of regional and higher–level mechanisms, the effects of which are closely interrelated. The functional state of the circulation and the respiration is continuously monitored by receptors at various places in the cardiovascular system and in brain structures, such as the ventral surface of the lower brain stem (Langhorst et al. 1975). This complex functioning is coordinated by the ANS (Grönlund et al. 1991). There is some evidence that the dynamics of the cardiorespiratory operation and their interrelationship incorporate not only linear but also nonlinear properties (Hoyer et al. 1995, Zwiener et al. 1995).

During the transition period from intrauterine to extrauterine life as well as during the first weeks of the neonatal life the cardiorespiratory coordination is essentially developed (Heymann et al. 1981). But there is a high risk of critical disturbances of this fundamental process in the neonatal period.

The maturation of the two different sleep states: active sleep (rapide eye movement, REM) and quite sleep (non rapide eye movement, non-REM) occurs during the third trimenon of the intrauterine life. After birth, the patterns of RM and HRF represent main parameters by which both sleep states can be differentiated (Prechtl 1974). Their dynamics and interaction reflect the individual state of development of the autonomic coordination. They were analysed in the first parts of each of the investigations presented.

Systemic hypoxia (lack of an adequate oxygen supply) is a frequently occuring external event which leads to strong changes in the autonomic coordination of the cardiorespiratory control. Hypoxia influences the short-term autonomic activity within a few hundred miliseconds after the onset of critical oxygen lack in the arterial blood (Morgan et al. 1995). However, long-term effects also occur by a change of the ANS balance between the sympathetic and parasympathetic tone. In order to differentiate the influences of both components of the ANS, a chemical disruption of the parasympathetic (vagal) efference was done using a cholinergic blockade. Therefore we investigated the cardiorespiratory dynamics during hypoxia and cholinergic blockade in comparison to general anesthesia as an experimental reference state.

Although, there are some species-specific differences in the dynamics of autonomic maturation, the main components of developmental processes are common for the mammals and of special importance for studies of normal and disturbed brain development in early extrauterine life (Buckley 1986). By means of corresponding investigations of a newborn piglet, pathogenetically relevant states could be produced while monitoring the key functions of the ANS.

2.2 Data Recording and Preprocessing

By means of an appropriate experimental design and an extensive monitoring (for details see Zwiener et al. 1996) the following states of a spontaneously breathing newborn piglet were classified or produced, respectively:

- REM sleep versus non-REM sleep
- hypoxia and cholinergic blockade versus general anesthesia.

In each state our multichannel recording equipment sampled several signals simulteneously, however, here we study first of all only two of them: the electrocardiogram (ECG) and the respiratory movement (RM). We have recorded data of several piglets of mixed breed, both sexes, a mass of 1.6 ± 0.35 kg, and at the age of 2–3 days. Nevertheless, for this pilot study the data of only one representative piglet were analysed.

For ECG recording the piglets were instrumented by sticking electrodes (5 mm in diameter) on every fore and hind leg, and for RM recording (impedance respirography) on both sides of the chest wall. The ECG was digitized at a sampling rate of 2048 Hz, and the RM at 128 Hz.

For heart rate calculation, the individual R waves, with the R wave peak as the fiducial point, were sequentially recognized, and thus the series of RR intervals $T_1, T_2, T_3, \ldots, T_n$ were obtained as a function of the beat number. This series constitutes the tachogram, its reciprocal represents the instantaneous heart rate fluctuation (HRF) used in this paper. The resulting calculation precision of about 0.49 ms corresponds to a frequency fluctuation of about 0.04% of the mean heart rate of 50 per minute. The recorded data were checked for artefacts and trends by continuous play back, and disturbed intervals were rejected.

One problem of the data preprocessing is that we have two different kinds of data which are the non-equidistant series of heart beats (QRS-complexes) and the continuous time series of RM. However, we need a synchronous presentation of these signals for their joint analysis. Therefore we used different kinds of common data sets. Each of these data sets consists of synchronous measurements of RM and HRF:

Non-equidistant (heart beat related): At each heart beat the instantaneous heart rate is calculated for the preceding RR interval. Simultaneously (at each of the heart beats) a sample of the RM is taken.

Equidistant, 128 Hz resampled: Here the instantaneous heart rate values are written into the corresponding preceding RR intervals at a rate of 2048 Hz. Then, the resulting "stepped" heart rate time series is resampled at 128 Hz. RM is sampled simultaneously.

Equidistant, low pass filtered, 8 Hz resampled: The data of the preceding procedure were 4 Hz low-pass filtered in order to emphasize the physiologically essential slow fluctuations of interest. Then the extend of data was reduced by the 8 Hz resampling of RM and HRF which is in accordance with the sampling theorem.

Preprocessing of the HRF for miogram analysis: The fast algorithm for the generalized mutual information analysis (see Pompe 1996) requires nearly continuous data. However, the equidistant and at 128 Hz resampled HRF data sometimes attain only 20 or even less different values within periods of about 15 s – HRF forms a staircase signal. To overcome this problem the staircase was transformed to a stepwise linear signal by linear interpolation of the values at the steps of the staircase.

3 Results

Figure 1 shows the time series RM and HRF which are switching between REM and non-REM sleep. The sleep state pre-classification in the top of the figure was performed independently by means of the measured eye movement EM and the electrocorticogram. The signals RM and HRF during general anesthesia, hypoxia, and cholinergic blockade are represented in Fig. 2.

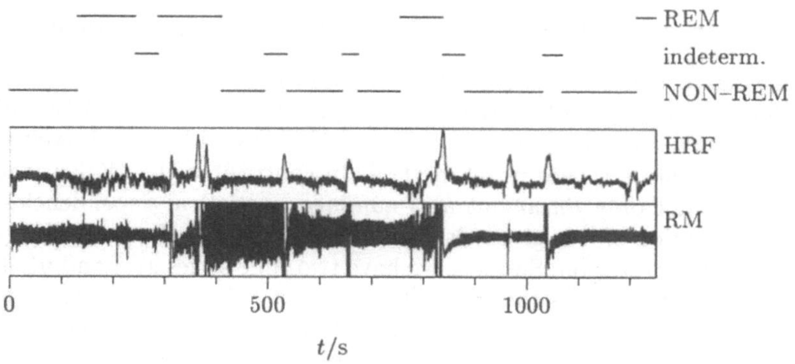

Fig. 1. Heart rate fluctuations HRF and respiratory movement RM of a piglet during different sleep states

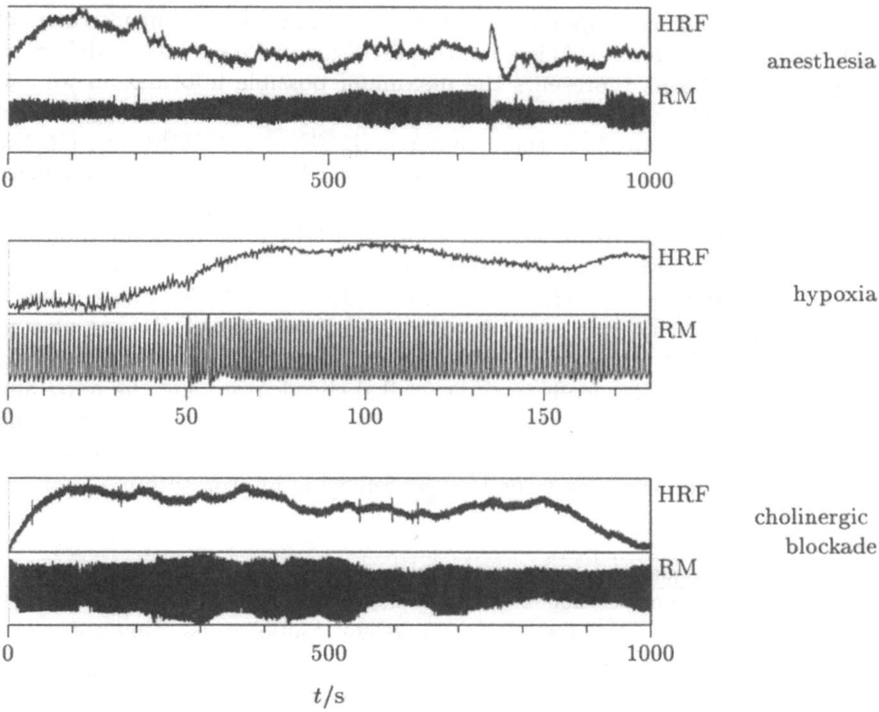

Fig. 2. Simulteneously recorded heart rate fluctuations HRF and respiratory movement RM of a piglet during different physiological states

3.1 Correlograms and Miograms for Measuring Dependencies

We use now the correlation function (CORF) and the so-called generalized mutual information function (GMIF) to investigate dependencies of HRF and RM of the piglet data in Figs. 1 and 2. The CORF is well-known, however, the GMIF represents a rather novel method. In contrast to the CORF, the GMIF describes not only *linear* but also *nonlinear* statistical dependencies. A detailed description of it is given elsewhere (Pompe 1996). Here we shortly summarize its interpretation: Suppose we have two time series $x(t)$ and $y(t+\tau)$, then GMIF(τ) represents the information we get from the quantity $x(t)$ on the other $y(t+\tau)$, and vice versa. This information is understood as a mean over all instants t, for fixed time lag τ. In general we have

$$0 \leq \mathrm{GMIF}(\tau) \leq -\log_2 \varepsilon \quad \text{for each } \tau \ .$$

The lower bound is attained iff there are no statistical dependencies within the relative measuring precision ε $(0 < \varepsilon \ll 1)$. The upper bound is attained iff there is a deterministic relation within precision ε. In the following we always work with $\varepsilon = 0.05$ corresponding to a partitioning of the amplitude range of each signal with 20 bins (this is a relative precision of 5%). Thus, $\mathrm{GMIF}(\tau) = -\log_2 0.05 \approx 4.3$ bit represents the maximum possible information we can get. For $y(t+\tau) = \mathrm{RM}(t+\tau)$ and $x(t) = \mathrm{RM}(t)$, e.g., we investigate auto dependencies. Here we have the symmetry $\mathrm{GMIF}(\tau) = \mathrm{GMIF}(-\tau)$. On the other hand, for $y(t+\tau) = \mathrm{RM}(t+\tau)$ and $x(t) = \mathrm{HRF}(t)$ cross dependencies are considered. Here $\mathrm{GMIF}(\tau) \neq \mathrm{GMIF}(-\tau)$ holds, in general.

Due to a fast algorithm we can apply the GMIF method for a sliding data window analysis with moderate computational efforts. Here we work with a sampling rate of 128 Hz, a window length of about 15 s, and a window shift of 2–4 s. For each data window the CORF and the GMIF are plotted vertically, encoded by a gray scale. Horizontally the starting time of the data window runs. This leads to the so-called correlogram and the miogram, respectively. Mutual information does not discriminate between positive and negative correlation. That is why we compare the miograms with the *squared* correlogram. Then the gray scale has the following interpretation:

gray level	squared correlogram	miogram
white	no correlation	no dependencies
gray	linear dependencies	(possibly nonlinear) dependencies
black	maximal correlation	one quantity determines the other within precision ε

Suppose the miogram detects dependencies (gray or black) and the squared correlogram says that there are no correlations (white), then the dependencies must be purely nonlinear.

Sleep States. Figure 3 shows the squared auto and cross correlograms of the time series of Fig. 1. First consider the correlogram of Fig. 3 b). The nearly horizontal (most dark) line for $\tau = 1.2 - 2\,\mathrm{s}$ reflects the variation of the breathing period of the piglet during the different sleep states. The line around $\tau = 0.6 - 1\,\mathrm{s}$ corresponds to negative correlations at half the breathing period. The other (most dark) lines represent multiples of the breathing period. From them we can better read off small variations.

In the first non-REM phase ($t = 0 - 75\,\mathrm{s}$) the breathing period develops from nearly 2 s down to 1.4 s and then up to 2 s again. This behavior is repeated once more for $t = 75 - 130\,\mathrm{s}$. During later non-REM periods (e.g. for $t = 380 - 520\,\mathrm{s}$) we observe again this down and up of the breathing period, but now in much shorter intervals $\Delta t = 20 - 30\,\mathrm{s}$.

In the first REM phase ($t = 140 - 380\,\mathrm{s}$) the breathing period starts with the much larger value of 2.8 s at $t > 140\,\mathrm{s}$ and it decreases down to 1.9 s at $t = 350\,\mathrm{s}$. Then it increases again. For time lags τ up to 7 s, correlations decrease more rapidly during the REM than during the non-REM sleep. This becomes obvious from the curve at the top of the correlogram representing the mean of CORF(τ) over all τ at each instant t. It could be considered as an average measure of linear relations in the signal at instant t.

The shape of the squared correlogram of RM is resembled in that of HRF shown in Fig. 3a). This reflects a correlation between the respiratory movement and the heart rate fluctuations (respiratory sinus arrhythmia). The cross correlogram of RM and HRF in Fig. 3c) verifies this correlations. It is remarkable that the cross correlogram attains its first maximum at $\tau \approx 0$ during non-REM sleep but at $\tau \approx -0.2\,\mathrm{s}$ during REM sleep. This means that the coupling from RM to HRF is somewhat (more) delayed for REM sleep.

The same behavior of the breathing period is detected by the miograms (Fig. 4). However, they have a higher mean gray level because they indicate nonlinear dependencies in regions where correlations (linear dependencies) vanish. Another striking difference between the correlograms and the miograms is that in the later the cross dependencies between RM and HRF have nearly the same integral strength (averaged over all τ) for REM and non-REM sleep (compare the curve at the top of Fig. 3c) with that of Fig. 4c)). This gives us a hint that the RM/HRF coupling becomes more complex during REM sleep.

Finally it should be mentioned that the vertical dark stripes in the miograms are artefacts. They occur at humps in the time series of Fig. 1 indicating a sudden increase in the heart rate. Such behaviour is physiologically normal. However, they represent instationarities in the data windows used for the sliding window analysis. For such situations the GMIF method is not applicable any longer – instationarities in the data produce a positive bias of the GMIF estimator (Pompe 1996).

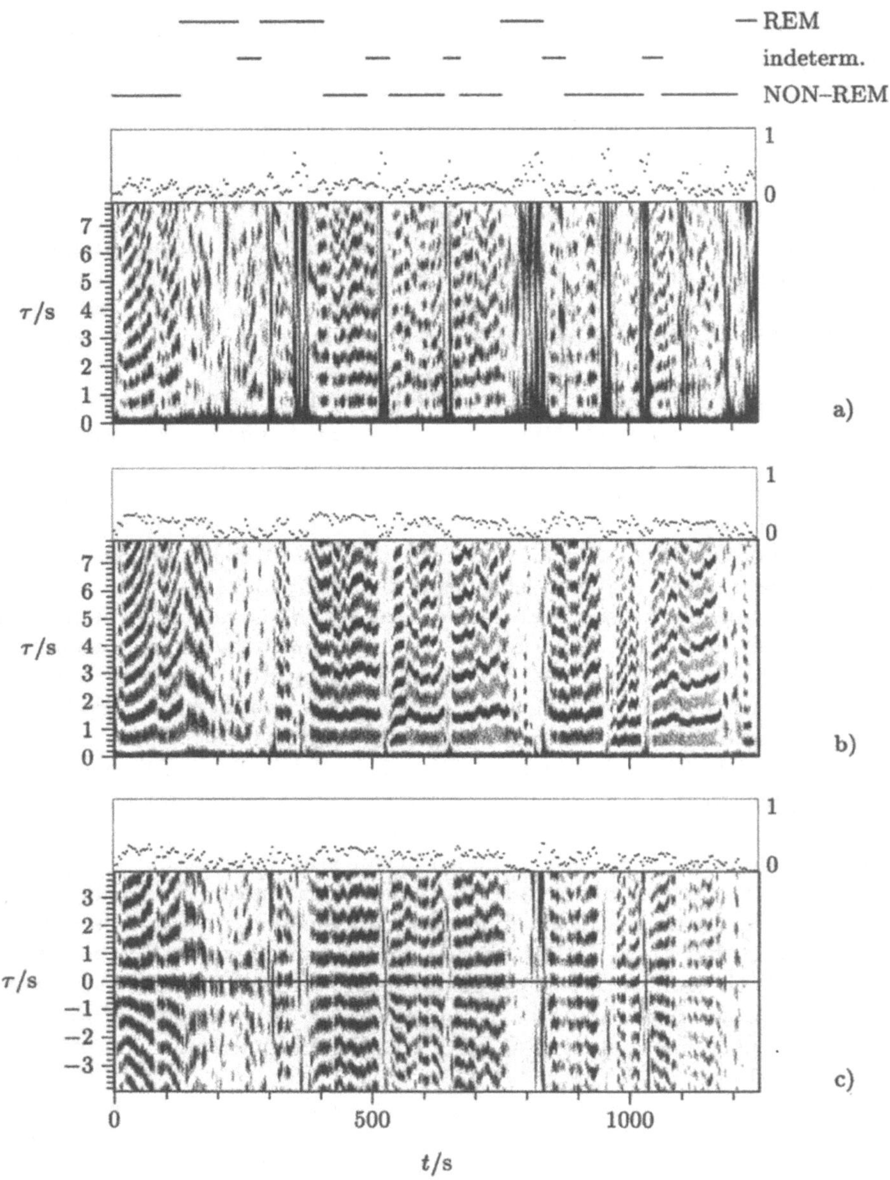

Fig. 3. Squared correlograms of piglet data during different sleep states: a) squared auto correlogram of heart rate fluctuations HRF, b) squared auto correlogram of respiratory movement RM, c) squared cross correlogram of HRF(t) and RM($t+\tau$). The curves at the top of each correlogram represent the mean over all time lags τ

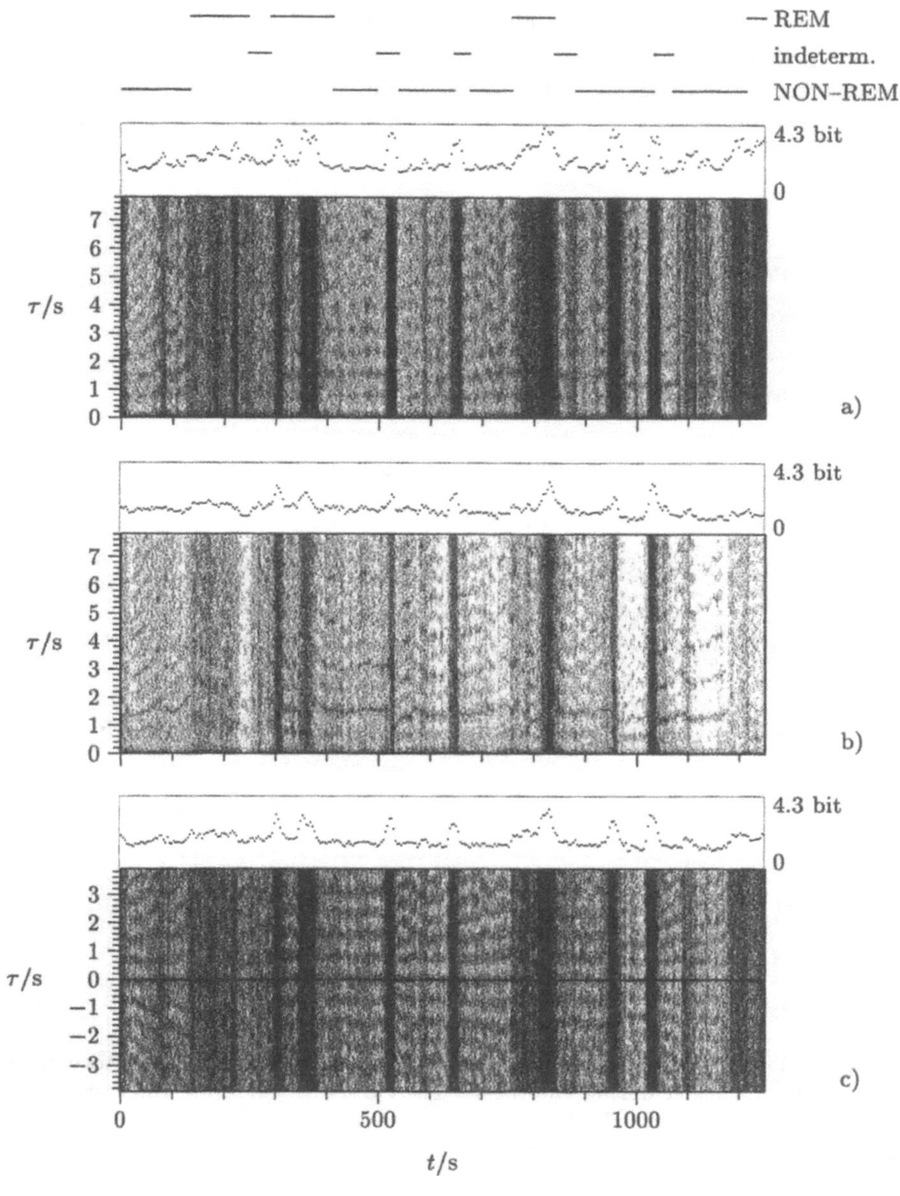

Fig. 4. Miograms of piglet data during different sleep states: a) auto miogram of heart rate fluctuations HRF, b) auto miogram of respiratory movement RM, c) cross miogram of HRF(t) and RM($t + \tau$). The curves at the top of each miogram represent the mean over all time lags τ

Anesthesia, Hypoxia, and Cholinergic Blockade. The same investigations as described above were done for the data in Fig. 2. The squared auto correlogram and miogram of the piglet during anesthesia are shown in Fig. 5. The breathing period varies now in the small range of $1.3 - 1.7\,\mathrm{s}$. There are

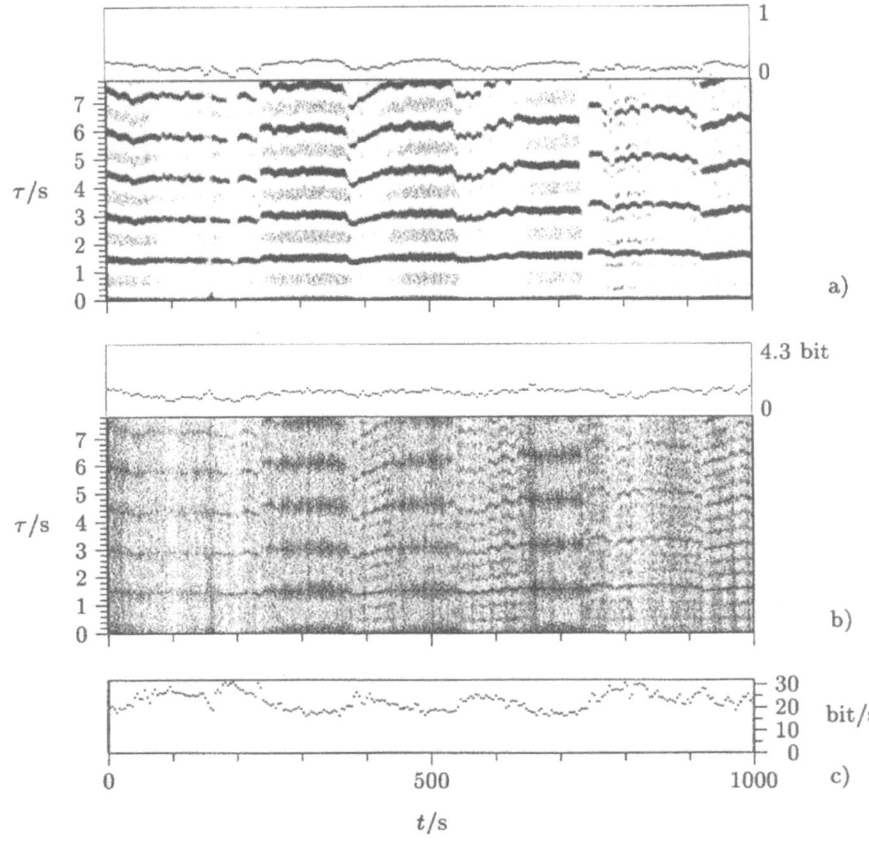

Fig. 5. Squared auto correlograms a) and auto miogram b) of the respiratory movement RM of a piglet during anesthesia. The curves at the top of the correlogram and miogram represent the corresponding means over all time lags τ. Curve c) represents the decay of GMIF forming the miogram for $\tau = 0 \ldots 8\,\mathrm{ms}$

rather stationary phases, e. g., for $t = 240 - 380\,\mathrm{s}$ and $t = 440 - 540\,\mathrm{s}$. They are interrupted by somewhat more active phases, e.g., around $t = 400\,\mathrm{s}$ and $600\,\mathrm{s}$. In contrast to the correlogram, the miogram detects more clearly an additional characteristic time at $1/3$ of the breathing period during more active phases.

Figure 5 c shows the initial decay of the GMIF forming the miogram of Fig. 5 b. It represents the rate of information production (on the base of a "first order memory"). There are arguments (e.g., Herzel, this volume; Palus, this volume;Pompe 1996) that this curve qualitatively resembles the development of the largest Lyapunov exponent (metric entropy), presuming that there is an underlying (chaotic) dynamic system. The later assumption cannot be reliably proven for our piglet data. Nevertheless, higher values of the initial decay of the GMIF indicate a more complex (irregular) behavior. A coupling between HRF and RM was detected, however, the linear coupling was found to be rather weak.

Some results for hypoxia are shown in Fig. 6. Here the breathing period varies slowly from 1.2 to 1.5 s. The dynamic activities are rather poor. Similarily to anesthesia, there was detected a coupling between HRF and RM, but the linear coupling was found to be weak. The initial decay of the GMIF forming the miogram of Fig. 6b is nearly constant at (17 ± 2) bit/s (not shown in the figure).

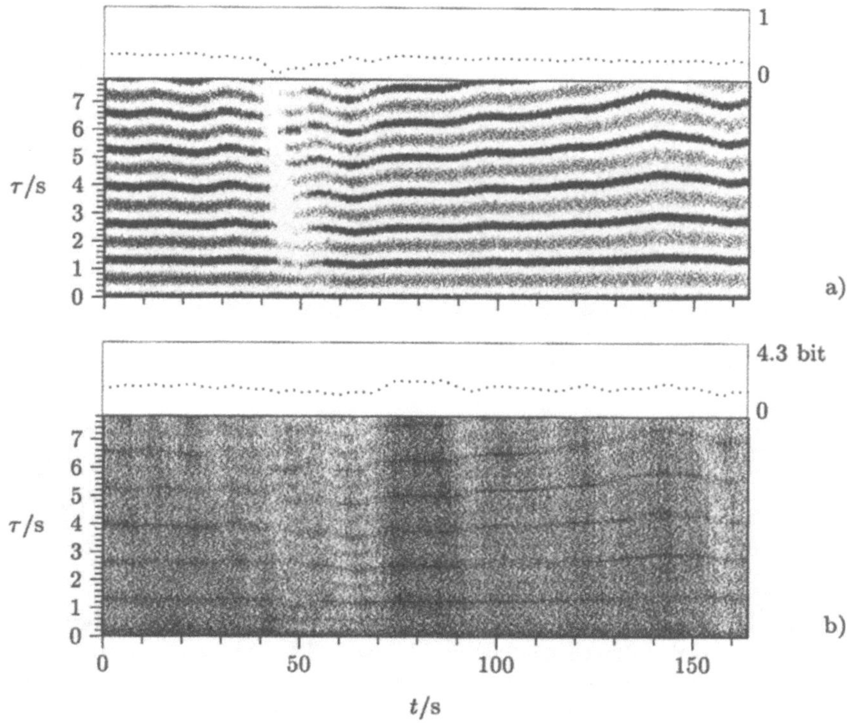

Fig. 6. Squared auto correlograms a) and auto miogram b) of the respiratory movement RM of a piglet during hypoxia. The curves at the top of the correlogram and miogram represent the corresponding means over all time lags τ

Finally, Fig. 7 shows results for cholinergic blockade. In the first part, up to 600 s, the breathing period varies rather actively in the range 1.4 – 2.1 s. This is coupled with a higher mean HRF and larger amplitude of RM (see both lower time series in Fig. 2). For $t = 600 - 900\,\mathrm{s}$ the breathing period varies only in the small range 1.5 – 1.6 s. This is coupled with a lower mean HRF and smaller amplitude of RM. Both, the cross correlogram and miogram indicate that the coupling between HRF and RM is almost vanishing.

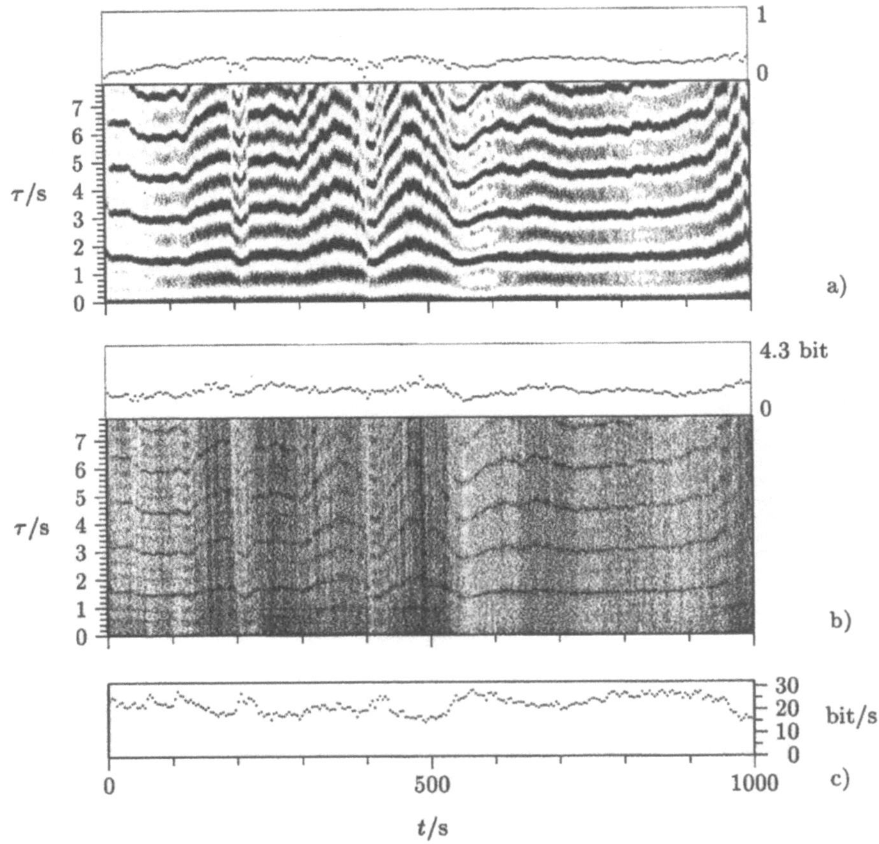

Fig. 7. Squared auto correlograms a) and auto miogram b) of the respiratory movement RM of a piglet during cholinergic blockade. The curves at the top of the correlogram and miogram represent the corresponding means over all time lags τ. Curve c) represents the decay of GMIF forming the miogram for $\tau = 0 \ldots 8\,\mathrm{ms}$

3.2 Probing Nonlinearities by Using Redundancies and Surrogate Data

The redundancy (R) – in the case of two variables also known as mutual information – was used to investigate the cardiorespiratory system. The univariate version was applied when the dynamic properties and the nonlinearities of the individual time series of RM and HRF were studied. The bivariate version, in particular, was applied when the dynamic relations between RM and HRF within their coordinated behavior were investigated. Then the mutual information from the scrutinized (original) data and the mean mutual information from the surrogates, as well as the test statistics are plotted as functions of the time lag. Detailes can be found elsewhere (Paluš 1996). We used the low-pass filtered and 8 Hz resampled data of RM and HRF.

Sleep States. The estimated redundancy of a REM sleep interval and a non-REM sleep interval (preclassified data sets of 1024 samples) are shown in Fig. 8. In all signals (RM, HRF, and RM/HRF) significant nonlinearites could be indicated. The significance parameter

$$S = \frac{\mathrm{R_{orig.data}} - \mathrm{meanR_{surrogates}})}{\mathrm{SD_{surrogates}}}$$

was $S > 20$ in some regions of all of these figures. Since also nonstationary data produce different properties in the surrogate data the signals were differentiated in order to decrease trends. Furthermore, the quality of the surrogates was checked by the linear R (LR). Since we could show that the LR did not differ between the original data and the surrogates, the significant decrease of R clearly indicates nonlinearities in the original data.

By means of the investigated R of the original and surrogate data the amount of nonlinearities and the time related pattern could be estimated with regard to different functional states of the cardiorespiratory system. The R measures all dependencies (linear and nonlinear). In the surrogates only the linear dependencies are preserved, so similarities to the original data are usually not remarkable, but the differences. If the oscillations are regular enough they can be detected also by linear tools (such as autocorrelation) and are reflected in the surrogates. But the process itself is nonlinear. That means that the oscillations of RM during non-REM sleep, e. g., are nonlinear. This oscillatory process is not regular (with zero entropy rate) but has positive entropy rate due to some dynamic noise or chaos.

An essential influence of RM to HRF is the well-known respiratory sinus arrhythmia. This transfer property is clearly to be seen by comparison of the estimated R of RM, HRF, and RM/HRF during non-REM sleep. The estimated R of RM and HRF as well as that of RM/HRF are higher during non-REM sleep than during REM sleep. This result corresponds to the higher regularity and the smaller amount of influences and minor level of information processing in the brain during non-REM sleep.

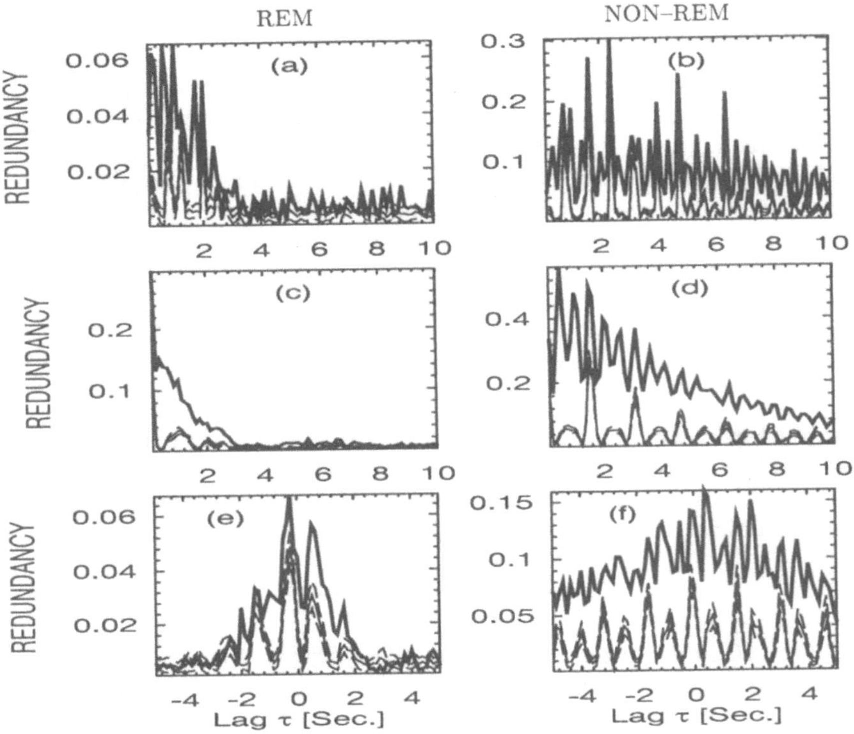

Fig. 8. Redundancies R in dependency on the time lag τ for the different sleep phases of the piglet data in Fig. 1 — REM: $t = 144 \ldots 272\,\mathrm{s}$, NON–REM: $t = 410 \ldots 538\,\mathrm{s}$. Fat solid line: original data, thin lines: surrogate data (mean $\pm$ standard deviation). a) and b): HRF, c) and d): RM, e) and f): NON–REM

Anesthesia, Hypoxia, and Cholinergic Blockade. The estimated R of selected data sets (1024 samples) of general anesthesia, hypoxia, and cholinergic blockade are shown in Fig. 9. Significant nonlinearities were found in RM, HRF, and RM/HRF during general anesthesia and hypoxia. However during cholinergic blockade HRF and RM/HRF became linear. During general anesthesia the similarities between the R of RM of the original and the surrogate data are remarkable. The significant differences between the R of the original data and its surrogates however clearly indicate nonlinear oscillations.

Hypoxia effected a slightly decreased R in the HRF but a clear increased R in the RM in comparison with general anesthesia. It can be concluded that this – in this situation more effective – nonlinear behavior is based on the changes in the autonomic coordination of the cardiorespiratory control. A stronger coordination with regard to the oxygen supply can be identified by

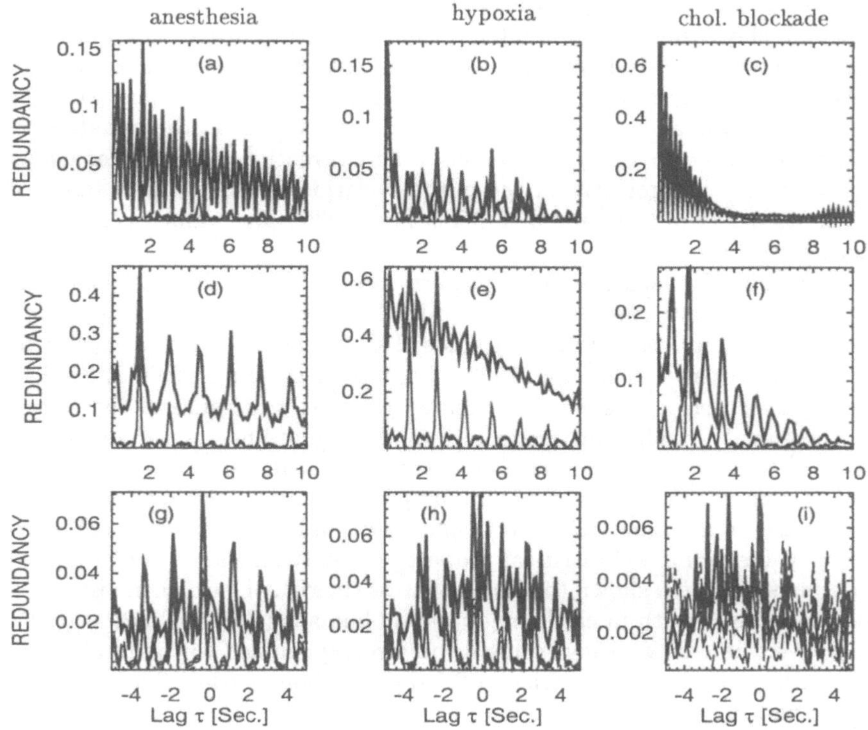

Fig. 9. Redundancies R in dependency on the time lag τ for the piglet data in Fig. 2. Fat solid line: original data, thin lines: surrogate data (mean $\pm$ standard deviation). a), b), and c): HRF, d), e), and f): RM, g), h), and i): RM–HRF

the increased R of both the original RM and in the surrogates.

During cholinergic blockade the R of HRF was dramatically increased. However, the R of RM/HRF was about zero. Under the additional consideration of the R of RM it can be concluded that during cholinergic blockade the couplings between RM and HRF disapeared. The approximatly constant heart rate corresponds to this result. In this case the coordination to the RM became meaningless. The disappearance of nonlinearites is a further essential indicator of the reduced autonomic coordination.

3.3 Window Pattern Entropy

It has been shown by many medical groups that 3-dimensional Poincaré plots of RR intervals are a useful tool for the analysis of heart rate variability. It was shown that both global 24-hour images of the trajectory of RR intervals reconstructed by means of the Takens delay coordinates and such images within a short time window (integer time counted in heart beats) are useful for the qualitative study of heart rate variability in humans (Żebrowski et al. 1995a). The images were observed with the Takens delay equal to 2 heart beats. This value was chosen by trial and error based on the quality of the images obtained in three dimensions. The conventional method of defining the delay at the first zero of the autocorrelation function failed due to the nonstationarity of the 24-hour ECG studied. The same value of the delay is used here throughout.

A complexity measure called pattern entropy (PE) was devised as a practical way of quantifying the delay coordinate images of RR intervals. A definition and a comparison with other kinds of static entropies are given elsewhere (Żebrowski et al. 1994, Paluš 1993). Because of certain details of its definition, pattern entropy should be treated more as a measure of the ordering of the signal (number of frequencies in the signal and its variance) rather than as a measure of predictibility. Note, that *pattern entropy has the peculiar property that it will be large for processes which are well ordered and stationary in the time while small otherwise* – quite the opposite to Shannon entropy, for example. Statistical properties of pattern entropy of 24-hour RR interval series were shown to allow to assess the risk of cardiac arrest and the risk of sudden death in cardiomyopathy in humans (Żebrowski et al. 1995b). Moreover, pattern entropy seems to be directly coupled to the level of norepineprine (correlation coefficient −0.74) on cardiomypathy (Popławska et al. 1994) and in normals.

The window pattern entropy (WPE) discussed in this section was calculated within a 100 beat window (about 40 s of real time for the piglet data discussed below). This window was slid along the given data series every 1 beat so that this static entropy could be monitored as it changed with the time. This technique of using a static entropy to analyze the dynamics of nonstationary processes, designed for the study of the nonstationary Holter ECG recordings of humans, turned out to be advantages also for the study of time varying processes such as the sleep dynamics in animals.

In the investigation presented here, the inverse of the instantaneous heart rate HRF is used, which is the series of RR intervals. Since the data are nonstationary by assumption, it is important to note that the histograms which are needed to calculate entropy were found using a fixed bin width called fixed-W approach. For the heart rate variability data a bin width of 15 ms (8% of the data range) was assumed. There was no prior experience with RM data and it was found that for this data the bin width that gave the most reliable values of pattern entropy was a very narrow 2% of the data range. Note that for Holter ECG of humans, where the analog signal is A/D converted at 128 Hz, the level of the sampling error precludes the use of such small bin widths. Here, the data was originaly sampled at 2048 Hz which allowed the use of much narrower bins.

The technique of the sliding, narrow time window yields a WPE which is a function of the time – each value corresponding to a different data window. It should be remembered, however, that this complexity measure is a form of static entropy. As such its value should not be directly compared with the redundancy R (Sect. 3.2) and the generalized mutual information GMIF (Sect. 3.1), which are dynamic entropies and yield information on how information about the system is gained (or lost) as the states of the system evolve. Nevertheless, several similarities but also essential differences between the results of WPE on one side and GMIF and R on the other side were found.

Sleep States. The behavior of WPE is not as clearly related to the sleep states as, e. g., CORF and GMIF (see Fig. 10). The WPE of HRF appears to

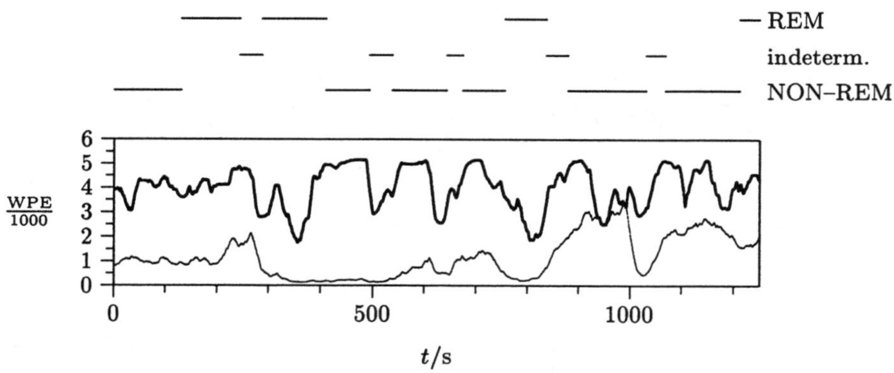

Fig. 10. Window pattern entropy WPE of the piglet data in Fig. 1. Thick line: WPE of RR, thin line: WPE of RM

be sensitive to characteristics such as the fast transients in the heart rate (indetermined intervals) and others. Nevertheless, the WPE of RM is inversely correlated to the variance of RM. Note, however, that the use of WPE with RM is a novel application of pattern entropy and some improvements for this kind of data may be sought.

Anesthesia, Hypoxia, and Cholinergic Blockade. During general anesthesia high amplitudes of WPE of RM and HRF were found (see Fig. 11) which are typical for the high ordering (high average WPE – low number of frequencies in the signal). The similar pattern of the WPE of RM and HRF indicate the couplings between these quantities.

During hypoxia the WPE of HRF is clearly smaller than during general anesthesia. Also here the similar pattern of the WPE indicate the couplings between RM and HRF.

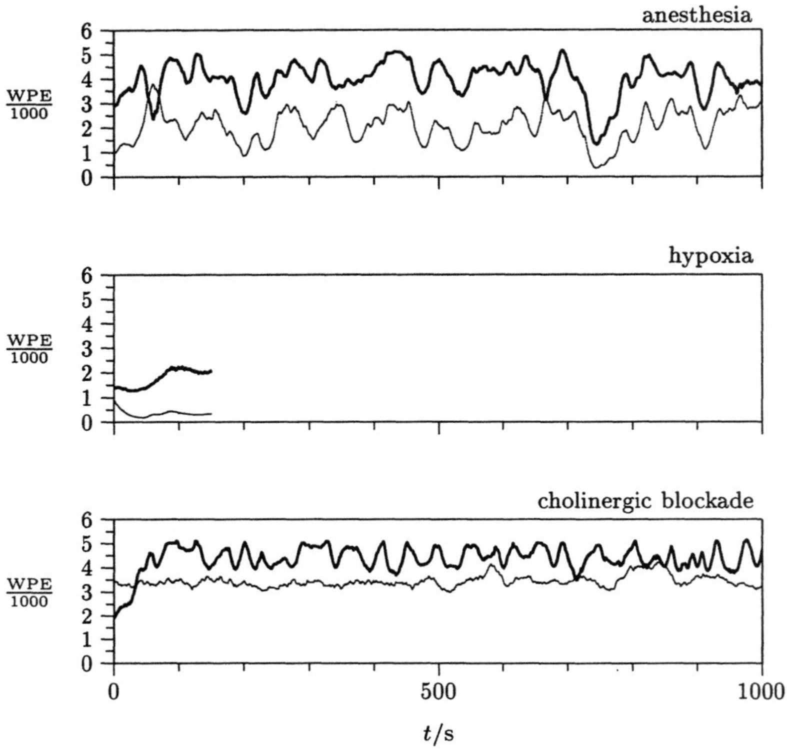

Fig. 11. Window pattern entropy WPE of the piglet data in Fig. 1. Thick line: WPE of RR, thin line: WPE of RM

The reduced coordination during cholinergic blockade could also be confirmed by the WPE estimated. The large average value of WPE of HRF corresponds to the approximately constant heart rate. The WPE of RM, however, shown even less variance and its pattern is not correlated to the WPE of HRF.

The WPE curves found represent the different behavior between non-REM sleep and REM sleep as well as between general anesthesia, hypoxia, and cholinergic blockade. The parameters chosen (delay parameter $\tau = 2$, window length = 100 beats) were sufficient to describe essential dynamics of the cardiorespiratory system by WPE as a function of the time.

3.4 Phase Synchronization in Cardiorespiratory Interactions

In this subsection we study the phase locking between the respiratory and cardiac systems. This approach is based on the recently demonstrated effect of phase synchronization of chaotic oscillators. As it was shown elsewhere

(Rosenblum et al. 1996, Pikovsky et al. 1996), if the coupling between two chaotic, or a chaotic and a periodic oscillators is strong enough, their phases are entrained, while the amplitudes can remain chaotic and independent. A weaker type of synchronization has been also demonstrated, where the frequencies are entrained, while the phase difference exhibits a random-walk type motion. The phase difference between two signals can be determined by means of the Hilbert transform. We emphasize that this technique can be used to study the relationships in *non-stationary* bivariate data (Rosenblum et al., this volume).

We present here examples of calculation of the phase difference between RM and HRF for the changing sleep states and during anesthesia. The sampling rate amounts to 8 Hz. In order to eliminate low-frequency trends, the moving average computed over the 45 point window was subtracted from the original data. This window length has been chosen by trial – its variation in the range 20 – 150 points does not practically effect the results.

Sleep States. The results for the sleep states are shown in Fig. 12. For a

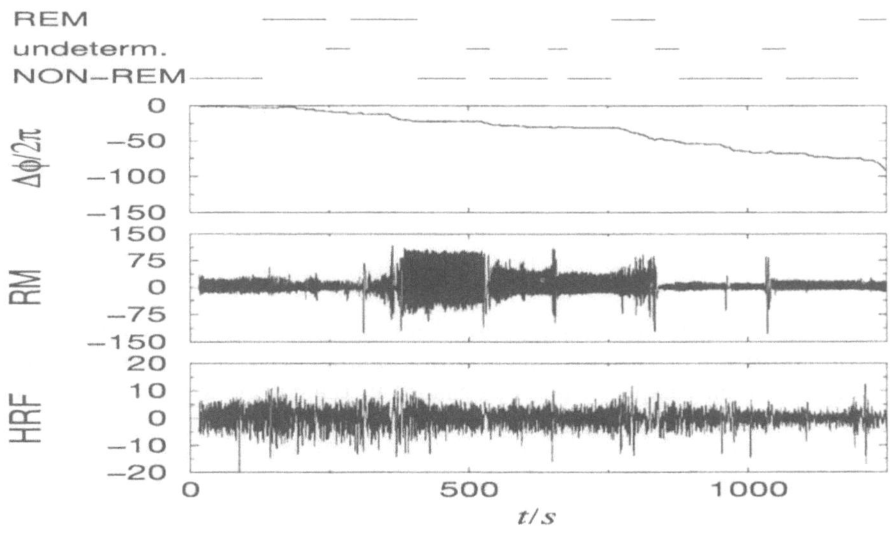

Fig. 12. Relative phase $\Delta\phi$ between RM and HRF during REM and NON–REM sleep. RM and HRF are pre–processed (see text for details) in order to eleminate low–frequency trends

better view one time interval is enlarged in Fig. 13. During non-REM sleep, phase synchronization of RM and HRF was found. This is clearly indicated by the constant (with exception of 2π jumps) phase shift between these processes.

During REM sleep, the alternating intervals of synchronous and non-synchronous behaviour are found. Due to a large number of 2π jumps, the synchronous epochs may be identified as intervals of frequency entrainment rather than of phase locking. The reason for this behavior of the relative phase is probably the disturbing influence of the EM and higher brain activity.

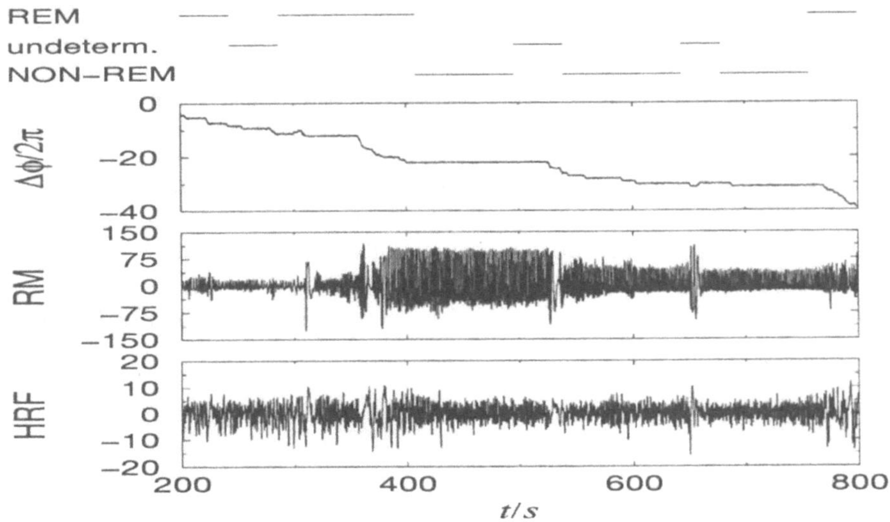

Fig. 13. Zoomed part of Fig. 12

General Anesthesia. During general anesthesia essential afferent pathways to the cortex are blocked and the resulting efferent activity is correspondingly reduced. We may expect that such change in the control system may essentially tell on cardiorespiratory interaction.

The results for general anesthesia are shown in Fig. 14. Epochs of approximate phase locking of different order such as 1 : 1 (a), 2 : 1 (b), or 3 : 2 (c) were found between periods of non-synchronized behavior of RM and HRF.

4 Discussion

The objective of the presented paper was the evaluation of new techniques of nonlinear systems analysis with regard to the study of the cardiorespiratory coordination. The questions were:

- Are there nonlinearities in RM and HRF?
- How can they be estimated?

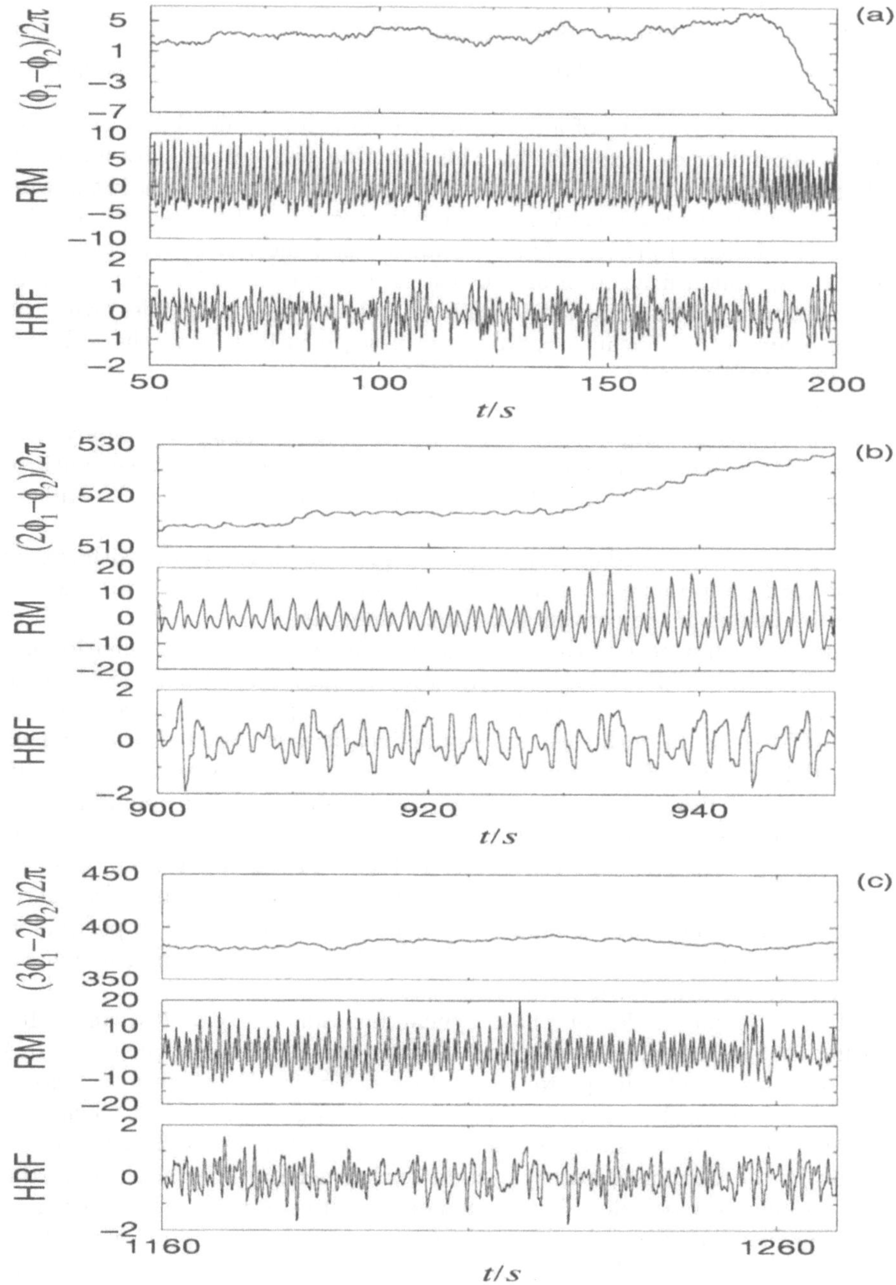

Fig. 14. Phase relation between RM and HRF during general anesthesia. a): part with approximately 1 : 1 entrainment, b): part with approximately 2 : 1 entrainment, c): part with approximately 3 : 2 entrainment. Slow trends are removed from RM and HRF

- How can non-stationarities be considered?
- How can nonlinear interrelations between RM and HRF be investigated?

An essential property of a dynamic process is its predictability or its dynamic dependencies, respectively. This question was adressed by means of different entropy measures: The short-term mutual information and the window pattern entropy were advantageous with regard to the non-stationarities of the physiological process. The redundancy and the test of surrogate data enablėd the quantification of nonlinearities. Furthermore, the comparison of the short-term mutual information (linear and nonlinear properties) with the correlograms (only linear properties) gave some insights into the different dynamic organization.

Another essential property of the ANS is the cardiorespiratory coordination which was investigated by means of the corresponding joint analysis of RM and HRF dynamics. Additionally to the joint-entropy measured synchronizations between RM and HRF were searched for by means of the Hilbert transform.

In the presented medical investigations some essential results with regard to the cardiorespiratory coordination were found which can basically not be found by means of the conventional linear analysis.

One medical aspect of the investigations dealt with the coordination between RM and HRF during different sleep stages. During non-REM sleep, there is a lower level of physiological information processing in comparison with REM sleep. In both sleep states nonlinearities were found in all signals (RM, HRF,RM/HRF). However, the linear dependencies were dominating in RM and HRF during non-REM sleep. In opposite, the corresponding nonlinear dependencies were dominating during REM sleep. The linear interrelations between RM and HRF were reduced during REM sleep, but the corresponding nonlinear inter-relations were increased. By means of these results the importance of the nonlinear behavior can be shown. In particular, if the linear dependencies become weak or disappear, the nonlinear dependencies are essential for medical diagnosis.

Two different modes of phase synchronization between RM and HRF during the sleep periods were found. During non-REM sleep, phases of RM and HRF are practically permanently locked. During REM sleep, epochs of frequency locking alternate with short intervals of non-synchronous behavior.

Another medical aspect of the investigations was the different coordination in the cardiorespiratory system during general anesthesia, hypoxia, and cholinergic blockade. Significant nonlinearities were found during general anesthesia and hypoxia. During general anesthesia approximate phase locking of different order such as 1:1, 2:1, and 3:2 were found alternating with non-synchronous periods.

During hypoxia the autonomic coordination of the cardiorespiratory system is greatly changed in order to compensate for the lack of oxygen. In this situation the system behavior is dominated by the organized RM and its coupling to HRF. The surrogate data indicate that the changes due to hypoxia are much stronger in the nonlinear dependencies than in the linear dependencies.

During the cholinergic blockade the parasympathetic efferences, representing an essential pathway for interaction in the ANS, are disrupted. In this situation the nonlinearities of HRF and, in particular, the couplings between RM and HRF, disappeared. It can be concluded that the parasympathetic efferences are essential for the nonlinear coordination within the ANS.

Also by means of these results can be confirmed that the nonlinear multivariate signal analysis gives new insights in the complex functioning of the cardiorespiratory system which can basically not be found by means of the conventional linear methods. Possible improvement of diagnostic applications in comparison to linear methods could be shown in the presented pilot study.

5 Acknowledgments

We are gratefull to B. Walter for supporting the animal experiments and to T. Schreiber for helpful suggestions in the data preprocesssing. This work was supported by grants: BMBF 01 IN 202 F/8 (D. Hoyer), BMBF ZZ 904 TP5 (R. Bauer), DFG Po442/2-1 (B. Pompe), Academy of Sciences of the Czech Republic No. 230404 (M. Paluš), KBN 2 P30B 20 09 (J. J. Żebrowski), DFG Zw47-1/3 (U. Zwiener). M. Rosenblum acknowleges financial support from the Alexander von Humboldt Foundation.

References

Buckley N. M. (1986): Maturation of circulatory system in three mammalian models of human development. Comp. Biochem. Physiol. A**83**, 1–7

Grönlund J. U., Kalli S. T., Siimes A. S. I., Sydänmaa M., Antila K. J., Välimäki I. A. T. (1991): Do beta-adrenergic blockade and sleep state affect cardiorespiratory control in neonatal lambs? kultivariate autoregressive modeling approach. Pediatric Research **29**, 272–277

Heymann M. A., Iwamoto H. S., Rudolph A. M. (1981): Factors affacting changes in the neonatal systemic circulation. Ann. Rev. Physiol. **43**, 371–383

Hoyer D., Wagner H., Zwiener U. (1995): Phase portraits of transfer systems – nonlinear transfer properties between respiratory movements and heart rate fluctuations. Biomedizinische Technik **40**, 316–321

Hoyer D., Schmidt K., Zwiener U., Bauer R. (1996a): Characterization of complex heart rate dynamics and their pharmacological disorders by nonlinear prediction and special data transformations. Cardiovascular Research **31**, 434–440

Hoyer D., Schmidt K., Bauer R., Zwiener U., Köhler M., Lüthke B., Eiselt M. (1996b): Complexity and predictability of heart rate variability and respiratory movements in the rabbit. IEEE Eng. Med. Biol. (accepted)

Kurths J., Voss A., Saparin P., Witt A., Kleiner H. J., Wessel N. (1995): Quantitative analysis of heart rate variability. Chaos **5**, 88–94

Langhorst P., Stroh–Werz M., Dittmar K., Camerer H. (1975): Facultative coupling of reticular neuronal activity with peripheral cardiovascular and central cortical rhythms. Brain Research **87**, 407–418

Morgan B. J., Crabtree D. C., Palta M., Skatrud, J. B. (1995): Combined hypoxia and hypercapnia evokes long-lasting sympathetic activation in humans. J. Appl. Physiol. **79**, 205–213

Paluš M. (1993): Coarse-grained entropy rates for characterization of complex time. Physica D **93**, 64–77

Paluš M. (1996): Detecting nonlinearity in multivariate time series. Phys. Lett. A **213**, 138–147

Pompe B. (1996): Ranking and Entropy Estimation in Nonlinear Time Series Analysis. (in this volume)

Popławska W., Grzybowski J., Żebrowski J. J., Baranowski R., Bilińska Z. T., Pszona B. (1994): Pattern entropy – nonlinear noninvasive estimation of adrenergic drive in patients with dilated cardiomyopathy. J. Amb. Monitoring **7** 80

Prechtl H. F. R. (1974): The behavioural states of the newborn infant (a review). Brain Res. **76**, 1304–1311

Pikovsky A. S., Rosenblum M. G., Kurths J. (1996): Synchronization in a population of globally coupled chaotic oscillators. Europhysics Letters **34**, 165–170

Rosenblum M. G., Pikovsky A. S., Kurths J. (1996): Phase synchronization of chaotic oscillators. Phys. Rev. Lett. **76**, 1804–1807

Żebrowski J. J., Popławska W., Baranowski R. (1994): Entropy, pattern entropy and related Mmthods for the analysis of data on the time intervals between heart beats from 24 hour electrocardiograms. Phys. Rev. E **50**, 4187–4205

Żebrowski J. J., Popławska W., Baranowski R. (1995): Pattern entropy a tool for nonlinear dynamical analysis of a biological nonstationary system: the human heart. Acta Phys. Pol. **26**, 1055–1064

Żebrowski J. J., Popławska W., Baranowski R., Buchner T. (1995): Properties of 24-hour holter recordings of RR-intervals: a nonlinear quantitative method through pattern entropy of 3-D return maps. Computers in Cardiology, 177–180

Zwiener U., Hoyer D., Lüthke B., Schmidt K., Bauer R. (1995): Relations between parameters of spectral power densities and deterministic chaos of heart rate variability. J. Autonom. Nerv. System **25**, 132–135

Zwiener U., Hoyer D., Bauer R., Lüthke B., Walter B., Schmidt K., Hallmeyer S., Kratzsch B., Eiselt M. (1996): Deterministic-chaotic and periodic properties of heart rate and arterial pressure fluctuations and their mediation in piglets. Cardiovascular Research **31**, 455–465

Cardiorespiratory Synchronization

M. Schiek[1], F.R. Drepper[1], R. Engbert[2], H.-H. Abel[3], and K. Suder[1]

[1] Institut für Biologische Informationsverarbeitung, Arbeitsgruppe Kardiorespiratorische Rhythmen, Forschungszentrum Jülich GmbH, Jülich, Germany
[2] Institut für Theoretische Physik und Astrophysik, und Innovationskolleg "Formale Modelle kognitiver Komplexität", Universität Potsdam, Germany
[3] Abteilung für Kardioanästhesie, Städtisches Klinikum Braunschweig, Germany

Abstract. The complex behaviour of cardiorespiratory dynamics is shown to be related to the interaction between several physiological oscillators. This study is based on electrocardiogram and respiratory flow data obtained from 3 different subjects during paced breathing at 10 different pacing cycle lengths ranging from 5 s to 12 s. Two different methods ideally suited for the analysis of synchronization pattern of coupled oscillators are applied: 1. Symbolic dynamics based on symbol coding adapted for the detection of respiratory modulation of cardiac parasympathetic activity discloses two regimes of different synchronization behaviour within the frequency area corresponding to the Arnold tongue of 1:1 frequency-locking between respiratory flow and respiratory heartbeat variation (respiratory sinus arrhythmia). 2. The analysis of the phase shift between respiratory flow and respiratory sinus arrhythmia indicates that synchronization is not a static but a dynamic phenomenon. The observed dependence of the phase shift on respiratory cycle length shows large inter-individual variation. These findings turn out to be further hints for the existence of an additional central oscillator in the frequency range of respiration interacting with the central respiratory oscillator driving mechanical respiration.

1 Introduction

The understanding of oscillators and their mutual synchronization is a crucial problem in physiology (v. Holst 1937, Winfree 1980, Glass and Mackey 1988, Strogatz et al. 1992, Collins and Steward 1993). There is detailed insight into the rhythm generator of the human heart, the so-called 'integrate-and-fire' mechanism of the pacemaker cells of the sino-atrial node. But under certain conditions like pathological arrhythmia or external electrical stimulation even these pacemaker cells show a very complex dynamic behaviour (v.d. Pol and v.d. Mark 1928, Guevara et al. 1981). Many physiological oscillatory phenomena like electric cortical activity, circadian rhythm, hormonal cycles, menopausal hot flashes, walking, chewing or respiration are not based on the activity of pacemaker cells (v. Holst 1937, Wever 1979, Glass and Mackey 1988, Plesch et al. 1988, Pöppel et al. 1990, Kronenberg 1991, Richter et al. 1992). Therefore, the knowledge of their mechanisms of rhythm generation is rather limited. A characteristic property of some physiological oscillators is that, normally performed as an autonomic motor act, they can be voluntarily controlled. Examples are chewing, breathing or motor co-ordination like bimanual production of polyrhythms (Plesch et al. 1988,Engbert et al., this volume). Therefore, the frequency of voluntary performance can be chosen as control parameter. This property is an excellent prerequisite for the application of new concepts developed in the context of dynamic systems theory.

Among the rhythmic phenomena in physiology the subsystems of the cardiorespiratory system (heart, blood pressure control loop, sympathetic and parasympathetic cardiovascular neurons, and the respiratory network) have been studied for long time and deep insight exists in part of their functions (Richter et al. 1992, Schiek et al. 1995a, Seidel and Herzel 1995, Suder 1996). This knowledge is the basis for a physiological interpretation of the results obtained by time series analysis of cardiorespiratory data. However, there are still open questions of high clinical interest, especially about central respiratory dynamics (Schläfke 1988). One controversially discussed item is the existence of different central nervous oscillators in the frequency range of respiration and the relevance of their synchronization behaviour for explaining respiratory dysfunctions (Koepchen 1988).

In the present study we use the analysis of cardiorespiratory interaction to get information about the mechanism of central respiratory rhythm generation. In particular, we look for further evidence for the existence of an additional oscillator in the frequency range of respiration interacting with the central respiratory oscillator driving mechanical respiration. The experimental results are investigated by an analysis of symbolic dynamics and phase shifts. These methods are ideally suited to studying the dynamics of coupled oscillators.

2 Cardiorespiratory Interaction

2.1 Respiratory Sinus Arrhythmia

At rest during spontaneous breathing, heartbeat intervals decrease during inspiration and increase during expiration. This interaction between heart rate and respiration was already noted by Ludwig (1847). Since then the so-called respiratory sinus arrhythmia (RSA, heart rhythm is often called sinus rhythm because the electrical activity which initiates the rhythmic contraction of the heart originates at the sino-atrial node in the right atrium of the heart) has been the subject of many studies.

The RSA is mainly caused by central respiratory activity (Koepchen et al. 1961, Hildebrandt 1966). Hereby inspiratory activity is more important than expiratory activity (Davies and Neilson 1967, Hirsch and Bishop 1981) and respiratory movement is not required for its appearance (Koepchen et al. 1961, Hildebrandt 1966, Hirsch and Bishop 1981). RSA amplitude, defined as the difference between the longest and shortest heartbeat interval within the respiratory cycle, is a quantitative index of parasympathetic cardiac control (Eckberg 1983, Hayano et al. 1994a). The amplitude of RSA increases with respiratory cycle length (Angelone and Coulter 1964, Hildebrandt 1966, Davies and Neilson 1967, Hirsch and Bishop 1981, Eckberg 1983, Hayano et al. 1994a) and, to a smaller extend, with respiratory tidal volume (Eckholdt and Schubert 1975, Hirsch and Bishop 1981). Some authors found that RSA amplitude remained constant beyond respiratory cycle length about 10 s to 12 s (Hirsch and Bishop 1981), others even observed a decrease beyond this cycle length (Angelone and Coulter 1964). These findings often are interpreted as a resonance of RSA with the so-called Mayer wave, which is a nonrespiratory rhythmic blood pressure variation with a mean period of 10 s (Angelone and Coulter 1964, Hildebrandt 1966, Koepchen 1984).

Mathematical models of the blood pressure control loop are able to reproduce the dependence of RSA amplitude on respiratory cycle length due to the band-pass property of the control loop. The band-pass property is determined by the delay time within the sympathetic part of the loop (DeBoer et al. 1987, Schiek 1994, Seidel and Herzel 1995, Schiek et al. 1995b). Therefore, it has been suggested that this band-pass property generates the Mayer wave or amplifies this autonomic cardiovascular rhythm generated by an intrinsic oscillator of a period of 10 s in the autonomous nervous system (Koepchen 1984, DeBoer et al. 1987, Seidel and Herzel 1995).

Phase shifts between respiration and heart rate are less variable between different individuals than RSA amplitude. Phase shift is not constant, but varies with respiratory cycle length (Angelone and Coulter 1964, Davies and Neilson 1967, Kelman and Wann 1970, Luczak and Raschke 1975, Eckberg 1983). The reason for this relationship is not clear. Angelone and Coulter interpreted their findings as a proof for resonance within the cardiorespiratory system, whereas Davis and Neilson claimed that the dependence of the phase shift on respiratory cycle length can be explained by a constant time delay between the onset of inspiration and the peak of the resulting heart rate variation.

2.2 Physiological Background

In healthy subjects heart rate is determined by the activity of the pacemaker cells of the sino-atrial node located in the right atrium (Schmidt and Thews 1995, Glass et al. 1991). The rhythmic activity of these cells is generated by an 'integrate-and-fire' mechanism: each cycle consists of a time interval of slow depolarisation and an action potential. The velocity of slow depolarisation i.e. the frequency of these pacemaker cells mainly depends on their innervation by parasympathetic and sympathetic neurons. Parasympathetic activity decreases, sympathetic activity increases the slow depolarisation of pacemaker cells and consequently the heart rate. On a time scale between 1 s and 30 s parasympathetic and sympathetic activity in humans at rest mainly is modulated by respiratory activity. This respiratory modulation is amplified by the blood pressure control-loop. The sensors of blood pressure control-loop are the baroreceptors which respond with an increase of firing rate to increase of atrial blood pressure and its time derivative. Baroreceptor activity is led to the cardiovascular and respiratory structures in the brainstem and increases parasympathetic and decreases sympathetic activity. The strength of this negative feedback is reduced during inspiration (Koepchen et al. 1961, Eckberg et al. 1980). To a minor extent respiratory movement also affects heart rate by modulating diastolic filling of the heart which leads to a corresponding modulation of pulse pressure and consequently to a respiratory modulation of heart rate via the blood pressure control loop.

The respiratory rhythm is generated within the brainstem not due to simple pacemaker cells (Richter et al. 1992, Schmidt and Thews 1995). Therefore, knowledge about central respiratory dynamics is much more limited compared to cardiac dynamics. The respiratory rhythm generator, as a part of the respiratory network, consists of different subgroups of neurons, which show a characteristic discharge pattern during different phases of the respiratory cycle. The respiratory rhythm is generated by a complex inhibitory and excitatory coupling between and within these neuronal subgroups (Richter et al. 1992). It has been shown in piglets that apnea produced by laryngeal stimulation (upper airway-induced apnea appear to be the major cause of apnea in newborns) are associated with an interruption of the oscillation within this network. The interruption is characterised by a prolonged activation of the socalled postinspiratory neurons and an inhibition of the other neuronal subgroups (Lawson et al. 1991). This leads to the assumption that the oscillation of this respiratory rhythm generator is directly associated with the oscillatory electrical phrenic discharge controlling inspiratory movements. There are several experimental results which support the hypothesis of the existence of an additional central oscillator in the frequency range of respiration:

- The frequency-locking ratios between blood pressure waves and electrical phrenic activity changed between 1:2, 1:3, and 1:4 in dogs and rabbits while respiratory movement and cardiac parasympathetic control were suppressed. Blood pressure waves also showed "phase-jumps" of 180° (Koepchen and Thurau 1958).
- The respiratory modulation of the blood pressure control loop continued during absence of phrenic discharge in dogs (Koepchen et al. 1961).
- Heart rate fluctuations in the former respiratory frequency range occurred during paced respiration at low frequencies (Schiek 1994).
- Phase shift 0° or 180° between respiration (flow or respiratory movement) before and after central apnea was observed in infants with sleep related apnea (Drepper et al. 1995).

3 Experimental Design

The cardiorespiratory data, ECG and respiratory flow (uncalibrated thermistor signal) with 1000 Hz sample rate, were obtained from three healthy

Table 1. Study protocol

Session	total duration	respiratory cycle length	number of cycles	duration
I	29′:45″	9.0 s	70	10′:30″
		5.5 s	70	6′:25″
		11.0 s	70	12′:50″
II	30′:20″	8.0 s	70	9:′20″
		6.0 s	70	7′:00″
		12.0 s	70	14′:00″
III	33′:15″	7.0 s	70	8′:10″
		5.0 s	70	5′:50″
		10.0 s	70	11′:40″
		6.5 s	70	7′:35″

men, aged between 29 and 35 years, without any history of cardiopulmonary diseases. All subjects did not smoke for more than 1 year. Because all subjects are members of the involved institutes, they were informed about the specific purpose of the study. To get a rough estimate of the intra-individual variation of the cardiorespiratory dynamics, after two weeks the experiment was repeated for one subject. From each subject data were obtained during paced breathing at 10 different cycle lengths (5, 5.5, 6, 6.5, 7, 8, 9, 10, 11 and 12 s), 70 cycles at each frequency. The whole experiment was split into three sessions of nearly equal duration (table 1). To check for transient phenomena pacing frequencies were mixed in a way that large frequency jumps were achieved. The sessions were performed at the same time of different days all within one week, 2–3 hours after the last meal or caffeinated beverage. When two sessions were measured on a single day, they were separated by a 45 minute break. All measurements were done in 45° head up tilt position. Volunteers were asked to breath in phase with a growing and vanishing light circle, so that they were able to predict the switching between the respiratory phases (expiration or inspiration). The ratio of expiration to inspiration length was kept constant (4:3).

Cardiac cycle length (RR interval) was determined as time interval between two successive ventricular depolarisations with an accuracy of about 1 ms. Each RR interval was numbered according to its position within the respiratory cycle. For technical reasons beginning of respiratory cycle was defined by the onset of expiration. Each heartbeat interval was assigned to the respiratory cycle covering more than half of the heart cycle.

Within all data sets only one irregular heartbeat was observed. Data corresponding to pacing with a cycle length of 6 s of experiment C-2 had to be excluded because the performance of paced breathing was insufficient. In all other data sets the accuracy of paced breathing was within 10%. In general, performance was better during low than high pacing frequencies.

4 Physiological Data Analysis

4.1 Symbolic Dynamics

The transformation of a time series into a symbol sequence is achieved by a coarse-graining of the phase space (Hao 1991, Beck and Schlögl 1993). This is done by dividing the phase space into a finite set of cells each associated with a different symbol. The easiest transformation is to divide a d-dimensional phase space into d-dimensional cubes of equal size (Wackerbauer et al. 1994). By partitioning into cells of different sizes and shapes one can focus on special dynamic properties of the observed system (Schwarz et al. 1994, Kurths et al. 1995). Depending on the specific question of the investigation different ways of partitioning the phase space can be used. The appropriate number of different symbols is limited by the length of the time series from which the symbol sequences are derived. The accuracy of the symbol description is refined more and more by introducing more symbols but consequently the statistical confidence level of the occurrence of the symbol drops in accordance with the number of possible symbols. With regard to a later interpretation of the statistical results of the symbol sequence it is useful to choose a symbolic coding which can easily be interpreted in a physiological way. If one observes a system with two pronounced time scales one can make use of this situation by subdividing the symbol sequence defined on the faster time scale into words corresponding to the slower time scale (?).

From Cardiorespiratory Data to Symbol Sequences

According to the two pronounced time scales within the physiological data (heart rate and respiratory rate) symbol definition is done in two steps. In a first step each heartbeat interval is mapped to the symbol 1 or 0, depending on whether its difference to the preceding heartbeat interval exceeds a certain level or not. Since RSA has a highly varying amplitude, the level s is locally adapted for each respiratory cycle to:

$$s = \langle dRR \rangle + b \cdot \mathrm{sd}(dRR) \; , \qquad (1)$$

where $\langle dRR \rangle$ denotes the mean of the RR interval differences dRR and $\mathrm{sd}(dRR)$ its standard deviation. The parameter b determines the percentage of the symbol 1, it is kept constant for all pacing frequencies of one

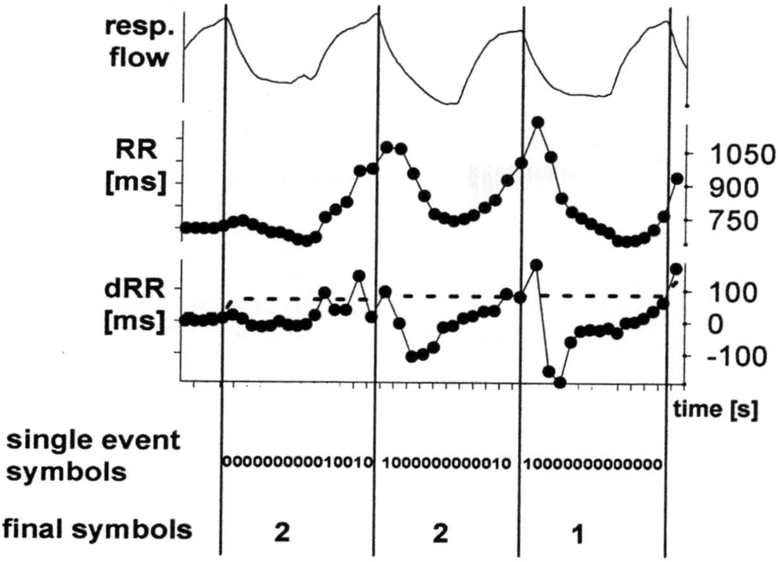

Fig. 1. Respiratory flow is measured by sampling the uncalibrated voltage of a pair of thermistors attached to nose and mouth (inspiration upwards). Heartbeat intervals (*RR*) are determined as time intervals between successive ventricular depolarisations (R peak in the ECG). *dRR* denotes the difference of a heartbeat interval to the preceding interval. The symbolic coding is explained in the text.

experiment. In a second step the number of transitions from 0 to 1 within each respiratory cycle is taken as the final symbol (Fig. 1).

The motivation from the physiological point of view for this symbol definition is the detection of changes of the autonomic innervation of the heart influenced by respiration. Sudden prolongation of heartbeat intervals (single event symbol 1) is caused by an increase in parasympathetic activity, which in contrast to sympathetic activity acts on a shorter time scale (Schmidt and Thews 1995). The final symbol represents the number of times of sudden rising cardiac parasympathetic activity within one respiratory cycle. Since synchronization of heart rate variability to respiration is mainly mediated by parasympathetic innervation the final symbol characterises cardiorespiratory synchronization.

Results

The symbol sequences can be used to visualize the specific dynamic properties focused on by the symbol definition described above. To visualize the dependence of the symbolic dynamics on the control parameter (pacing cycle length) we plot the final symbols, coded by grey scale levels, against respiratory cycle number (time) on the abscissa and pacing cycle length on the ordinate. To estimate the effect of the percentage of the single event symbol 1 determined by parameter b (1) we look at all symbol plots for percentages between 10% and 90% in steps of 2.5%.

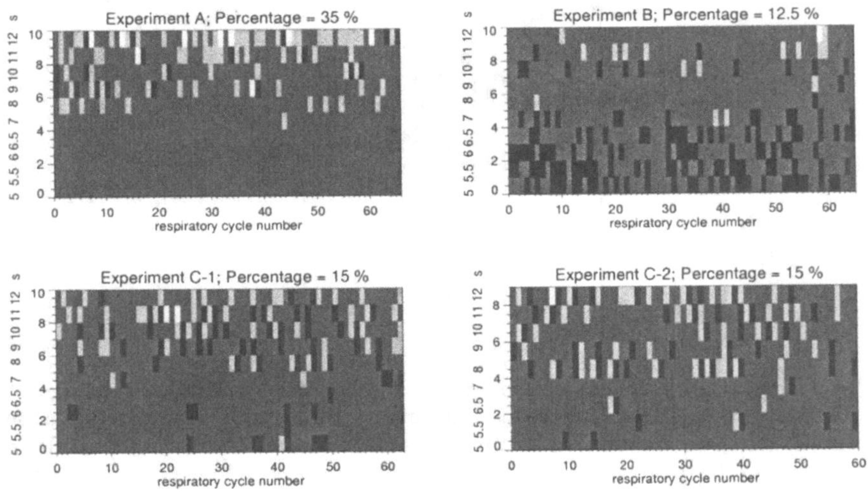

Fig. 2. Final symbols coded by grey scale levels (black denotes final symbol 0, white denotes final symbol 3): Respiratory cycle number (time) on the abscissa and pacing cycle length (control parameter) on the ordinate. Final symbols obtained by the symbol definitions show a qualitative change in symbol distribution at pacing cycle length 8 s.

The first important result is that there is no obvious transient nor instationarity within the symbol sequences for all experiments and all percentages, but there is a dependence of the symbol distribution on the control parameter. For longer pacing cycles, symbols with higher numbers are more frequent. This is a rather trivial result. However, it is remarkable that an abrupt change in symbol distribution occurrs around the pacing cycle length of 8 s in all experiments. To both sides of this critical value of the control parameter symbol distribution stays comparatively constant. In contrast to this the winding number (i.e. the number of heartbeats per respiratory cycle) shows an overall trend of constant increase, in particular, the winding numbers of all experiments at pacing cycle length of 8 s vary between 7.8 and 10.3 and show no plateau behaviour to either sides. The visibility of this finding varies with the chosen percentage.

For further analysis we select for each experiment the symbol definition showing most clearly the qualitative change in symbol distribution at pacing cycle length 8 s (Fig. 2). The χ^2test is used to quantify the visual impression. In particular, we calculate the probability of the assumption that two experimental distributions have been generated by the same process. The χ^2 probabilities for all combinations of control parameters are visualized in Fig. 3 by grey scale coding. For all experiments the dynamic behaviour of the symbol sequences corresponding to different control parameters is split into two areas, ranging approximately from 5 s to 7 s and from 8 s to 12 s. Within each area there is a high probability of similar symbol distribution and therefore of similar dynamics, whereas between the two areas there is a very low probability of similar dynamics. The distinctness of the probability distributions of the two areas is even more pronounced when the probability matrix is calculated for the distribution of words consisting of three successive symbols.

As can be seen in Fig. 3, in experiment A, B, and C-2 one finds a rather sharp transition between pacing cycle length 7 s and 8 s whereas in experiment C-1 a smoother transition is observed. Symbol sequences corresponding to pacing cycle length 10 s stand out in experiment A, B and C-2. In experiment A and C-1 one observes an additional transition beyond pacing cycle length 11 s.

In summary, the control parameter "pacing cycle length" determines the symbol sequences represented by its visible impact on the χ^2 probability. The session number or the position in time within one session exerts no obvious dependence on the symbol sequences. Furthermore, there is no indication of an influence of the preceding pacing frequency. Since symbol definition focuses on cardiorespiratory synchronization, the observed transitions between qualitatively different symbol distributions correspond to a qualitative change in synchronization behaviour. This shows that two regimes of different cardiorespiratory synchronization behaviour exist within the control parameter area between 5 s and 12 s. As will be shown in the following, these two regimes both belong to the Arnold tongue of 1:1 frequency-locking between respiratory flow and heartbeat intervals. The synchronization regimes stand out in Fig. 3 as distinct areas of high probability of similar dynamics.

The rather high χ^2 probability of similar dynamics of the time series corresponding to pacing cycle length 10 s and the ones corresponding to the short pacing cycle length in experiments A, B and C-2 leads to the assumption of an interaction between the so-called Mayer wave in heart rate fluctuations and respiration. However the sharpness of this phaenomenon makes this interpretation rather unlikely because the Mayer wave is known to be a broad frequency phenomena (Golenhofen and Hildebrandt 1958, Koepchen 1984). But the observed interaction tends to influence cardiorespiratory synchronization pattern on a restricted frequency band.

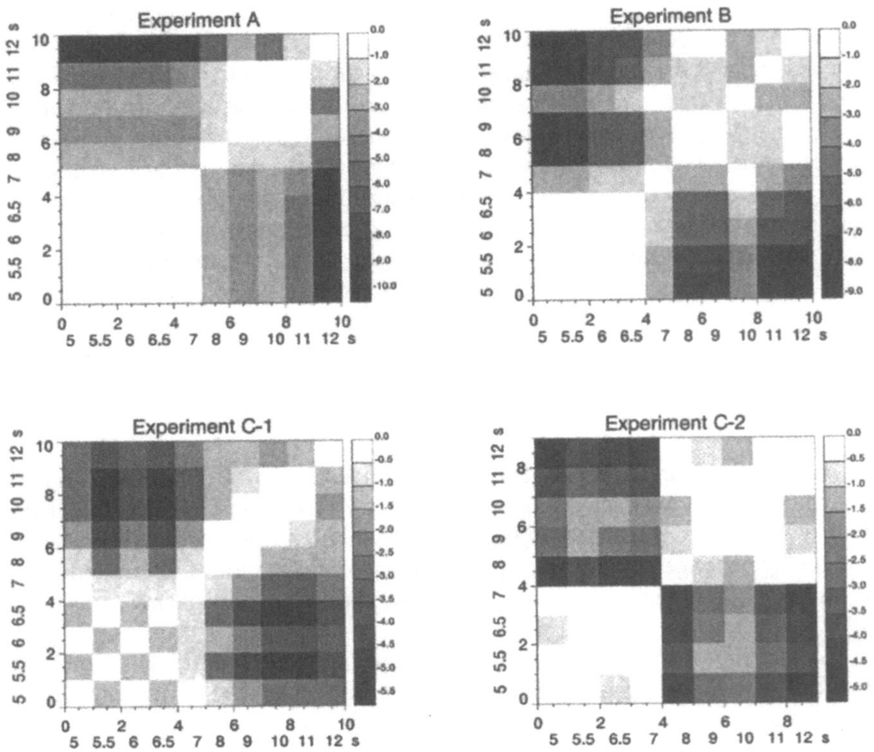

Fig. 3. Logarithm to the base of 10 of χ^2 probabilities for the distribution of the final symbols for all combinations of control parameters coded by gray scale levels. Note that time series corresponding to pacing cycle length 6 s in experiment C-2 is excluded because performance of paced breathing was insufficient.

4.2 Analysis of Phase Shifts

Method

A second method to investigate the dynamics of coupled oscillators is the analysis of phase shifts between signals coupled to different oscillators. In the context of cardiorespiratory interaction phase differences between mechanical respiration and the respiratory sinus arrhythmia (RSA) have already been investigated e.g. by Angelone and Coulter (1964) and by Eckberg (1983). The calculation of phase shifts between oscillatory phenomena with similar frequencies can be done in many different ways. Angelone and Coulter determined the time differences between respective maxima of the two cardiorespiratory signals, whereas Eckberg (1983) also discussed time differences between respective minima (more accurately: points in time where 5% of the relative change has occurred after the respective maxima or minima) . Furthermore Hayano et al. (1994b) used a complex demodulation technique based on the introduction of a reference oscillation with a frequency which lies in the centre of the relevant frequency band. However, their main focus was on instantaneous frequency changes.

In the following, two additional methods which lead to a definition of a phase shift at every point of time will be applied. The first is based on an oscillatory phase defined with the help of the so-called Hilbert transform (Panter 1965, ?). The second more imaginative method is outlined within the present context. Both methods start to transform each oscillatory signal x_1, x_2 to a rotation process, by introducing an additional related signal $\tilde{x}_1, \tilde{x}_2$, as a second co-ordinate in the plane often called the embedding plane. In both methods the additional signals $\tilde{x}_1, \tilde{x}_2$ are obtained from the corresponding originals by a kind of delay or time shift. The second method is based on a characteristic period length of the original time series. For such signals having a characteristic period length the Hilbert transform can be understood as a kind of delayed signal with a delay time corresponding to one quarter of the characteristic period length. Whereas for the Hilbert transform there is no need to pick out the dominant period length in advance the second method uses a predetermined time shift chosen as one quarter of the characteristic period length τ (in our case the period length of the paced respiration). Within this two-dimensional (embedding) space a rotation angle ϕ about the origin can be defined:

$$\phi(t) = \arctan \frac{\tilde{x}(t)}{x(t)}\,, \tag{2}$$

where $\tilde{x}(t)$ denotes the delayed signal which is either the discrete approximation to the Hilbert transform (3) or the fixed delay (4).

$$\tilde{x}(t) = \frac{1}{\pi} \sum_{i=1}^{\infty} \frac{x(t-i) - x(t+i)}{i}\,, \tag{3}$$

$$\tilde{x}(t) = x(t - \tau/4)\,. \tag{4}$$

The derivative with respect to time of the phase angle (2) can be interpreted as an instantaneous rotation frequency. Using the addition theorem of the arctan the phase difference between the two oscillatory signals can be written as

$$\phi_1 - \phi_2 = \arctan \frac{\tilde{x}_1 x_2 - \tilde{x}_2 x_1}{\tilde{x}_1 \tilde{x}_2 + x_1 x_2}\,. \tag{5}$$

In practice, however, it turns out that the rotation frequency derived from (2) often does not coincide with the intuitive concept of the oscillatory frequency of the original signals. Superimposed oscillations with a lower frequency may lead to an underestimation of the rotation speed (some of the rotations in the plane may not enclose the origin). Superimposed higher

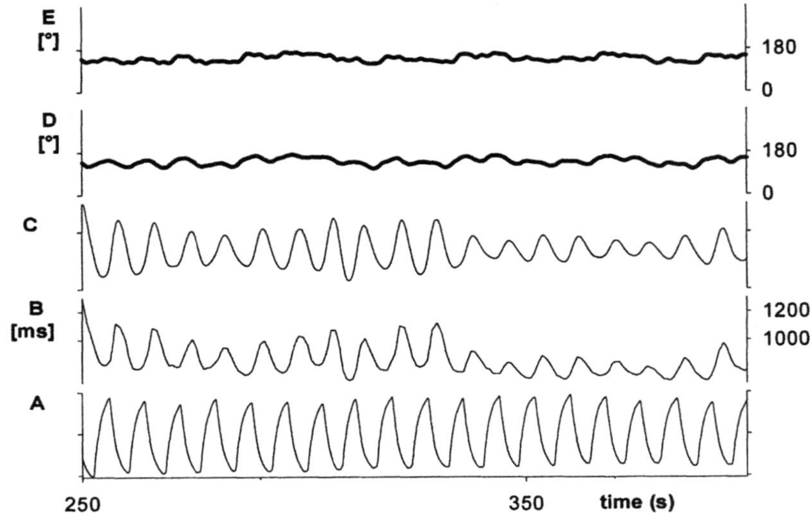

Fig. 4. Phase shifts (5) between two cardiorespiratory signals: (A) thermistor voltage as uncalibrated data corresponding to respiratory flow and (B) 10 Hz resampling of the linear interpolated nonequidistant time series of heartbeat interval lengths. (C): band-pass filtered heartbeat intervals as described in the text; (D) and (E): time series of instantaneous phase differences between heartbeat interval lengths and respiratory flow, (D) obtained by Hilbert transform (3) and (E) by fixed time delay (4).

frequency oscillations may lead to an overestimation of the rotation speed (additional small loops around the origin due to ubiquitous noise). To eliminate these disturbances the oscillatory signals first have to be treated by a band-pass filter eliminating the higher and lower frequencies. The simplest way to implement a phase neutral frequency filter is to use a moving average as the low-pass filter and the difference to a moving average as a high-pass filter. In our case the boundary frequency of the high-pass filter corresponds to the respiratory frequency and the boundary frequency of the low-pass filter is chosen four to five times larger. Thus the moving average width of the high-pass filter is equal to the oscillatory period. In addition, the phase differences are considered modulo 360°.

To apply these to methods to the cardiorespiratory data the non-equidistant time series of heartbeat intervals have to be converted by linear interpolation into equidistant time series with a sampling rate equal to the respiratory flow data.

Results

Figure 4 shows in its upper part the phase differences obtained by the Hilbert transform as (D) and those obtained via fixed time delay as (E). The comparison of the two differently defined instantaneous phase shifts shows a relatively good agreement. In particular, both phase shifts have the same average value and low frequency fluctuations around the mean tend to be synchronous. However, the instantaneous phase shifts obtained by the Hilbert transform display less high frequency fluctuations but additional oscillations in the frequency range of respiration. The high frequency fluctuations of the fixed time shift method have been partly eliminated by an additional low-pass filter adjusted to the heart rate. The additional fluctuations in the frequency band of respiration of the Hilbert transform lead to a higher variance of the fluctuations of the phase shift in (D). The absence of the high frequency component in (D) is due to a different filter characteristics of the Hilbert transform compared to the fixed delay.

Due to the nonlocal definition of the Hilbert transform in (3) the dynamics of the Hilbert phase contains contributions from the dynamics of the amplitudes, which are absent in the case of the instantaneous phase defined by fixed time delay. As has already been noted in the context of the analysis of symbolic dynamics, the respiratory sinus arrhythmia is characterized by highly variable amplitudes. To account for this, the symbols were defined independent of amplitude fluctuations. For the same reason the second definition of the instantaneous phase was chosen for the analysis of phase shifts characterizing cardiorespiratory synchronization. The comparative advantage of the Hilbert transform, to be less sensitive to frequency shifts, is of minor importance in the present study.

The finite variations around the mean in both graphs (Fig. 4 D, E) are characteristic for a stable 1:1 frequency-locking. Similar behaviour we find in all time series of phase shifts corresponding to all pacing cycle lengths of all experiments, except cycle length 5 s and 5.5 s in experiment B. In all other cases a clear 1:1 frequency-locking between respiratory flow and heart rate variation is found. The discretness in time of the heartbeat intervals is not reflected in the method presented. For this reason the present analysis is not able to detect phase-locking between heartbeat intervals and respiratory flow. Phase-locking in this sense can be detected e.g. by analysing the points in time of the heartbeats in relation to the latest onset of exspiration (Schiek 1994). Using this method, phase-locking of ratio 1:1 and 1:2 (i.e. the position of all heartbeats within the respiratory cycle repeats every or every second respiratory cycle) is sporadically observed for short time intervals up to one minute.

The fluctuations of the phase shifts around the mean partially seem to be non-random. The further investigation of the low frequency dynamics goes beyond the scope of the present study. Instead we focus on the question how the average phase shifts and their standard deviations depend on the res-

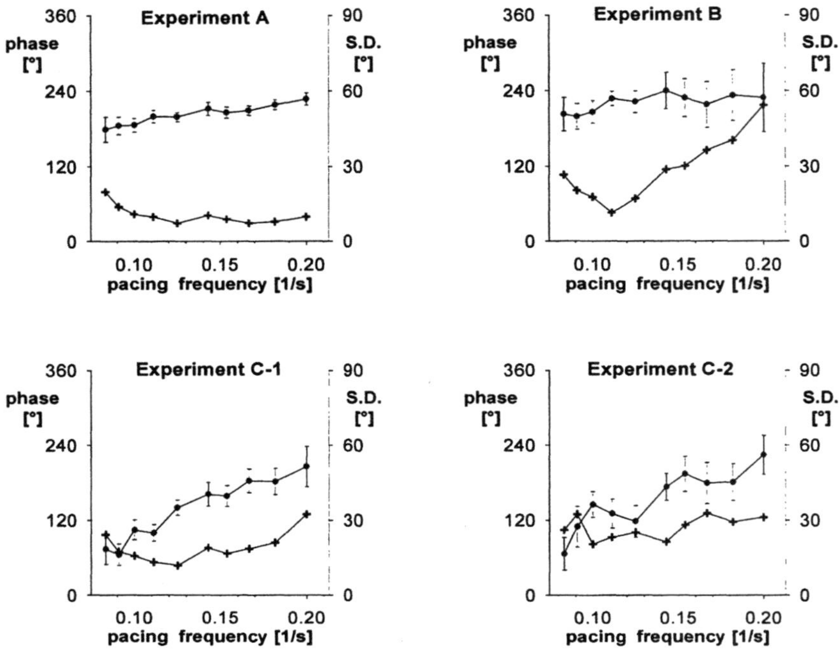

Fig. 5. Means (points) and standard deviations (crosses) of instantaneous phase shifts as computed by equations (4) and (5) for ten different speeds of paced respiration plotted on the abscissa (period lengths: 12, 11, 10, 9, 8, 7, 6.5, 6, 5.5, 5 s) and four different experiments (C-2 was obtained from the same subject two weeks after experiment C-1).

piratory frequency. As already observed by several authors (Angelone and Coulter 1964, Davies and Neilson 1967, Kelman and Wann 1970, Luczak and Raschke 1975, Eckberg 1983) the phase shifts have an over all tendency to increase with increasing respiratory frequency. However, it turns out that the strength of this dependence is markedly different for the three subjects involved in the study (Fig. 5). Whereas for subject C the frequency dependence is of the same order of magnitude as found in the literature, for subject A and B the variation is much lower. For these subjects the values coincide only for the higher frequencies. The apparent deviations from monotonicity for experiment C-2 turn out to be not significant. Measurements recorded on a single session show no obvious deviation from monotonicity.

However, the dependence of the standard deviation of the instantaneous phase differences on the pacing frequency is significant. Figure 5 suggests that the frequency coupling is more stable in the intermediate frequency range

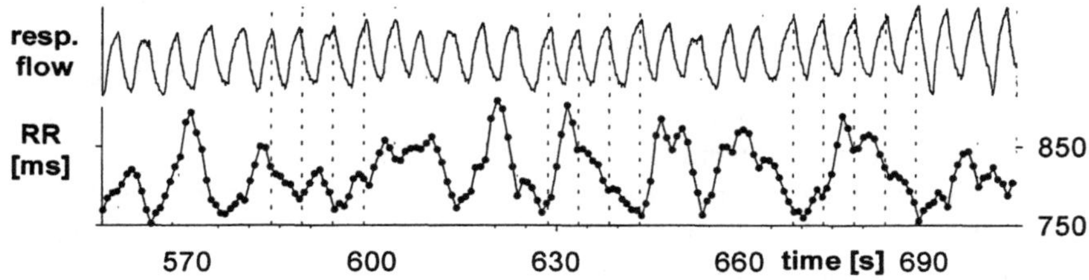

Fig. 6. Respiratory flow (uncalibrated thermistor signal) and heartbeat intervals (RR) corresponding to pacing cycle length 5 s in experiment B. In the time interval shown sudden transitions to different frequency-locking regimes 2:1, 3:1, and 4:1 between the two signals occur.

corresponding to respiratory period lengths of about 7 s to 9 s. To both sides the frequency synchronization has the tendency to become unstable. The steep increase of the standard deviation at the higher frequency end for subject B is due to transitions to different frequency-locking regimes 2:1, 3:1 and 4:1 (Fig. 6). In this case the RSA pattern repeats every second or third respiratory cycle. In all other cases the frequency-locking is stable. The general observation of an increase of the standard deviation at both ends of the frequency band together with the visual inspection of the time series of the instantaneous phase shifts leads to the conclusion that the observed cardiorespiratory synchronization is a dynamic one. In contrast to this, a static synchronization would be defined as a stable phase-locking.

5 Discussion

The analysis of phase shifts and symbolic dynamics prove to be powerful complementary tools for the investigation of cardiorespiratory synchronization.

The analysis of the phase shift between respiration and heart rate variation indicates that the whole range of the control parameter (pacing cycle length) corresponds to the Arnold tongue of 1:1 frequency-locking between respiratory flow and respiratory heartbeat variation (respiratory sinus arrhythmia, RSA), except for the higher frequency range in experiment B. Phase-locking between heartbeat intervals and respiratory flow is observed only sporadically for short time intervals up to 1 minute. The results obtained by the analysis of symbolic dynamics indicate that within the frequency area of 1:1 frequency-locking between respiratory flow and RSA there is a sudden switching between two qualitatively different synchronization regimes at pacing cycle length 8 s. The synchronization regimes stand out in Fig. 3 as distinct areas of high probability of similar dynamics. The standard deviation of the phase shifts in Fig. 5 shows a tendency of increasing at both ends of the range of pacing frequencies. This leads to the conclusion that the synchronization is not static but dynamic. The question whether the low frequency phase dynamics indicating dynamic synchronization is governed by stochastic jumps between different phase-locking ratios or by a nonlinear determinism has to be decided by further studies.

The dependence of the phase shift on respiratory cycle length in experiment C-1 and C-2 is in agreement with the findings in the literature (Angelone and Coulter 1964, Davies and Neilson 1967, Luczak and Raschke 1975, Kelman and Wann 1970, Eckberg 1983). Following Davies and Neilson this simply reflects a constant time delay between the onset of inspiration and the peak of the resulting heart rate variation. In experiment A and B, however, the variation of the phase shift over the control parameter range is much smaller, which means that the corresponding time delay between the two signals varies between 3 s and about 6 s, and, therefore, needs a more complex explanation, for example a frequency dependent delay. Following Angelone and Coulter (1964) and Hildebrandt (1966) the strong dependence of the phase shift on the respiratory cycle length as observed in experiment C-1 and C-2 is due to a resonance of the RSA with the Mayer wave, a non-respiratory rhythmic blood pressure variation with a mean period of 10 s. However, Angelone found a sharp decrease of the slope of the phase shift above 10 s, characteristic for a resonance. Such a kink in the slope is not seen in any of our experiments. Nevertheless, at pacing cycle length of 10 s the synchronization regime beyond 8 s shows a minor disturbance, expressed by a rather high χ^2 probability when compared to the synchronization regime at low pacing cycle length (Fig. 3). The interpretation of this disturbance as an interaction with the Mayer wave requires the assumption of the existance of an intrinsic 10 s oscillator in the central nervous system because of the sharpness of this phenomenon.

Time series corresponding to the values 5 s and 5.5 s of the control parameters in experiment B show a more complex frequency-locking than the 1:1 ratio observed in all other time series. Respiration and heart beat variation are predominantly 2:1 frequency-locked but also frequency-locking ratios of 3:1 and 4:1 are observed (Fig. 6). These findings are quite similar to those described by Koepchen and Thurau (1958) who analysed blood pressure waves in anaesthetized dogs and rabbits. After suppressing respiratory movements by medication of succinylcholin and interrupting cardiac parasympathetic control by cooling parasympathetic nerves they observed spontaneous switching between frequency-locking ratios 2:1, 3:1 and 4:1 of central respiratory activity and blood pressure waves. Additionally they observed "phase-jumps" of 180° of the blood pressure waves. (Similar "phase-jumps" of 180° occur within longer time intervals of 2:1 frequency-locking in the two time series mentioned above.) To explain their findings Koepchen and Thurau assumed a central oscillator in the vasomotoric centre which normally shows a 1:1 frequency-locking with respiration but can be desynchronized under certain conditions. The complex frequency-locking pattern at high pacing frequencies in experiment B suggests that desynchronization of central oscillators in the frequency range of respiration can be provoked by a sudden change from low to high paced breathing frequencies (Table 1).

The detection of two qualitatively different synchronization regimes which are both not characterised by stable phase-locking within the frequency area of the Arnold tongue corresponding to a 1:1 frequency-locking between respiration and RSA is in agreement with the hypothesis of the existence of two different central "respiratory" oscillators. Following this hypothesis, the well-known respiratory rhythm generator driving mechanical respiration and an additional oscillator in the frequency range of respiration show a 1:1 frequency-locking over the whole control parameter range. However, at pacing cycle length 8 s the synchronization between them shows a sudden change. This sudden change is not reflected by the means and standard deviations of instantaneous phase shifts (Fig. 5).

The interpretations of our results have to be seen as preliminary and should be confirmed by a larger study involving more volunteers and covering a wider range of paced breathing frequencies and additional methods as mentioned above.

Acknowledgements

The authors wish to thank Dr. M. G. Rosenblum for his expertise in the application of Hilbert transformation to experimental data and for valuable discussions.

References

Angelone A.A., Coulter N.A. (1964): Respiratory sinus arrhythmia: a frequency dependent phenomenon. J. Appl. Physiol. **19** (3), 479–482

Beck C., Schlögl, F. (1993): *Thermodynamics of chaotic systems.* Cambridge nonlinear science series **4** (Cambridge University Press)

DeBoer R.W., Karemaker J.M., Strackee J. (1987): Hemodynamic fluctuations and baroreflex sensitivity in humans: a beat-to-beat model. Am. J. Physiol. **253**, 680–689

Collins J.J., Steward I. (1993): Coupled nonlinear oscillators and the symmetries of animal gaits. J. Nonlin. Sci. **3**, 349–392

Davies C.T.M., Neilson J.M.M. (1967): Sinus arrhythmia in man at rest. J. Appl. Physiol. **22** (5), 947–955

Drepper F.R., Abel H.H., Geissler J., Hörnchen H., Schiek M. (1995): Phase behaviour at the onset of respiration after central apnea in infants. Pflügers Arch. - Eur. J. Physiol. **429** (6) [suppl], 128 [Abstract]

Eckberg D.L., Kifle Y.T., Roberts V.L. (1980): Phase relationship between normal human respiration and baroreflex responsiveness. J. Physiol. London **304**, 489–502

Eckberg D.L. (1983): Human sinus arrhythmia as an index of vagal cardiac outflow. J. Appl. Physiol. **54**, 961–966

Eckholdt K., Schubert E. (1975): Zum Einfluss der Atemtiefe auf die Sinusarrhythmie des Herzens. Acta biol. med. germ. **34**, 767–771

Glass L., Mackey M.C. (1988): *From Clocks to Chaos.*, Princeton University Press, Princetown, NJ

Glass L., Hunter P., MacCulloch A. (1991): *Theory of Heart.* Springer Berlin, Heidelberg

Golenhofen K., Hildebrandt G. (1958): Die Beziehung des Blutdruckrhythmus zu Atmung und peripherer Durchblutung. Pflügers Archiv **267**, 27–45

Guevara M.R., Glass L., Schrier A. (1981) Phase-locking, period-doubling bifurcations, and irregular dynamics in periodically stimulated cardiac cells. Science **214**, 1350–1353

Hao B.-L. (1991): Symbolic dynamics and characterization of complexity. Physica D **51**, 161

Hayano J., Mukai S., Sakakibara M., Okada A., Takata K., and Fujinami T., (1994a): Effects of respiratory interval on vagal modulation of heart rate. Am. J. Physiol. **267**, H33–H40

Hayano Taylor J.A., Mukai S., Okada A., Watanabe Y., Takata K., and Fujinami T., (1994b): Assessment of frequency shifts in R-R interval variability and respiration with complex demodulation. J. Appl. Physiol. **77** (6), 2879–2888

Hildebrandt G. (1966): Rhythmus und Regulation unter besonderer Berücksichtigung der Blutdruckregulation. Z. Gesamte Inn. Med. Ihre Grenzgeb. **22**, 206–213

Hirsch J.A., Bishop B. (1981): Respiratory sinus arrhythmia in humans: how breathing pattern modulates heart rate. Am. J. Physiol. **241**, H620–H629

v. Holst E. (1937): Vom Wesen der Ordnung im Zentralnervensystem. In *Zur Verhaltensphysiologie bei Tieren und Menschen, Band 1.*, Piper, 1969

Kelman G.R., Wann, K.T. (1970): Studies on sinus arrhythmia (Abstract). J. Physiol. London **213**, 59P–60P

Koepchen H.P., Thurau K. (1958): Untersuchungen über Zusammenhänge zwischen Blutdruckwellen und Ateminnervation. Pflügers Archiv **267**, 10–26

Koepchen H.P., Wagner P.-H., Lux H.D. (1961): Über die Zusammenhänge zwischen zentraler Erregbarkeit, reflektorischem Tonus und Atemrhythmus bei der nervösen Steuerung der Herzfrequenz. Pflügers Archiv **273**, 443–465

Koepchen H.P. (1984): History of studies and concepts of blood pressure waves. In *Mechanisms of Blood Pressure Waves.* Miyakawa K., Koepchen H.P., and Polosa C. Eds., Springer, Berlin, Heidelberg

Koepchen H.P. (1988): Regulation of respiration. Pflügers Archiv **408**, Suppl. 1, 23–25

Kronenberg F. (1991): Menopausal hot flashes: Randomness or rhythmicity. Chaos **3**, 271–278

Kurths J., Voss A., Witt A., Saparin P., Kleiner H.J., Wessel N. (1995): Quantitative analysis of heart rate variability. Chaos **5**, 88–94

Lawson E.E., Richter D.W., Czyzyk-Krzeska M.F., Bischoff A., Rudesill R.C. (1991): Respiratory neuronal activity during apnea and other breathing patterns induced by laryngeal stimulation. J. Appl. Physiol. **70** (6), 2742–2749

Luczak H., Raschke F. (1975): Regelungstheoretisches Kreislaufmodell zur Interpretation arbeitsphysiologischer und rhythmologischer Einflüsse auf die Momentanherzfrequenz: Arrhythmie. Biol. Cybernetics **18**, 1–13

Ludwig C. (1847): Beiträge zur Kenntniss des Einflusses der Respirationsbewegung auf den Blutlauf im Aortensystem. Arch. Anat. Physiol. **13**, 242–302

Panter P. (1965): *Modulation, Noise, and Spectral Analysis* (McGraw–Hill, New York)

Plesch O., Bishop B., McCall W.D.Jr. (1988): Comparsion of automatic and voluntary chewing patterns and performance. Exp. Neurol. **99** (2), 326–341

v.d. Pol B., v.d. Mark J. (1928): The heartbeat considered as a relaxation oscillation and an electrical model of the heart. Phil. Magazine **6**, 763

Pöppel E., Schill K., v. Steinbüchel N. (1990): Sensory Integration within temporally neutral system states: a hypothesis. Naturwissenschaften **77**, 89–91

Richter D.W., Ballany K., Schwarzacher S. (1992): Mechanisms of respiratory rhythm generation. Curr. Opin. Neurobiol. **2**, 788–793

Schiek M. (1994): Quantifizierung und Modellierung der respiratorischen Sinusarrhythmie. Berichte des Forschungszentrums Jülich; **2899**

Schiek M., Drepper F.R., Abel H.-H., Schläfke M.E., Koepchen H.P. (1995a): Nonlinear model of the respiratory sinus arrhythmia. Pflügers Arch. - Eur. J. Physiol. **429** (6) [suppl], 104 [Abstract]

Schiek M., Drepper F.R., Abel H.-H. (1995b): Mathematical model of respiratory sinus arrhythmia. Wien Med. Wochenschr. **145** (17–19), 492–4

Schläfke M.E. (1988): Plötzlicher Kindstod: Klinische Physiologie und Modelle. In *Der plötzliche Kindstod* Andler W., Schläfke M.E., Trowitzsch E. (Eds), Acron-Verlag, Berlin

Schmidt R.F., Thews G. (Eds.) (1995): *Physiologie des Menschen.* Springer, Berlin, Heidelberg

Schwarz U., Kurths J., Witt A., Benz A.O. (1994): Analysis of solar spike events by means of symbolic dynamics methods. Astron. & Astrophys. **277**, 215–224

Seidel H., Herzel H. (1995): Modelling heart rate variability due to respiration and baroreflex. In *Modelling the Dynamics of Biological Systems.* Springer, Berlin, Heidelberg

Strogatz S.H., Mirollo R.E., Matthews P.C. (1992): Coupled nonlinear oscillators below the synchronization threshold: relaxation by generalized Landau damping. Phys. Rev. Lett. **68**, 2730–2733

Suder K. (1996): Nichtlineare Analyse eines Modells der respiratorischen Sinusarrhythmie. Berichte des Forschungszentrums Jülich; **3213**

Wackerbauer R., Witt A., Atmanspacher H., Kurths J., Scheingraber H. (1994): A comparative classification of complexity measures. Chaos, Solitons & Fractals **4**, 133-173

Wever R.A. (1979): *The Circadian System of Man.* Springer, Berlin, Heidelberg

Winfree A. (1980): *The Geometry of Biological Time.* Springer, Berlin, Heidelberg

III EEG Analysis

EEG Signal Analysis by Continuous Wavelet Transform Techniques

Marco Ende[1], Alfred K. Louis[2], Peter Maass[1], and Gottfried Mayer-Kress[3]

[1] Institut für Mathematik, Universität Potsdam, D-14415 Potsdam, Germany
[2] Institut für Numerische Mathematik, Universität des Saarlandes, D-66041 Saarbrücken, Germany
[3] Ross Institute, East Hampton, NY 11937, U.S.A.

Abstract. Wavelet techniques are used to analyse EEG signals. In a first step the deviation of an expected power law determines the scale (frequency) at which some unexpected events happen. The positions of the local extrema of the wavelet transform are computed in a second step, this locates the instances at which these events take place.

1 Introduction

The basic goal of signal processing is to extract some desired information from a given set of measured data. Amongst the most powerful tools are time-frequency- or time-scale-representations of the signal. They can be obtained e.g. by Wigner-Ville-, Gabor- or wavelet transforms. Both types of representation aim at transforming and displaying the given data in such a way, that (in the case of a one-dimensional signal) a dominant value at (ω, b), resp. at (a, b), reflects the presence of a significant detail at time $t = b$ with local frequency ω, resp. with size a.

The present investigation intends to highlight the power of wavelet transforms for the detection of significant structures or unexpected events in EEG signals. The data under consideration were taken from the experiment described by Birbaumer et al. (1996), it consists of EEG measurements from various people who were exposed to different acoustic sequences. We will demonstrate the use of wavelet methods by analyzing the EEG measurement of five people which resulted from the sequence "periodic, melody".

2 Wavelet Transform

The continuous wavelet transform of a signal $f \in L^2(\mathbb{R})$ with respect to the wavelet ψ is defined by $(a, b \in \mathbb{R},\ a > 0)$:

$$L_\psi f(a,b) \ = \ a^{-1/2} \int_{\mathbb{R}} f(t)\, \psi\left(\frac{t-b}{a}\right)\, dt \ . \qquad (1)$$

Hence the signal f is "scanned" by a pattern ψ whose size varies with a, i.e. $L_\psi f$ is a time-scale-representation of f. By t we denote dimensionless time, where the basic time step $\tau = 1/256$s is chosen according to the sampling rate of the EEG signals.

The wavelet transform is invertible: f can be recovered from $L_\psi f$ with the help of a dual wavelet $\tilde{\psi}$ by

$$f(t) = \int_{\mathbb{R}} \int_{\mathbb{R}+} L_\psi f(a,b) a^{-1/2} \tilde{\psi}\left(\frac{t-b}{a}\right) \frac{da\, db}{a^2} \tag{2}$$

if ψ and $\tilde{\psi}$ satisfy the admissibility condition

$$1 = \int_{\mathbb{R}} \frac{|\widehat{\psi}(\eta)\widehat{\tilde{\psi}}(\eta)|}{|\eta|} \, d\eta \quad .$$

Here $\widehat{\psi}$ denotes the Fourier-transform of ψ.
This leaves a lot of freedom in choosing a wavelet ψ for a particular application. We tested the mexican hat wavelet, a sine-like wavelet and a Morlet-wavelet

$$\psi(t) = \pi^{-1/4} \left(\mathrm{e}^{-\mathrm{i}5t} - \mathrm{e}^{-25/2}\right) \mathrm{e}^{-t^2/2} \quad , \tag{3}$$

for analysing EEG signals. The best results were obtained with the Morlet-wavelet, which is particularly suited for detecting oscillating, but local, structures in a noisy background.

The computation of the wavelet transform requires the evaluation of the integral in the definition (1) for various parameters (a, b). This can be avoided by implementing the fast wavelet transform which computes $L_\psi f$ on the dyadic grid

$$\{(a,b) = (2^m, 2^m k) \mid m, k \in \mathbb{Z}\}$$

by recursive convolutions. But this requires some severe restrictions on the choice of ψ, moreover the knowledge of $L_\psi f$ on the dyadic grid is too coarse for a more subtle investigation. In particular the discretization of the parameter a by powers of two will not reliably capture significant structures on intermediate scales. Therefore we used the full continuous wavelet transform in the following experiments.

The investigation of EEG signals rests on two properties of the wavelet transform:

1. The asymptotic behaviour of $L_\psi f$ as $a \to 0$ reveals the local regularity of f. This property can either be used pointwise (b_0 fixed) or after averaging over some interval $b \in I = [b_{\min}, b_{\max}]$. I.e. we consider

$$n(a) = \int_I |L_\psi f(a,b)|^2 \, db \tag{4}$$

which, depending on the smoothness of f and ψ, obeys a power law for estimating the asymptotic decay rate (power law)

$$n(a) \sim a^s \quad .$$

Any deviation of the expected decay rate at a_0 reveals the presence of some significant structures, or unexpected events, of size a_0.

2. The wavelet transform at (a_0, b_0) is equivalent to the scalar product of f with the scaled and shifted wavelet $a^{-1/2}\psi\left(\frac{\cdot - b}{a}\right)$:

$$L_\psi f(a,b) = \left\langle f \, , \, a^{-1/2}\psi\left(\frac{\cdot - b}{a}\right) \right\rangle_{L^2(\mathbb{R})} \quad . \tag{5}$$

Hence computing the wavelet transform for $a \in (a_{\min}, a_{\max}]$ amounts to a matched pattern analysis performed simultaneously on the whole range of scales $(a_{\min}, a_{\max}]$. A local extremum of $L_\psi f$ at (a_0, b_0) therefore indicates a significant structure of size a_0 at time b_0. In the case of the Morlet-wavelet this can moreover be interpretate as the existence of a localized oscillation with frequency $\omega = 5/a_0$ at time $t = b_0$, hence in the following example, where EEG signals were sampled at a rate of 256Hz, a local maximum at a_0 corresponds to a physical frequency of $\Omega = \omega/\tau$. For more details on properties of wavelet transforms the reader is referred to either the textbooks Daubechies (1992), Louis et al. (1994) or to the series Chui (1992), where recent results on the theory and applications of wavelet techniques are published.

3 Analysis of EEG Traces

We selected five sets of data ("periodic, melody", Fz-electrode, first playing, test persons c01104–c01108) out of the measurements acquired in the experiment as described in Birbaumer et al. (1996). They were used for a first test of how continuous wavelet transform techniques perform in analysing EEG signals. The measurements from the electrodes Fz, Pz, Cz and the acoustic sequence "periodic, melody" were chosen, since they exhibit significant differences between the test persons. Typical sets of data are depicted in Fig. 1. For all five test persons the indicator function $n(a)$, for the definition see (4), was computed over the range $a \in (5, 20]$, see Fig. 2.

This already allows a clear discrimination between subjects who react to the music and those who do not react. The strongest reaction can be seen in the indicator function for person c01104, see Fig. 2. Moreover for some measurements one can observe the "octave effect": a reaction around $a \sim 15.5$ is mirrored by another, less pronounced maximum at $a \sim 7.8$, which resembles a doubling in frequency.

Primarily, our main aim is not primarily to extract physiolocigal information from the given data, we rather concentrate on demonstrating the use of

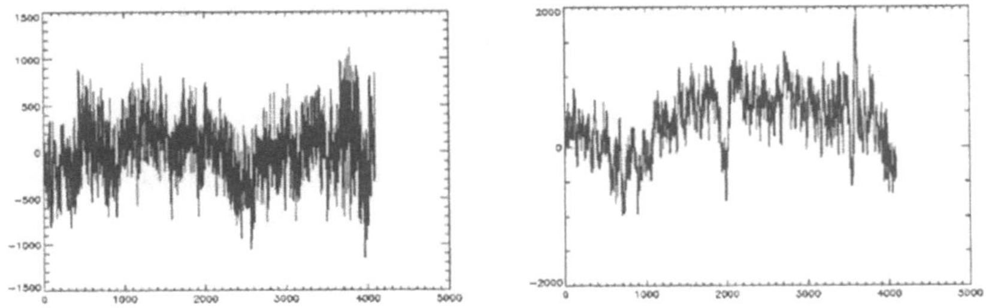

Fig. 1. EEG measurements, Fz electrodes, subjects c01104 and c01107.

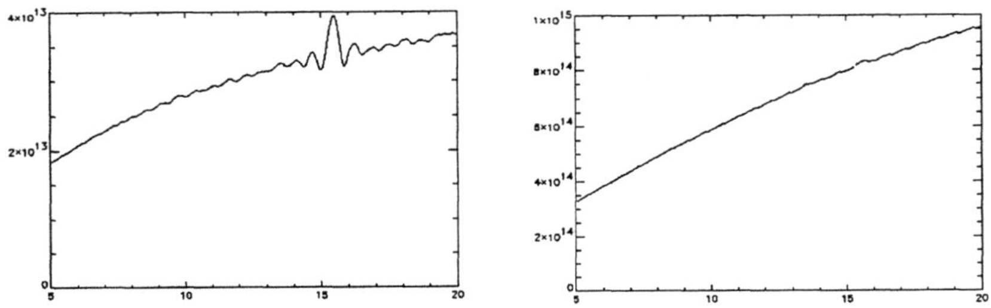

Fig. 2. The indicator function $n(a)$ for subjects c01104 and c01107. Subject c01104 reacts strongly near $a \sim 15.5$, subject c01107 shows no reaction.

wavelet methods for this particular application. Hence we did not perform a statistical verification of our findings nor did we aim at investigating the full set of data collected in the above mentioned experiment.

Therefore we proceeded by further investigating the EEG data of that person, who seemd to react strongest to the "periodic, melody" data. For this person we also examined the measurement corresponding to the central (Cz) and parietal (Pz) placement of the electrodes. For all of them one finds a critical scale around

$$a \sim 15.5 \quad .$$

Hence we display $L_\psi f$ with a fine discretization for $a \in (14, 18]$, see Fig. 3. According to (5) we search for local extrema which account for a localization in time of the significant reactions, Fig. 3. Hence simple thresholding was performed on three electrodes (Fz, Cz, Pz) of person c01104. The same compuations were done for all three electrodes (Fz, Cz, Pz). This allows to read off the critical frequencies and the instances at which the person reacted strongest:

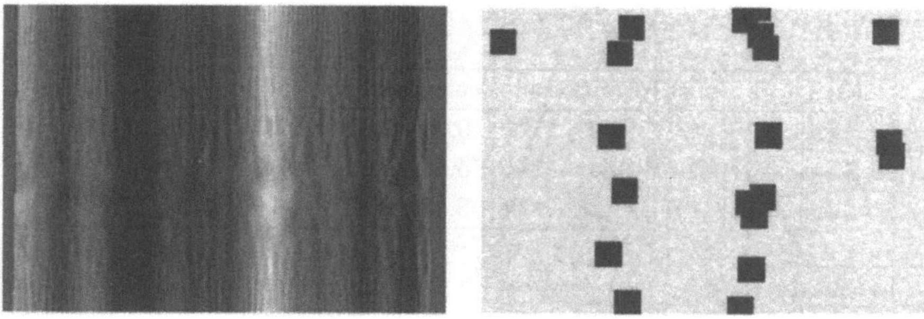

Fig. 3. The wavelet-transform of the Fz-measurement of subject c01104 zoomed to the intervall $a \in (14, 18]$ (left), the most significant events after thresholding are marked in the (b, a)-plane (right). We erased a certain region around each detected event since a local extremum spreads over some area.

Table 1. List of wavelet maxima for different electrodes

electrode Fz			electrode Cz			electrode Pz		
b	a	$\lvert\Re(L_\psi f(a,b))\rvert$	b	a	$\lvert\Re(L_\psi f(a,b))\rvert$	b	a	$\lvert\Re(L_\psi f(a,b))\rvert$
198.95	17.54	1.34e+05	187.25	16.97	2.34e+05	163.84	16.70	1.74e+05
1176.14	14.78	1.32e+05	210.65	17.55	2.47e+05	187.25	16.93	2.00e+05
1199.54	16.30	1.31e+05	234.06	18.00	2.60e+05	210.65	17.51	2.02e+05
1275.61	17.40	1.29e+05	263.31	17.74	2.32e+05	245.76	16.53	2.28e+05
1322.42	15.61	1.31e+05	795.80	17.38	2.37e+05	269.17	16.10	1.76e+05
1351.68	14.15	1.29e+05	819.20	15.32	2.13e+05	292.57	15.94	1.51e+05
1380.94	17.74	1.32e+05	854.31	17.58	2.39e+05	315.98	15.24	1.72e+05
2387.38	14.07	1.50e+05	877.72	16.14	2.14e+05	339.38	14.26	1.50e+05
2422.49	17.83	1.68e+05	901.12	17.21	2.08e+05	362.79	17.38	1.59e+05
2463.45	15.46	1.74e+05	1427.75	14.74	2.12e+05	719.73	16.29	1.51e+05
2486.85	14.59	1.85e+05	1451.16	17.36	2.16e+05	795.80	17.34	1.99e+05
2510.26	15.29	1.67e+05	1474.56	17.86	2.29e+05	1427.75	17.82	1.67e+05
2539.51	18.00	1.69e+05	1497.97	17.03	2.14e+05	1509.67	17.22	1.58e+05
2562.92	17.64	1.58e+05	1527.23	16.36	2.23e+05	1562.33	17.04	1.66e+05
2586.32	15.53	1.49e+05	1562.33	17.68	2.22e+05	1609.15	18.00	1.74e+05
2609.73	17.47	1.31e+05	1585.74	15.25	2.08e+05	3124.65	16.88	1.55e+05
2638.99	16.33	1.32e+05	1609.15	18.00	2.38e+05	3200.71	16.64	1.74e+05
3709.78	17.69	1.28e+05	3785.84	16.38	2.23e+05	3815.10	15.28	1.68e+05
3744.89	16.24	1.37e+05	3815.10	14.59	2.30e+05	3861.91	16.02	1.61e+05
3768.29	16.07	1.48e+05	3856.06	17.08	2.25e+05	3908.72	17.43	1.78e+05

Taking into that the basic time step is $\tau = 1/256$s we combine those maxima of the wavelet transform which fall in the same intervall of length 0.25 seconds. This yields the following list of extraordinary events in the time series Fz, Cz, Pz, see Table 2.

Table 2. Time instances (in seconds) of events for three different electrodes

Critical events for different electrodes									
Fz	0.78	4.64	5.07	5.34	9.39	9.67	10.01	10.25	14.67
Cz	0.87	3.15	3.43	5.62	5.86	6.19	14.92		
Pz	0.73	1.05	1.33	2.96	5.58	6.00	6.29	12.35	15.18

One should note that the significant events can be well localized in time, however the corresponding frequencies/scales vary mostly between $a = 14.5$ and $a = 17.5$. This variation is strong enough so that these points would not have been detected with a single scale analysis of f. Hence one really needs the full multi-scale resolution of the continuous wavelet transform in order to reliably detect all significant structures.

The following Fig. 4 shows a small portion of the Pz measurement, which contains one of the detected events. Next to it we display the wavelet transform of this portion for a fixed parameter $a_0 = 16.5$. One observes a local maximum of $L_\psi f(16.5, b)$ near $b = 245$. This corresponds to a local oscillation (period $\sim$ 20) at time t = 245, which stemms from the almost equidistant maxima in the data at time $t = 245$.

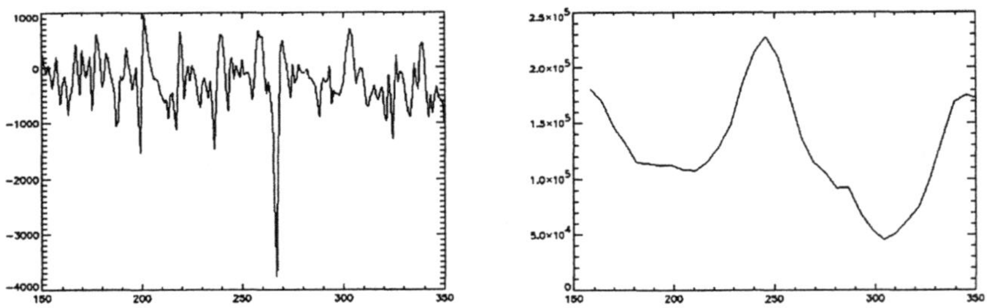

Fig. 4. The figure on the left displays a portion of the Pz-measurement, which contains an unexpected event. The figure on the right shows the wavelet transform of this section, the local maximum locates the unexpected event.

4 Algorithm

The software code used in this investigation can be obtained by sending an email to mende@rz.uni-potsdam.de.

The basic outline of the algorithm is as follows.

1. The discrete data set is interpolated using either linear or cubic spline interpolation.
2. The evaluation of the integrals (1), which define the continuous wavelet transform, is done by Simpson's rule.
3. For the Morlet-wavelet one may either display the real-part, the imaginary part or the modulus of $L_\psi f$. We used the real part in our computations.
4. In Figs. 4–6 we used an equidistant discretization with steplength $h_a = 1/125$ and $h_b = 4096/700$, which corresponds to $h_b/256\text{s} = 0.023\text{s}$. The thresholding parameter for Figs. 4–6 was set to a value between 65% and 80% of the global maximum.

References

Birbaumer N., Lutzenberger W., Rau H., Mayer-Kress G., Choi I., Braun C. (1996): *Perception of Music and Dimensional Complexity of Brain Activity*, Int. J. Bifurcation and Chaos 6 (2), 267–278

Chui C. (Ed.) (1992–96): Series on *Wavelet Analysis and its Applications, Vols. I–V*, Academic Press, New York

Daubechies I. (1992): *Ten Lectures on Wavelets*, CBMS, SIAM Publ., Philadelphia

Louis A.K., Maaß P., Rieder A. (1994): *Wavelets: Eine Einführung in Theorie und Anwendungen*, Teubner Verlag, Studienbücher Mathematik. English translation by Wiley, New York

Relativity of Dynamical Measures in Study of the Human Brain

Local Lyapunov Exponent as a Tool for Comparative Studies

Zbigniew J. Kowalik[1,4], Thomas Elbert[2], Larry E. Roberts[3] and Manfried Hoke[4]

[1] Department of Neurology, University of Düsseldorf, Germany
[2] Department of Psychology, University of Konstanz, Germany
[3] Department of Psychology, McMaster University, Canada
[4] Institute of Experimental Audiology, University of Münster, Germany

Abstract. The deterministic character of cortical processes in the human brain invites an assessment of nonlinear methods of analysis and exploration of their applicability. Because brain processes are frequently nonstationary, questions arise as to the validity of formal assumptions. Here we present a simple algorithm for estimation of the local divergence measure (Mayer-Kress 1994, Kowalik and Elbert 1994) based on the local Lyapunov exponent (Wolff, 1992) that can be used for nonstationary data.[1]

1 Introduction

Nonlinear methods are applicable to psychology, medicine, economy and politics where measured phenomena (like those of physics) frequently generate complicated patterns of data in time and space. These patterns appear to be deterministic because they can be described by certain control rules, and also nonlinear because unexpected outputs are sometimes observed. Biomedical and physiological mechanisms that generate chaotic dynamics are also candidates for nonlinear analysis. We have approximately three types of dynamical measures which can practically be applied:

- The dimension of the topological structure of reconstructed attractors (which measures the complexity of trajectories drawn in phase space);
- Entropy (which describes the homogenity of this space);
- Lyapunov exponents (which characterize the tendency of a system to change its dynamical state or more simply the chaoticity or unpredictability of the system).

There are two main problems of analysis of biomedical data, the influence of noise and nonstationarity. While the first problem is general for all real

[1] See also comments in (Nicolis et al. 1983).

systems, the second problem is of special concern when dealing with living organisms. Biological and psychological development involves transitions on a wide range of time scales that appear to be of a nonstationary character.

Frequently physiological experiments are performed in such a way that during the period of measurement a quasistationary state is reached. However, even when this happens (and it may not always happen), the dynamical state of the original measurement may not be replicated during repetitions of an experiment. The problem that now appears is, should we use nonlinear systems theory if some of the condition(s) required for mathematical description are violated, and if so, how do we do so?

The simplest solution seems to be to repeatedly measure a system at a fixed set of parameters. However, because the dynamics of the system may not be reproduced, the meaning of replication is uncertain. Perhaps in some specific cases when a series of experiments is very well controlled one can expect that dynamics will be repeated, but in general this may not be true. Another approach is to trace the evolution of system dynamics by using a running window (time-scanning) technique. While this method may not solve the problem of comparative studies, a marriage of scanning and replication methods may give us valuable information if we interpret their results carefully. Our presentation describes some simple algorithms for the description of nonstationary dynamical systems and applies these algorithms to simulated data. Applications to real data are described elsewhere (Skinner et al. this volume; Herzel et al. this volume).

2 Nonstationary Nature of Brain Signals

The physical nature of data used for dynamical studies should be considered. For example, EEG (electroencephalography) and MEG (magnetoencephalography) data are generated by flows of electrical charges consequent on neural activity in the cortical structures of the brain. Although MEG and EEG thus measure different physical aspects of the same activity, frequently they give different information about a system's dynamics. The brain has at least two subsystems governing its activity. First, several basic systems are automatic and repetitive and are responsible for maintaining life functions. Examples are autonomic regulation, diurnal rhythms, and homeostatic processes that support energy expenditure. The second subsystem of the brain supports an individual's decision processes. The physical basis of this subsystem is only partly known. Nevertheless, we can try to separate the deterministic and noisy contents of data that we gather to measure the output of this second subsystem. Moreover, we can try to extract from the measured output signals the different dynamical variables which describe the behavior of system.

An interpretation of the genesis of the deterministic part of system performance is difficult if one uses global measures of brain activity in form of EEG/MEG. Some useful information can be gained through comparison of experimental and control conditions. However, determinism cannot be identified by surrogate data tests (Theiler et al. 1992) such as those in which the phase of signals (imaginary component) arising from different dynamics is randomly shuffled. Because such tests discard information about the point in time at which system dynamics changed, surrogate data tests cannot detect nonlinearity in nonstationary patterns. One can speak about the deterministic parts only in the relative manner by comparing dynamical measures between different generators (measured persons) or between different time-intervals. A measure that we use to characterize nonstationary systems is presented in the next section.

3 The Local Lyapunov Exponent

There are many published methods in the scientific literature which can be applied for the estimation of the largest positive Lyapunov exponent. Calculation of these exponents in real systems makes sense only for short segments of a time series, owing to the senstitive dependence of chaotic systems on initial conditions. All of these methods assume that the experimentally investigated system is stationary. Moreover, high-dimensional systems require a large number of points. Two very well known algorithms are the method of Wolf et al. (1985) in which the separation between nearby pairs of points is estimated as a function of time (iteration), and the method of Eckmann and Ruelle (1985), in which the Jacobian at each chosen point is estimated through a least-squares technique. Both methods work well for low-dimensional systems although the first one seems to be more robust while the second exhibits strong dependence on the embedding parameters. A running window technique combined with the algorithms of Wolf and collegues gives a useful picture of nonstationarity in the data (Iaseimidis and Sackellares 1991; Elbert et al. 1994; Elbert et al. 1997) but is very time consuming and unsuitable for on-line analyses. A modification of this method for nonstationary data (the "chaoticity measure" of Kowalik and Elbert (1994) improved the speed of estimation (30–100 times faster depending on parameters) but has problems of interpretation when nearly-periodic time series are studied.

The local properties of an attractor can be investigated by observing system dynamics over a short period of time. One measure which directly characterizes the short-term-predictability of a dynamical system is the local Lyapunov exponent. Wolff (1992) estimated this measure by averaging the following quantities:

$$\lambda_{i,m} = \frac{1}{m}\frac{1}{n_i}\sum_{j\in S_j} \log\left|\frac{X_{i+m} - X_{j+m}}{X_i - X_j}\right| \tag{1}$$

where[2]

[2] Originally, Wolff's expression did not contain a term representing experimental noise. One is added here to reflect the noise in biological systems.

$i = 1, ..., n - m$; $S_i = \{j : \text{Noise} < |X_i - X_j| \leq V\}$, $n_i = \#(S_i); m \in N$ and $V >$ Noise.

One of the most important advantages of this formulation is its simplicity and speed of estimation. Its weakness is the large number of parameters, although this feature also means that the method is sensitive to changes in parameter values. Because of these properties the method may be useful for the on-line measurement and/or for comparative studies. Deterministic mechanisms of physiological processes allow us to assume that at least in the short term the processes that we measure exhibit a quasistationary character. If this assumption is accepted, the observation of dynamical measures in short time-slices characterizes their momentary state, and the observation of the temporal evolution of such measures provides information about the development of the physiological system. Combining the local divergence measure described in (1) above with a windowing technique gives us a useful tool for measurement of global system dynamics.

4 Numerical Algorithm

A practical estimation of the local *LE* by the above mentioned technique can be performed in the following steps:

1 Set lag m and small V ($<$ 10% of the attractor's "diameter") and the width of the window (128–2048 points).
2 Fix the index i to the first point in the set ($n_i = 1$).
3 Find a set (S_1) of indexes for which distance between the running point and the reference point will be smaller than V.
4 For all indexes from the set S_I calculate distances to all running points (X_i, X_j) and ($X_{i+m} - X_{j+m}$).
5 Avoid 0 (omit if $X_{i+m} - X_{j+m} = 0$).
6 Sum logarithms to get $\lambda_{i,m}$.
7 Shift a window STEP-points forwards. If not the end of datafile then goto #2.
8 If ready, estimate the window-average (Λ) and its standard deviation (σ). The width of the averaging window must be lower than the width of the λ-scanning window.

5 Simulations

The method was tested for both real data and numerical simulations. Here we show some results and dependencies of the local *LE* on different parameters in cases of the baker's transformation and Duffing's oscillator.

5.1 Baker's Map

The baker's transformation is given by the map:

$$x_{n+1} = \begin{Bmatrix} \beta x_n & : & y_n <= \alpha \\ 0.5 + \beta x_n & : & y_n > \alpha \end{Bmatrix} \tag{2}$$

$$y_{n+1} = \begin{Bmatrix} y_n/\alpha & : & y_n <= \alpha \\ \frac{1}{1-\alpha}(y_n - \alpha) & : & y_n > \alpha \end{Bmatrix} \tag{3}$$

To enable a comparison with other results we fixed the parameter $\beta = 0.25$. Figure 1 presents a dependence of local $LE(t)$ on three parameters. The generated time series was 4096 points long.

Figure 1A shows patterns of the $LE(t)$ function for different values of the parameter α of the baker's transformation. The dependence of the local LE on α is qualitatively in agreement with a theoretical tendency: The larger the value of α, the larger value of the LE, assuming that α is less than 0.5. There is a rapid increase of the LE for small α and the values saturate for α close to 0.5. The value of the local LE is dependent on the lag parameter m and on the locality (vicinity) parameter V (1), and on the assumed level of noise (grid parameter). The dependence of the local LE on the grid parameter is shown in Fig. 1B and on the vicinity parameter in Fig. 1 C. In both figures one observes a saturation effect. The smallest value of the Noise parameter, i.e. 2, produces the upper curve in Fig. 1B. There is almost no change in the average value of the local LE for Noise = 8. When the level of noise approximates the locality parameter (here $V = 40$) the corresponding values of the local LE go to zero. Thus, the natural limit of the grid parameter ($2 \leq$ Noise $< V$) fixes a range of available local LE-values. The range of available values of the local LE also depends on the vicinity parameter V when the grid parameter is fixed (here Noise = 8). This effect is shown in (Fig. 1C) where the local LE lies between 0.1 and 0.2 for $V = 30 - 65$. The upper pattern shown there corresponds to the most local estimation ($V = 30$). Increasing the vicinity parameter shifted the pattern of local LE's down to some saturation level.

5.2 Duffing's Oscillator

The double-well symmetrical Duffing oscillator is frequently used for modeling in the biological sciences due to its relatively simple form (see, for instance, Srinivasan and Nunez 1993). In the Duffing equation, there is a cubic nonlinear term in addition to the linear elastic one:

$$\ddot{x} + \gamma\, \dot{x} - ax + bx^3 = F_0 \sin \Omega t \tag{4}$$

The dynamics of the Duffing system can be manipulated by changing parameter values (γ, a and/or b), as well as by changing the amplitude of external

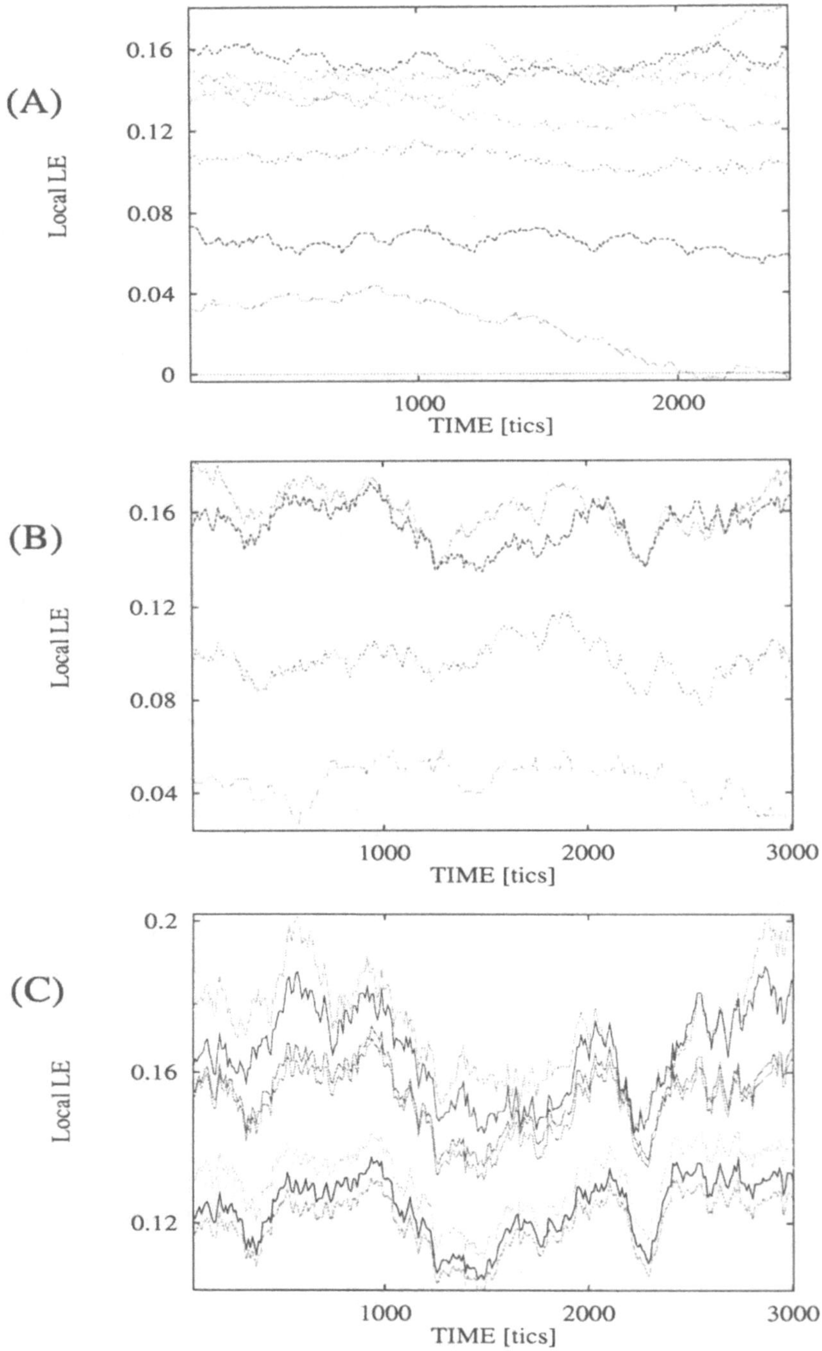

Fig. 1. Local *LE* vs time for different parameter $\alpha = 0.1, 0.15, ..., 0.499$ of the baker's transformation (A); different grid parameters Noise $= 2, 8, 12, 20$ (B), and different vicinity values $V = 30, 35, ..., 65$ (C).

driving force F_0. The Duffing equation is particularly interesting because it can generate what is termed intermittent behavior. The so-called *crisis induced intermittency* may represent a naturally occurring nonstationarity in physical systems which is similar to that of living organisms.

Let us observe a temporal evolution of such a system during a transition between two different dynamical states by mean of its local LE (Fig. 2). In

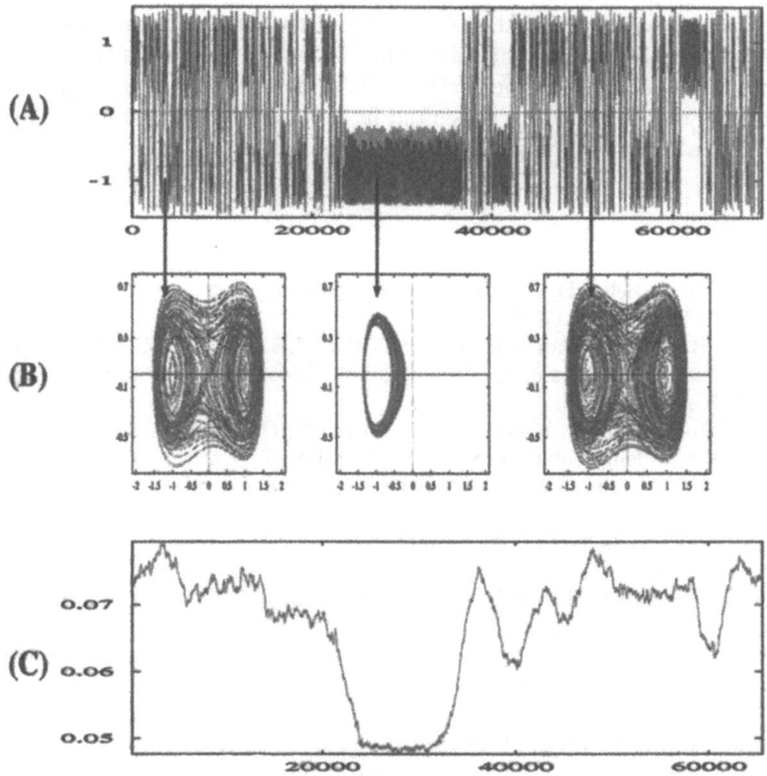

Fig. 2. Local LE *vs* time in the case of crisis induced intermittency in the Duffing oscillator (4). $\gamma = 0.045, a = 0.5, b = 0.5, F_0 = 0.1, (\Omega/2\pi) = 0.1417$. (A) The generated signal shows a distinct transition to a quasiperiodic state in the center of figure and returns to a chaotic mode of motion. (B) Three attractors corresponding to the three time domains indicated by arrows. (C) An evolution of the local LE obtained for the whole time series at following parameters: lag = 10, step = 10, width of window = 2048 points, vicinity $V = 40$, and grid parameter Noise = 8.

this case the state transition is highly visible without any dynamical measure. Nevertheless, it is noteworthy that there are subtle temporal fluctuations in the local LE that reflect small changes in the system's dynamics.

6 Discussion

Brain processes produce naturally nonstationary but deterministic electric (or magnetic) patterns. The investigation of such processes by the tools of traditional nonlinear dynamics raises interpretational problems because the asumption of stationarity is violated. Applying the local LE estimated in a running window can be used to characterize momentarily the dynamical state of investigated cortical structures. We presented here an algorithm for calculation of the local LE that can be used to detect changes between dynamical states. Moreover, this algorithm is faster than classical largest LE methods (LLE) and can be performed even during experiment, on-line. Some applications to real data are presented in other contributions in this volume (Skinner et al. this volume; Herzel et al. this volume).

References

Eckmann, J.-P., Ruelle, D. (1985): Ergodic theory of chaos and strange attractors. Rev. Mod. Phys. **57**, 617–655.

Elbert, T., Ray, W.J., Kowalik, Z.J., Skinner, J.E., Graf, K.E., Birbaumer, N. (1994): Chaos and physiology. Deterministic chaos in excitable cell assemblies. Physiol. Rev. **74**, 1–47.

Elbert, T., Rockstroh, B., Kowalik, Z. J., Hoke, M. (1997): Clinical applications of EEG analyses using tools derived from non-linear system theory. In Angelieri, F., Butler, S.R., Giaquinto, S., Majkowski, J. *Analysis of the Electrical Activity of the Brain.* Wiley, London, pp. 45–67.

Iasemidis, L.D., Sackellares, J.C. (1991): The evolution with time of the spatial distribution of the largest Lyapunov exponent on the human epileptic cortex. In Duke, Pritchard, W. (1991): *Measuring Chaos in the Human Brain.* World Scientific, Singapore, pp. 49–82.

Kowalik, Z.J., Elbert, T. (1994): Changes of chaoticness in spontaneous EEG/MEG, Integrative Physiological and Behavioral Science, **29**, 268–280.

Mayer-Kress, G. (1994): Localized measures for non-stationary time-series of physiological data, Integrative Physiological and Behavioral Science, **29**, 203.

Nicolis, J. Mayer-Kress, G., Haubs, G. (1983): Non-uniform chaotic dynamics with implications to information processing, Z. Naturforsch. **38a**, 1157–1169.

Srinivasan, R., Nunez, P.L. (1993): Neocortical dynamics, EEG standing waves and chaos, In Jansen, B.H., Brandt, M.E. *Nonlinear Dynamical Analysis for the EEG.* World Scientific, Singapore, pp. 310–355.

Theiler, J., Galdrikian, B., Longtin, A., Eubank, S., Farmer, J.D. (1992): Using surrogate data to detect nonlinearity in time series, In Casdagli E.M., Eubank, S. *Nonlinear Modeling and Forecasting.* Addison-Wesley, Reading, MA, pp. 163–188).

Wolf, A., Swift, J.B., Swinney, H.L., Vastano, J.A. (1985): Determining Lyapunov exponents from a time series. Physica, D **16**, 285–317.

Wolff R.C.L. (1992): Local Lyapunov exponents: Looking closely at chaos. J. R. Stat. Soc. B **54**, 353–371.

Classification of EEG Signals From a Music Perception Experiment Using Empirical Dynamical Models

Michael Kremliovsky and James Kadtke

Institute for Pure and Applied Physical Science
University of California at San Diego
La Jolla, CA 92093-0360, USA

Abstract. In this paper we describe a data classification method as applied to a set of EEG signals provided to the international workshop on Nonlinear Techniques in Physiological Time Series Analysis (Dresden, 1995). This method is derived from the theory of nonlinear dynamics, and consists of a classification processing chain which utilizes estimated sets of nonlinear ODEs as a data model. The investigation consisted of a blind analysis of the EEG recordings from a single human subject, taken while the subject was exposed to music segments of varying and controlled dynamical complexity. From this analysis, we find firstly that there appears to be dynamical similarity between many of the sensors, but that most sensors do not provide significant classification capability. However, at least one sensor is seen to demonstrate a strong classification performance, and this was used to generate a relative classification scheme for the data. This scheme shows significant distinction between certain musical samples, and from this we infer several general conclusions about the sensitivity of the particular subject to the musical data classes.

1 Introduction

This paper describes a preliminary dynamical systems analysis of EEG data provided to the Nonlinear Techniques in Physiological Time Series Analysis Workshop (Dresden, 1995), which is described in Birbaumer, et al. (1994). The experiment consists of recordings of EEG data from a human subject at nine different locations on the head, while the subject experiences perception of various artificially generated music segments. The music segments themselves were devised to have several particular characteristics, e.g. segments consisted of either melody, rhythm, or melody+rhythm, and were modulated according to a periodic, chaotic, or stochastic temporal evolution. EEG recordings were available for all of these combinatorial classes, simultaneously for each of nine sensors located at distinct areas of the subjects head. The principal aim of this experiment was to determine how the brain responds to different levels of musical complexity, whether the response was localized within the brain, and whether this could be quantitifed using phase trajectories.

In effect, this experimental configuration loosely defines a resonant system, in that we examine the output (EEG recordings) induced by a particular class of input or driving (musical segment) through a nonlinear system or filter (the subjects brain). One hypothesis for the type of perception presented in Birbaumer, et al. (1994) is that the grouped cell assemblies (localized regions) in the brains of the subjects individually respond to particular classes of input music, depending on previous musical experience, training, or perhaps intelligence. The complexity of the EEG response may then be accounted for by the level of entrainment between the individual cell assemblies and the musical input, and also between the cell assemblies themselves. This hypothesis and others was examined quantitatively by Birbaumer, et al. (1994), by looking for underlying dynamical complexity in the EEG signals as measured by the correlation integral of Grassberger-Procaccia. A hypothesis here is that the dimensionality of a heuristic underlying dynamical model is related to the number of individually active cell assemblies. The results of these analyses and other standard statistical correlation measures were described by Birbaumer, et al. (1994) across the spectrum of the various human subjects involved.

Here, we describe an analysis of a subset of this data using an alternate dynamical characterization method, which attempts to quantify signal structure by fitting empirical dynamical systems (sets of coupled ordinary differential equations, ODEs) to the data. To perform classification, we reformulate the general modeling procedure into that of an observational inverse problem, which distinguishes the EEG signals which appear to be dynamically related, or not so. Algorithmically, we define a classification chain by generating a feature space from the model coefficients, which yields statistical estimates of the dynamical relations between various signals (Kadtke (1995), Kadtke and Kremliovsky (1996a), Kadtke and Kremliovsky (1996b)). We also note one other important difference in our analysis, that being that we had available to us only the EEG recordings of a single individual, rather than the 18 subjects described by Birbaumer, et al. (1994). Hence, aside from poorer statistics, the possibility exists that our analysis may be peculiar to the traits of this individual, and our results constitute only a characterization of the perceptional classification capabilities of this particular brain.

We would also like to stress that our analysis was performed in a "blind" fashion, in the sense that no a priori information was used to generate the classification schemes. All results and conclusions were derived only from the data at hand. Since the analysis described here is far from exhaustive, we welcome any additional fundamental insight that may confirm or negate our conclusions. We also point out that since no a priori dynamical model can be postulated for the brain/sensory system, our classification schemes are relative; that is, they cannot be directly connected to any physical characteristics at this time.

In the remaining sections of this paper, we describe briefly the classification algorithm, the data characteristics, the data reduction method, and finally we draw generalizations and conclusions.

2 The Dynamical Classification Method

To classify the EEG signals described above, we use a technique derived from dynamical systems theory, described by Kadtke (1995), and Kadtke and Kremliovsky (1997). The reader is referred there for more detail. Here, we only give a brief qualitative summary of the analysis method, which suffices to understand the main conceptual elements.

For this technique, we assume the hypothesis that signal "structure" equates to a deterministic relationship which is generated by a low-dimensional dynamical system. In this sense, we equate determinism with local smoothness, similar to Salvino and Cawley (1994). To quantify this determinism, we use dynamical models (sets of nonlinear, coupled ODEs) which have been discussed extensively in the literature (Crutchfield and McNamara (1987), Cremers and Hübler (1987), Kadtke and Brush (1994)). The models are used to extract dynamical information from data, in what may be termed a "dynamical inverse problem". The advantages of this type of modeling include a very compact representation of nonlinear correlations to arbitrary order, good noise averaging characteristics, minimal data requirements, and ease of analytic manipulation. For data classification applications, we embed the global modeling method in a data processing chain which performs classification based on statistical properties from an ensemble of observations (Kadtke and Kremliovsky (1997)).

Under this hypothesis, we note that a lack of structure (i.e. the null hypothesis) can mean either that the system is truly random, or that it is a high-dimensional dynamical system, or that corrupting noise components are large and obscure the dynamics, or that the dynamical model chosen is an exceptionally bad representation of the data. That is, we cannot claim to capture any possible dynamical relationship that can exist in the data. We only aim to capture sufficient dynamical information to distinguish between distinct signal classes of interest for the problem at hand. Because the model can be quite generally defined, however, there is large flexibility in tailoring the model form to extract particular types of structure.

To begin, we first outline the modeling process. Here, we assume that any observed data structure of interest is originally generated by a physical system, which has a state space consisting of state variables $\mathbf{Z}(t) = \{z_1(t), z_2(t), z_3(t), \ldots\}$ at time t, and which evolve according to the dynamical rule $d\mathbf{Z}(t)/dt = \mathbf{F}[\mathbf{Z}(t)]$ for a stationary (autonomous) system. That is, the generating system can be described at least approximately as though it were a relatively low-dimensional dynamical system. We note that non-stationary data can often be described using explicitly time-dependent models (non-autonomous ODEs) of the dynamics. To use this formulation for the analysis of observed data, one invokes the well-known theorem of Takens to define a time-delayed reconstruction of the data: from a measured scalar signal $x(t)$ of some physical observable, we form the time-delay reconstructed state-space vectors $\mathbf{X}(t) = \{x(t), x(t-\tau), \ldots, x(t-(D-1)\tau)\}$ for some embedding dimension D, and some delay time τ. The importance of this construct is that it is known to preserve topological properties of the dynamics under relatively loose restrictions. Intuitively, this step constructs from the scalar data $x(t)$ an empirical dynamical state-space representation with potentially D active degrees-of-freedom, which is isomorphic to the original phase space for sufficient D. Hence, one can postulate a new dynamical model for the data, $d\mathbf{X}(t)/dt = \mathbf{F}[\mathbf{X}(t)]$, of dimension D, which can be used to compactly describe the time evolution of $x(t)$. Our aim here is to perform classification of observed signals using this form of the empirically determined reconstructed dynamical model, if indeed a dynamical model proves relevant to as a data description.

In using this framework for a classification scheme, we also point out a fundamental difference between a strictly modeling application (e.g., for prediction) and a classification application, where in the latter we desire only to distinguish a few known data classes. In a classification application, we do not seek an exact model for the data's temporal evolution; we seek only to find an empirical model which provides distinct, repeatable, and robust classification of the relevant dynamical behaviors, within the observation time scales in question. In this sense, a simple and "inappropriate" dynamical model can often provide adequate (or even superior) classification performance to a full, exact model of the dynamics. This is because we typically attempt to compare data classes with widely different dynamical properties, using only a single general model form. Although use of the inexact models poses several fundamental questions concerning uniqueness, in practice one can typically define a satisfactory hypothesis-testing structure to provide adequate data class distinction.

To construct the data classification algorithm in detail, we first assume a specific form for the empirical dynamical model which will quantify signal information. The form of the model (i.e. the model basis set) is typically dictated by expected signal characteristics, as well as expected noise levels, observation times, etc. For the analysis described in this paper, we use only a simple polynomial approximation for the dynamical model, with D independent variables which are the Takens time-delayed scalars. For example, the general form for each of the D possible dynamical equations can be written as

$$\begin{aligned} F[\mathbf{X}(t)] &= a_0 + a_1 x(t) + a_2 x(t-\tau) + \ldots + \\ &+ a_m x^2(t) + a_n x^2(t-\tau) + \ldots + a_q x(t) x(t-\tau) + \ldots \quad . \end{aligned} \tag{1}$$

This expansion includes all possible cross-terms up to the specified order P. The choice of D and P are based on classification performance requirements, which as mentioned are typically different from those criteria used for a purely modeling application. Intuitively, one can understand these basis terms as approximately capturing nonlinear correlations of some order (quadratic, cubic, etc.) inherent in the data.

An important point in choosing the model structure is that it must remain fixed throughout the entire analysis, in order that varying data observations may be compared in an unbiased way. Hence, care must be taken to define a model which is sensitive to the peculiarities of all signal classes of interest, to provide adequate class distinction. In this sense, the fixed model structure may be considered as a "dynamical filter", through which the universe of data observations are all viewed.

To use the specific form for the empirical model for classification purposes, we have developed a classification processing chain in analogy with detection theory. This scheme is graphically summarized in Fig. 2. Briefly, we can outline this analysis procedure as follows:

1. A fixed measurement window is designed to sample the scalar data, which consists of a sequence of data points of length L_w, a particular sampling rate, and may include other factors such as pre-filtering and data precision. These parameters are typically chosen empirically, based on time scales of the physical system, although fundamental information can be used as well. The data set is then divided into an ensemble of separate observations, using these window parameters; if the data is a continuous signal, we utilize a sliding window which is moved through a data set by L_s points to produce a set of quasi-independent observations indexed by α.
2. We use the dynamical model $\dot{\mathbf{X}}(t) = \mathbf{F}(\mathbf{X}(t))$ described above with a specific basis set (e.g. polynomials), the particular embedding dimension, order, and delay time to extract the dynamical information. This is done by first constructing the Takens time-delayed vectors for each data observation, and then estimating the model parameters for each observation, by least-squares fitting of the model to each window α. This produces a set of k estimated model coefficients $\mathbf{A}^\alpha = (a_1, a_2, \ldots, a_k)$. The fixed observation window, plus the fixed dynamical model, can together be thought of as the *dynamical filter*; i.e. all data observations will be viewed through this standard quantifier for all time.
3. From the k estimated coefficients for all $\mathbf{A}^\alpha$, we create a k-dimensional vector space $\mathcal{A}$, with each axis corresponding to a particular model coefficient a_j. Hence, each individual observation $\mathbf{A}^\alpha$ will lie at a single point in this vector space, and an ensemble of observations, $1 \leq \alpha \leq m$, will in general consist of a distribution of m distinct points in this vector space. In effect, this space now defines a metric, whereby the "closeness" of two different data observations can be measured. In analogy with detection theory, this space is the feature space of the classifier. Note that each point of this space defines a particular set of dynamical relations.
4. Since the dynamical filter is fixed, the information corresponding to different types of dynamical behavior is contained in the feature space distributions. Hence, we aim to utilize this space to perform characterization and classification. In particular, the statistics of the distributions of points

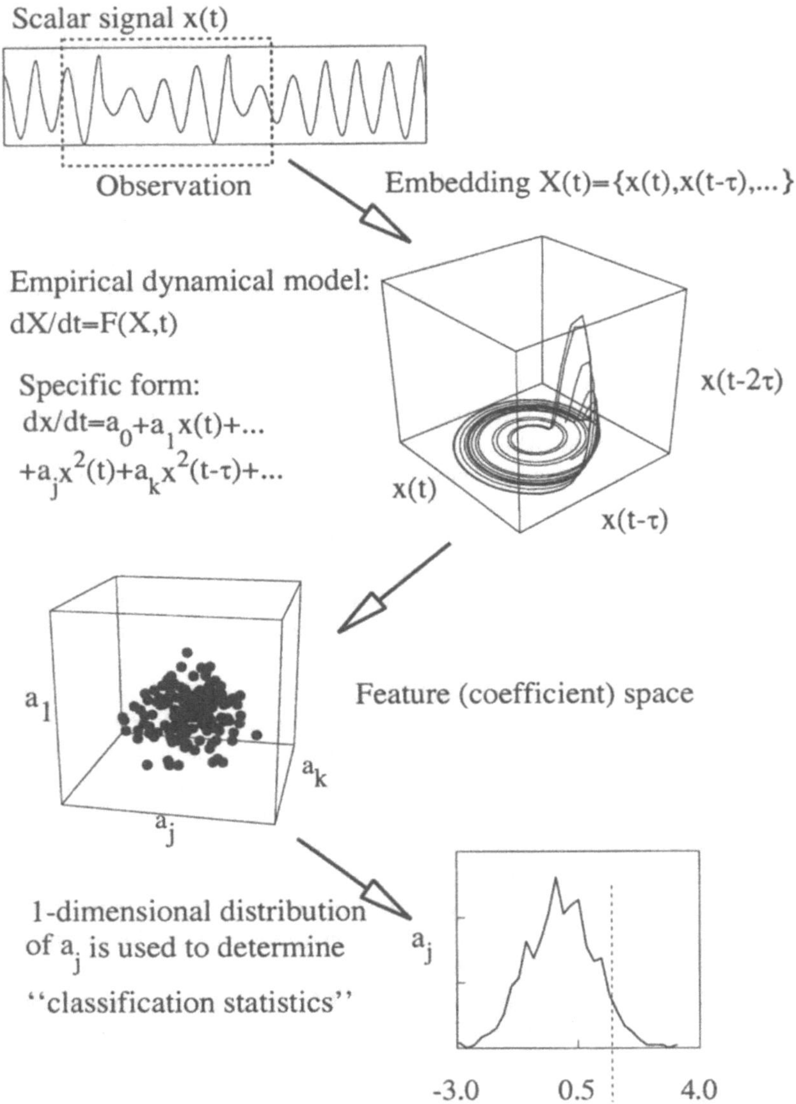

Fig. 1. An idealized schematic of the dynamical classification processing chain.

in $\mathcal{A}$ (e.g. for noisy data) can be used to estimate the probabilities that a data observation belongs to a particular dynamical class, or to a pure noise distribution. Many sophisticated methods can be brought to bear to partition this feature space and calculate classification probabilities, such as Neyman–Pearson criteria, kernel density estimation, and neural network methods. For the purposes of this paper, we use a crude but effective scheme: we estimate the probability of an observation belonging to a particular signal class by measuring its distance from the centroid of the feature distribution for that class, in units rescaled to the standard deviation of that distribution. Hence, one can state approximate classification probabilities under a Gaussian distribution approximation.

5. To provide classification of an arbitrary signal into a particular signal class, we typically assume that each class has been observed previously, and we have a good representation of its signal distribution. Alternately, if one can postulate an a priori model for a signal based on first principals, one may generate it's feature space distributions artificially. In this sense, we assume that we have built a library of dynamical classes, to which arbitrary signals are compared. We note, however, that if the application is purely one of detection, we need only generate the feature distribution for an artificial randomly distributed data set, since signal structure is defined to be any distribution which is statistically distinct from this "null-hypothesis" distribution.

In addition to these steps, it is necessary to ensure that the classification algorithm gives self-consistent results, and several criteria can be defined to check this. Most importantly, we require that observed data which consists of purely independently distributed noise must produce a model estimate which contains no information on average. Hence, the estimated model coefficient distributions must lie around the origin of $\mathcal{A}$ for the algorithm parameters chosen (colored noise may, however, lie away from zero). A pure "detector" can then be defined using this fact, by deriving a threshold away from this noise distribution that separates "structure" from "noise", and defines the hypothesis testing criteria. In a similar fashion, we may examine the statistical information in various individual coefficients for the relevant signal classes of interest, to determine which coefficients may be chosen to construct a robust classifier (i.e. reduce the dimensionality of the feature space $\mathcal{A}$). These can then be used to define a multi-hypothesis testing structure, using a variety of methods to partition the space. For further details, the reader is referred to Kadtke and Kremliovsky (1997).

To interpret the results of these classification analyses, we note that one may consider the extracted information on two levels. Firstly, without respect to any dynamical systems interpretation, one may view the ODE formulation as capturing nonlinear correlations of the data set, as would higher-order spectral measures such as bi-spectra, tri-spectra, etc. However the dynamical model also contains correlations not typically used, such as correlations between the signal and its derivative. Hence, this method yields classification features which provide additional signal characterization beyond typical linear measures (e.g. spectral), and which are often quite useful in the low-noise regime. Clearly, however, cross-terms exist in the dynamical model expansion which are not easily interpreted in this framework.

The second interpretation of the classification analysis is in terms of a fundamental dynamical representation of the data. Hence, we assume that some component of the data was originally generated by a dynamical system, and that we attempt to match primitive dynamical behaviors with those of previously observed data classes, via a low-order approximation. One may think of this more directly as matching simple phase space flow topologies to previously observed data, via templates of flow topology expressed as an expansion in the model basis set. A variety of such simple flow topologies can be easily generated as examples, although we do not discuss this here for brevity.

In the remainder of this paper, we apply the above method to the analysis of the EEG data discussed in the introduction.

3 Description of the Data

The data sets analyzed in this paper were identical to that described by Birbaumer, et al. (1994). The data files in question consisted of digitized time series of EEG recordings, recorded at 256 Hz sampling frequency, taken simultaneously from nine separate Ag/AgCl sensors located at different positions on a subjects head. The nomenclature for the separate sensors consisted of the following: F3, F4, Fz, C3, C4, Cz, P3, P4, Pz. Although individual data files were each identified with its sensor name of origin, the position on the head and relative position between sensors was unknown to us. For our analysis, only the recordings from a single patient were available, unlike the analysis of Birbaumer, et al. (1994). Each data file was approximately 4000 points in length. The data was apparently down-sampled by the experimentors to 128 Hz; the resulting signal-to-noise ratio of the available data was fairly high. Examples of the time series are shown in the figures of the next section.

In addition to the different sensors, the data segments were classified according to the type of musical stimulus (input) that was applied to the patient to produce the EEG signals (output). Firstly, the musical stimuli were divided by composition: these consisted of music segments of rhythm only (identified by "R"), melody only ("M"), and both rhythm and melody ("MR"). Secondly, each of these musical structures was identified by the type of temporal modulation for the rhythm or melody: music modulation could be periodic ("P"), chaotic ("C"), or stochastic ("S"). One segment of data realization was available for each combination of all characteristics. In all, there were (3 modulations) $\times$ (3 musical structures) $\times$ (9 sensors) $=$ 81 data segments available. These data files were mostly stationary over their entire length, except for short "glitches" that were typically removed from the analysis (the windows containing discontinuities were dropped during runs).

We emphasize that during our dynamical classification analysis, no a priori information was assumed about the experimental system, beyond what was available from the data files themselves. Hence, our analysis was relatively "blind", since we did not wish to introduce any bias from the preceeding analyses. We quote information from Birbaumer, et al. (1994) for reference only, to make contact with the original experimental framework. In this sense, our classification scheme is only relative, since no fundamental dynamical properties about the data have been inferred as yet.

In the next section, we describe the analysis of this data using the dynamical classification algorithm.

4 Description of the Dynamical Classification Analysis

To proceed with the analysis, we first examined the data to choose rough guidelines for the parameter values of the dynamical model and algorithm. The primary consideration is the data sampling rate relative to the apparent characteristic times of the data features. Here, we typically observed that there were roughly 5 to 6 samples per characteristic period, over the ensemble of data sets. This sampling rate is, in fact, quite low for the purpose of model analyses such as ours; the effect of low sampling rates is to significantly reduce the amount of information that can be extracted from the data, because the estimate of the signal derivative becomes poor. Hence, we are restricted in our choice of parameters for our algorithm, and construct a model accordingly. On the positive side, the relative noise level seemed quite low, so a reasonable analysis can still be performed. Based upon these considerations, we chose a dynamical model with a polynomial expansion which is of dimension $D = 3$, and order $P = 2$; i.e. utilizing scalar variables $\{x(t), x(t-\tau), x(t-2\tau)\}$, up to quadratic order. The choice of the delay time τ is typically about a quarter characteristic cycle, so we are restricted to at most $\tau = 2$. We perform classification using only one equation of the system of ODEs (Kadtke (1995)), which for $D = 3$, $P = 2$ yields 10 coefficients to be estimated.

Window lengths used to construct the data observations were chosen as the largest time before any noticeable effects of non-stationarity were typically manifested in the data. Here, we chose between 80 to 200 point windows, which corresponded roughly to between 15 and 40 characteristic cycles over most of the signal segments. Windows were slided by half their length through the data files to produce a set of quasi-independent observations. Generally, these parameters generated between 50 and 75 observations to define the observation ensemble, per data file. Using these rough guidelines, the parameter values were later "tuned" to provide improved performance during particular stages of the analysis, as described below. However, final algorithm parameters were all near these rough values, and the qualitative classification conclusions were robust to variations in the specific parameter values.

To define the data analysis, it was necessary to develop a non-standard, data driven approach. Our general goal is to determine if obvious differences in dynamical structure are exhibited by the different data sets, for model parameters as yet to be determined. In a generic classification problem, one would usually be given rough indications of the general signal classes, and have previous observations of each class to define a feature space library. Here, no prior information about the signal classes exist, and the number of independent classes is potentially huge (81). Therefore, it was necessary to define a multi-step analysis procedure, to reduce the complexity of the classification space. In this case, some detailed classification is likely lost, although a more intuitive framework for understanding the data is gained.

To perform a detailed classification analysis for this data, it would be necessary to generate an optimization search over the several relevant algorithm parameters, based on 81 different data sets, measured via a 10-dimensional feature space, and satisfying several different performance criteria. Although straightforward, the effort involved in this analysis is far beyond the scope of the preliminary analysis desired here. Instead, we define a series of steps designed to reduce the complexity of this search, which however yielding only approximate classification assessment. Essentially, we proceed by taking slices through the space at constant parameter values, and observing general properties of the correlations of the sensors and data. To do so, we firstly used a fixed set of preliminary algorithm parameters to analyze the performance of all electrodes over all data sets. This allowed us to categorize the different sensors into equivalence classes which were dynamically correlated, and also to choose the sensor or group of sensors with the best performance. Secondly, using the best sensor, we attempted to optimize its classification performance over all data sets by searching the available algorithm parameters. Thirdly, using the best sensor with optimal algorithm parameters, we again generated the refined classification space for the data, and drew conclusions about the dynamical relationships. Note that for this experiment, we do not have a well-defined classification goal with which to tailor our algorithm parameters, hence we aim only to provide maximal separation of signal classes which appear qualitatively distinct. These steps are discussed in more detail below.

Step 1: Sensor Correlation Analysis. In the first stage of the analysis, we seek to reduce the dimension of the classification parameter space by determining if any correlations exist between the various detectors. That is, we attempt to determine rough sensor groupings (e.g. equivalence classes), so that possibly the performance of only a few representative sensors need be investigated. We also seek to determine the best sensor(s) in terms of performance. Note that here, sensor correlation refers to dynamical similarity. To accomplish this, we first chose a fixed set of preliminary algorithm parameters which lie in the ranges discussed above, and generated the coefficient (feature) space distributions for each sensor, over all data sets. Examples of the data time series, and resulting feature space distributions, are shown in Fig. 4, for some representative music structure categories. To index the data series, we introduce the notation "music composition (music modulation)", hence Fig. 4 represents the data files R(S) (i.e. rhythm-only modulated stochastically), M(S) (i.e. melody-only modulated stochastically), MR(C) (i.e. melody and rhythm modulated chaotically) and so on. In each pairwise comparison, we show the time series and the feature space distributions, to give rough indications of the typical size, shape and separation of distributions. Also, note that the feature space plots in Fig. 4 represent only 3-dimensional projections (in this case of the linear coefficients) of the 10-dimensional feature space of model coefficients.

To determine more rigorously the actual groupings of the sensor distributions in the 10-dimensional feature space, we calculate a table of all the distribution's vector separations. This is generated by a) calculating the mean (centroid) μ and standard deviation s of each sensor distribution, for all data sets; b) calculating the Euclidean distances between all centroids in 10-dimensional space; and c) scaling those distances by the sum of standard deviations Q for pairwise elements (i.e. $Q \equiv \sigma_1 + \sigma_2$) to provide a length scale roughly in statistical significance (under a Gaussian hypothesis). Sensor groupings can then be understood by generating a cross-correlation table, whose elements are these rescaled distances. This scale can be intuitively understood as follows: for two distributions with e.g. $\sigma_1 = \sigma_2 = \sigma$, a value of $Q = 1$ indicates the centroids are 2σ apart, and the distributions statistically "just touch". Alternatively, a hyperplane separating the two-distributions would thus produce approximately a 1/3 chance of false classification (due to the distribution tails) under a Gaussian approximation. In practice, distributions are typically more highly grouped, hence resulting false classification probabilities are significantly less. Note that the actual Euclidean distances are arbitrary, hence only distances normalized by distributional moments have physical relevance (i.e. we perform only relative classification).

The overall sensor groupings were investigated by generating such "cross-correlation" tables for a particular set of code parameters, over all available data sets. Three such representative tables (Tables 1–3) are shown, for music data sets corresponding to: MR(C), MR(S), and M(C). Such tables were also generated for a few other representative sets of algorithm parameters, to check robustness. General, qualitative trends were deduced by inspection over all data sets. Based on this analysis, several observations concerning the dynamical correlation of the different sensors could be made:

- The sensors seem to fall into three distinct groupings, with the third group (C3, F3, P3, F4, Fz) showing some division into two sub-groups: (C3, F3, P3, F4) and (F4, Fz). This large group shows little or no classification capability over most of the data sets (see also Fig. 3 below).

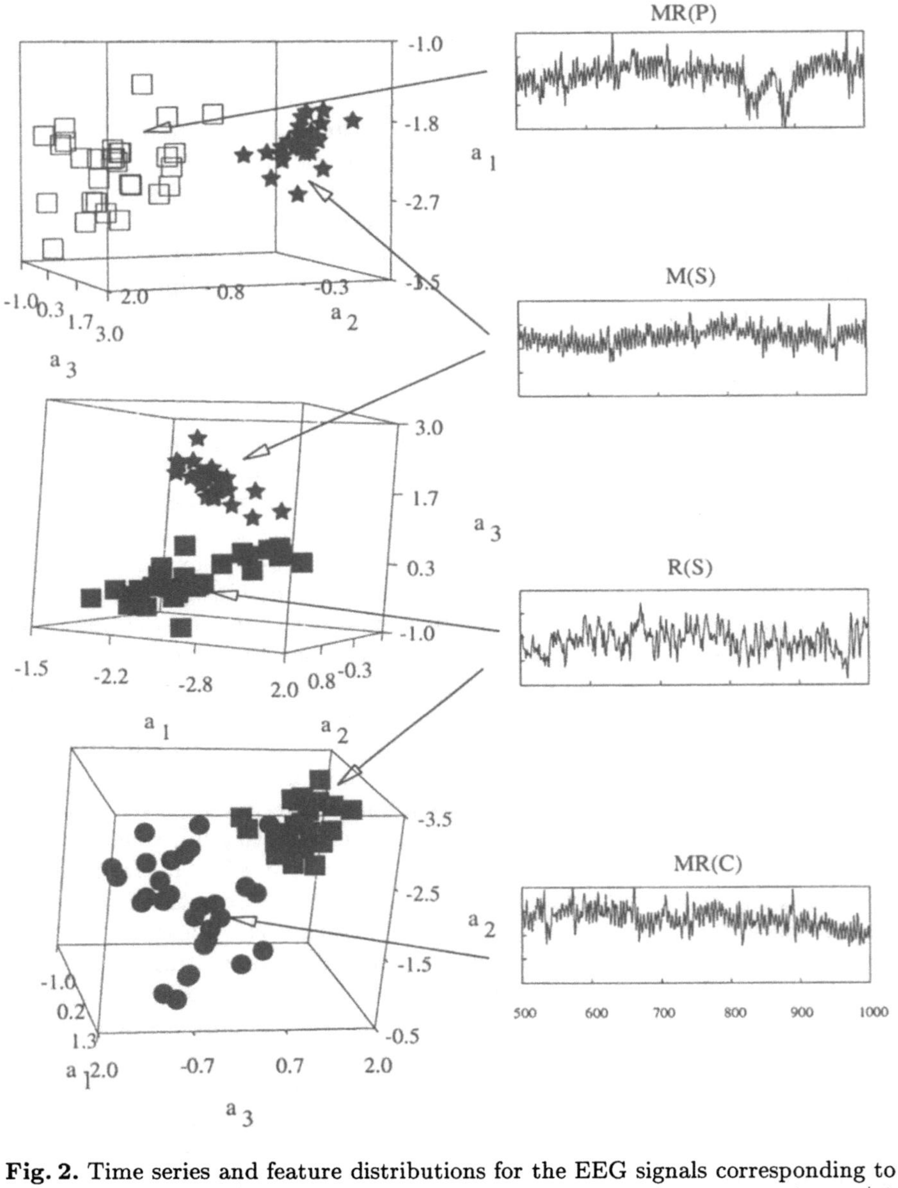

Fig. 2. Time series and feature distributions for the EEG signals corresponding to input music compositions (from top to bottom): (a) melody and rhythm (periodic) and melody (stochastic); (b) melody-only (stochastic) and rhythm-only (stochastic); (c) rhythm-only (stochastic) and melody and rhythm (chaotic).

- The second grouping (C4, Pz) shows weak to moderate classification performance, but only for a few of the data sets.
- The first grouping (Cz, P4) shows moderate to strong classification performance over a wide range of the data sets. Cz is clearly the best sensor of all investigated, with P4's performance weaker but still significant. Both Cz and P4 also show mutual consistency in their classification conclusions.
- Cz, and to a lesser extent P4, show surprising significance in their nonlinear coefficients for many signal classes. Most other sensors show very little.
- The majority of electrodes show some sensitivity to some particular type of music segment, while remaining insensitive to all others.
- Doubly-periodic music segments (i.e. MR(P)) exhibit almost no distinction for any of the sensors.

Table 1. Examples of cross-correlation tables (see also Tables 2 and 3) showing rescaled distances between feature distributions, indicating relative dynamical correlation. Tables show the distances between features generated by different sensors, for a single data file melody and rhythm (chaotic)

Sensor	C4	Cz	F3	F4	Fz	P3	P4	Pz
C3	0.27	0.54	0.18	0.25	0.35	0.20	0.23	0.26
C4		0.50	0.30	0.24	0.21	0.19	0.35	0.38
Cz			0.50	0.39	0.42	0.43	0.55	0.59
F3				0.34	0.40	0.13	0.16	0.19
F4					0.17	0.19	0.41	0.41
Fz						0.33	0.52	0.57
P3							0.19	0.19
P4								0.07

Using the above specific observations, two main conclusions can be deduced from this analysis. Firstly, the sensors show general, robust grouping into three (and possibly four) distinct classes; this grouping can be most compactly described by a diagram showing the general location of the distributions of individual sensors over the data sets, as indicated in Fig. 3. In this figure, the Euclidean distances of the indicated sensor symbols are schematic only, but reflect the relative correlations of the sensors, as measured by their ability to distinguish the different musical structures. Secondly, the Cz sensor

Table 2. Same as Table 1 for melody and rhythm (stochastic).

Sensor	C4	Cz	F3	F4	Fz	P3	P4	Pz
C3	0.24	0.81	0.09	0.19	0.33	0.16	0.27	0.28
C4		0.68	0.20	0.20	0.30	0.16	0.37	0.35
Cz			0.79	0.66	0.58	0.78	0.84	0.83
F3				0.20	0.38	0.13	0.24	0.24
F4					0.18	0.27	0.45	0.44
Fz						0.45	0.69	0.68
P3							0.19	0.18
P4								0.05

Table 3. Same as Table 1 for melody-only (chaotic).

Sensor	C4	Cz	F3	F4	Fz	P3	P4	Pz
C3	0.20	0.12	0.13	0.30	0.38	0.17	0.97	0.40
C4		0.15	0.21	0.18	0.26	0.30	0.77	0.27
Cz			0.11	0.30	0.34	0.19	0.88	0.37
F3				0.36	0.42	0.18	0.89	0.42
F4					0.12	0.19	0.69	0.19
Fz						0.14	0.67	0.28
P3							0.62	0.30
P4								0.70

clearly shows the best classification capability for this experiment, followed by P4, which is moderately well-correlated with Cz. Based on these conclusions, we restricted our future analysis to determining the optimal classification properties of the Cz and P4 sensors.

Step 2: Sensor Parameter Optimization. From the analysis of the above section, the Cz and P4 sensors were determined to have the best classification capability using the preliminary algorithm parameters. In this step, we aimed to optimize the performance of these two sensors, by performing an algorithm parameter search. To do so, it was necessary first to define a performance criteria. In a generic classification problem, one would typically be given previous observations of the relevant signal classes, and the task would

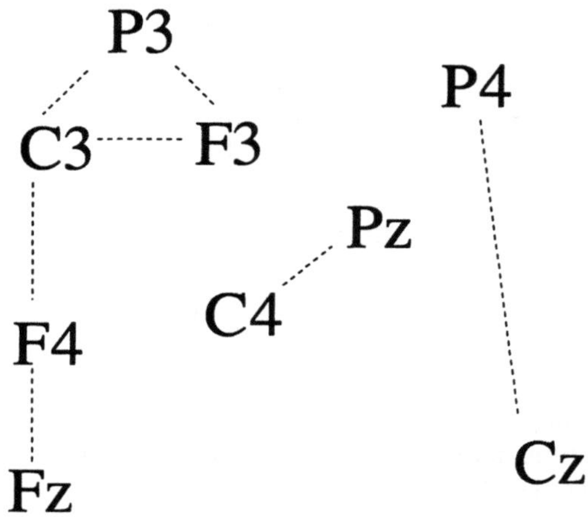

Fig. 3. Schematic of the approximate groupings of sensors, based on average separation of coefficient distributions in feature space. Euclidean distances in the schematic reflect a qualitative measure of dynamical correlation projected from a higher dimensional space. Dotted lines connect sensors with similar dynamical response.

be to generate optimum separation of the known class distributions in the feature space (and simultaneously to produce minimal separation between signals belonging to the same class). Here, we have no prior knowledge of the class structure for the data set, so we chose as our performance criterion the optimal separation of all feature distributions for all data sets. In general, this does not ensure a proper classification structure, however no other obvious and un-biased criteria can be formulated without prior information about the system.

The optimal separation criterion can be expressed in the following way: assuming that the centroid μ of the distribution of each data set is the vertex of a 10-dimensional N-gon, we seek to find algorithm parameters which produce the largest volume of the N-gon (see Fig. 4). To perform the optimization, we simply vary the possible algorithm parameters (window length, smoothing parameter, etc.) and observe this volume. The optimal parameters are then chosen as the ones that produce the largest volume, which can be estimated using the two-point distances between each centroid.

This procedure was used to determine the optimal parameter values for the Cz and P4 sensors for the data analysis step. These values were as follows: $L_w = 100$, $D = 3$, $P = 2$, $\tau = 2$, and the smoothed derivative was calculated over 5 points. In general, we found the classification performance to be robust to relatively wide ranges of these parameters, hence the sensor optimization was actually not critical. In the next section, we describe the data classification analysis using these parameters.

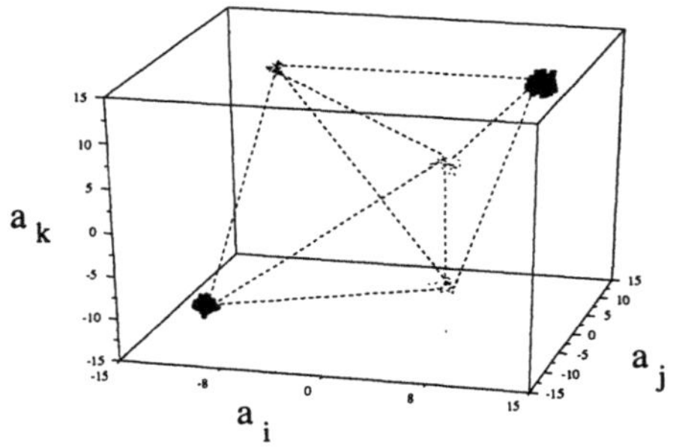

Fig. 4. Schematic indicating the optimization criterion for Cz and P4 sensors. Centroids of feature distributions form the vertices of a 10-dimensional N-gon, whose volume is maximized over the algorithm parameters.

Step 3: Sensor Data Classification. In this last step, we attempted to perform classification of the EEG signals based on the optimized sensor classification parameters. Here, we used the optimized parameters for the Cz and P4 sensors to generate the respective feature space distributions over all data sets. Note that at this point, we have only nine data sets remaining to analyze per sensor, hence we have dramatically reduced the parameter space to be searched. Using the renormalized distance between centroids of the data feature distributions, we generated a feature cross-correlation measure exactly as described in the previous section. The results of this analysis is shown in Tables 4 and 5, which give the matrices of the cross-correlation measures for the Cz and P4 sensor, respectively. All relevant classification structure under the Gaussian distribution assumption may be deduced from these tables.

Based on the two correlation matrices, several generic observations could be made about the apparent classification structure derived from the EEG signals. These can be summarized as follows:

1. The Cz sensor shows at least some distinction between the majority of all data sets, and in some cases shows large distinction. P4 shows quantitatively less sensitivity, however its qualitative conclusions largely correlate with Cz.
2. Cz is the most affected by stochastic melody, M(S), which shows the largest distinction of any data set or electrode from the average of all data sets.
3. For the Cz electrode, the largest single quantitative distinction is between stochastic melody, M(S), and any of the pure rhythms.
4. Data consisting of pure rhythm generally stands away from any data which includes melody, in both Cz and P4.

Table 4. Cross-correlation tables for the Cz sensor, over all data types. Values indicate the statistical separation of feature distributions for different music classes, under a Gaussian hypothesis. Parameter values used are: $L_w = 100$, $D = 3$, $P = 2$, $\tau = 2$

Sample	MR(P)	MR(S)	M(C)	M(P)	M(S)	R(C)	R(P)	R(S)
MR(C)	0.12	0.19	0.60	0.15	0.22	0.61	0.70	0.65
MR(P)		0.16	0.70	0.09	0.16	0.73	0.85	0.77
MR(S)			0.85	0.20	0.16	0.88	1.04	0.94
M(C)				0.77	1.01	0.09	0.14	0.12
M(P)					0.20	0.80	0.91	0.86
M(S)						1.02	1.23	1.14
R(C)							0.21	0.17
R(P)								0.22

Table 5. Same as Table 4 for the P4 sensor.

Sample	MR(P)	MR(S)	M(C)	M(P)	M(S)	R(C)	R(P)	R(S)
MR(C)	0.09	0.12	0.87	0.16	0.21	0.95	0.35	0.69
MR(P)		0.10	0.81	0.12	0.15	0.89	0.33	0.67
MR(S)			0.87	0.09	0.16	0.97	0.36	0.72
M(C)				0.88	0.90	0.18	0.13	0.15
M(P)					0.14	0.96	0.36	0.72
M(S)						0.94	0.40	1.73
R(C)							0.12	0.13
R(P)								0.09

5. Chaotic melody affects P4 strongly, and to a lesser degree Cz.
6. Chaotic rhythm also affects P4 strongly.

From the specific observations listed above, two general conclusions can likely be drawn about the classification structure of Cz and P4. Firstly, any music segment which includes melody can generally be distinguished strongly from any segment which does not. Secondly, this distinction is also sensitive to the structure of the melody itself, with stochastic melody providing the greatest distinction, followed by chaotic, with periodic melody showing significantly less distinction. In the next section, we draw general conclusions and implications of these results.

5 Conclusions

Although the EEG data sets represent a well-defined experimental structure with good data characteristics, it is difficult to draw any fundamental conclusions without a more detailed understanding of the physical system, the affects of experimental parameters, and a more exhaustive and representative data base. Hence, the conclusions here are preliminary and entirely data-driven, and their physical significance is as yet unclear. Within these caveats, however, we may state the following: firstly, the fact that only two sensors show well-defined classification capability (i.e. sensitivity) over a spectrum of data sets would seem to indicate some localization of processing activity in the brain of this particular subject. Secondly, the obvious distinction between sounds with melody, versus sounds without, would imply that some additional processing activity is occuring for melodic components. Thirdly, the sensitivity of the melodic processing activity is at least moderately related to the complexity of the melody, being most strongly affected by stochastic structure, and least affected by periodic.

With respect to these conclusions, however, we must point out again that the data observations are narrow with respect to the space of experimental parameters available. Specifically, this data represents observations of (i) a single subject, at (ii) a single data-taking session, with (iii) one contiguous data measurement per musical composition, and with (iv) pre-defined (low) sampling rates and data filtering. Although the most obvious problem is the lack of multiple observations over extended time, how the other experimental factors could affect the results is unknown to us. In addition, we have at this time no physiological interpretation or alternate quantitative analyses to confer or repudiate these conclusions.

In addition to the experimental factors, several aspects affecting the modeling process could be improved. These include the use of detection-theoretic thresholding methods to generate classification probabilities, via integrals over the feature distributions, and a search over more general dynamical model types, to determine if performance improvements can be made. With respect to the latter point, we note that the conclusions of the above analysis depend to a large degree on the ability to characterize the data by the low-dimensional polynomial model. Hence, reduced sensitivity of the classification scheme to particular data types can in practicality be due to the inappropriateness of the specific dynamical model. For example, the apparent lack of sensitivity in the frontal lobes may be due to their response dimensionality being somewhat higher than three (but still low-dimensional). Although we have typically found that the classification properties are relatively robust to model dimensions which are only approximately correct, this factor must certainly still be considered.

Finally, given that the conclusions of the above analysis eventually prove physiologically significant, we speculate that there exists the possibility of developing a practical classification tool for a variety of bio-medical applications, e.g. a warning monitor for the onset of epilepsy. Such a monitor could be made specific to particular patients by tailoring the algorithm parameters and dynamical model. Since the analysis of data (after the signal classification library has been generated) is computationally very efficient, such a classification scheme could easily be implemented on current dedicated processing chips to provide a physically small, real-time monitor.

Acknowledgements

The authors would like to acknowledge very useful discussions with Mario Inchiosa, Adi Bulsara, Gottfried Mayer-Kress, and Nikolai Makarenko. J.K. wishes to acknowledge support through UCSD by a grant from the Office of Naval Research, No. N00014-96-1-0056. M.K. and J.K. also wish to acknowledge support for this project through the RTA Corporation by various grants from the ARO and ONR.

References

Birbaumer N., Lutzenberger H., Rau H., Mayer-Kress G., Choi I., and Braun C. (1994): CCSR Tech. Report 94-28, preprint.

Cremers J., Hübler A. (1981): Construction of differential equations from experimental data. Z. Naturforsch. **42A**(8), 797–802.

Crutchfield J.P., McNamara B.S. (1987): Equations of motion from a data series. Complex Systems **1**(3), 417–452.

Kadtke J. (1995): Classification of highly noisy signals using global dynamical models. Phys. Lett. A **203**, 196–202.

Kadtke J., Brush J. (1994): Global modeling of chaotic time series with applications to signal processing. AIP Conf. Proc. **296**, AIP Press, New York, 205–230.

Kadtke J., Kremliovsky M. (1996a): Signal classification using global dynamical models. Part I. Theory. AIP Conf. Proc. **375**, AIP Press, New York, 189–202.

Kadtke J., Kremliovsky M. (1996b): Classifying complex, deterministic signals. In *Predictability of Complex Dynamical Systems*. Springer, Berlin, Heidelberg, 78–102.

Kadtke J., Kremliovsky M. (1997): Estimating statistics for detecting determinism using global dynamical models. Phys. Lett. A **229**(2), 97–106.

Salvino L.W, Cawley R. (1994): Smoothness implies determinism: a method to detect it in time series. Phys. Rev. Lett. **73**(8), 1091–1094.

IV Perception and Motor Control

Nonlinear Analysis of Perceptual-Motor Coupling in the Development of Postural Control

Steven M. Boker[1], Thomas Schreiber[2], Bernd Pompe[3], and Bennett I. Bertenthal[4]

[1] University of Notre Dame, Department of Psychology, Notre Dame, IN 46556, USA
[2] Department of Physics, University of Wuppertal, D-42097 Wuppertal, Germany
[3] Department of Physics, Ernst-Moritz-Andt-University, F.-L. Jahn Str. 16, D-17487 Greifswald, Germany
[4] University of Virginia, Department of Psychology, Charlottesville, VA 22903, USA

Abstract. The maintenance of balance while sitting or standing requires a control mechanism which can maintain upright posture as well as adapt quickly and flexibly to changes in the environment. Some sort of dynamical control must link visual, auditory, vestibular, and proprioceptive perceptual input to the motoric responses required to activate appropriate muscle groups in order to maintain balance. This dynamical control mechanism needs to use perceptual input to predict the future state of posture with respect to the environment if adaptive balance is to be maintained under changing conditions. These constraints suggest that a purely stochastic random-walk postural control system is unlikely, although others have been unable to reject a linear stochastic model for postural control of quiet standing.

The data presented in this chapter are drawn from an experiment that measures center of pressure in a sample of sitting infants who are exposed to a "moving room" stimulus paradigm. Three categories of analyses are applied to these data: mutual information, false nearest neighbors and surrogate data tests. These techniques will be applied to ask whether the center of pressure in sitting infants' postural control can be modeled as a linear system, whether there is a developmental change in the strength and direction of the coupling of that postural control to visual stimuli and whether any developmental change observed in this coupling carries an accompanying reduction in noise in center of pressure.

1 Introduction

Postural control in either sitting or standing involves controlling musculature in such a way that balance can be maintained under varying environmental conditions. A control mechanism must exist which maps visual, auditory, vestibular, and proprioceptive perceptual input onto the appropriate muscle groups. This control mechanism must take into account both changes in the environment and changes in the state of the postural system in order to quickly and flexibly adapt to new conditions. For example, if the apparently solid rock on which one is standing were to suddenly begin to move, then one's posture must respond immediately to this perturbation of the environment in order to remain standing.

Since there is a time delay between perception of a change in the environment and the possible adaptive control which can be applied in order to respond to that change, it would be sensible for the control mechanism to attempt to anticipate the changes in the environment in order to make prospective adaptation that will result in a close temporal match between perception and action. This can be illustrated by thinking about how one catches a ball; the hand is extended so that the ball and the hand will arrive at the same point at the same time.

For these reasons we are inclined to think of postural control as a dynamical system; a system that must integrate input from several perceptual modalities in order to maintain an equilibrium. However, this equilibrium is not a fixed point; even when we attempt to stand motionless we sway slightly. This motion during quiet standing has been studied by Collins & De Luca (1994) who have concluded that the postural control system can be modeled as a linear stochastic system. Collins & De Luca (1993, 1995) also argue that the posture utilizes two linear control mechanisms: an open-loop system over short time scales and a closed-loop system over longer time scales.

Vision and proprioception are two principal inputs to the postural control system (Howard 1986). This can be demonstrated using a *moving room* experimental paradigm (Lishman & Lee 1973). The moving room, or "Phantom Swing" as it was called in the 19th century, effectively dissociates the input from proprioception and vision. In this experimental paradigm, the subject stands in a room whose walls and ceiling are not fixed to the floor, but instead are mounted on wheels. When the experimenter moves the room, the subject receives visual input indicating self motion while the proprioceptive input indicates no self motion. A slowly oscillating room produces a significant swaying response in subjects, including infants (Bertenthal, 1990).

This work examines data from a moving room experiment performed on a group of infants (Bertenthal et al. 1996). These infants were selected to have ages which straddled the average age of onset of self-supported sitting: 6.5 months (Bayley 1969). The data from this experiment holds interest for examination using dynamical systems methods since changes in the dynamics of the postural control mechanism can be studied simultaneously with the coupling between perception and action. By studying developmental changes in the coupling between visual perception and postural action we hope to better understand the nature of postural control system.

2 Methods

Forty infants participated in the study, 10 in each of 4 age groups: 5, 7, 9 and 13 months. The infants were tested in a moving room designed as a 1.2 m

$\times$ 1.2 m $\times$ 2.1 m open-ended enclosure as shown in Fig. 1. The walls and ceiling of the enclosure were constructed of fiberboard covered with green and white vertically striped cloth and mounted on small rubber wheels that rolled on tracks fixed to the floor. The floor was padded and covered with white muslin cloth. Two flourescent lights mounted at the top of the two side walls illuminated the moving room. A small window in the middle of the front wall of the moving room provided a view to a small electronically activated toy dog that was used to fix the gaze of the infant at the beginning of each trial. A potentiometer was attached to one wall of the moving room such that the position of the room could be measured by the voltage drop through the potentiometer. The position of the room was sampled at 50 Hz and converted to a digital time series using an 8 bit A/D converter thereby creating a time series $R = \{r_1, r_2, r_3, \ldots, r_N\}$ representing the room movement.

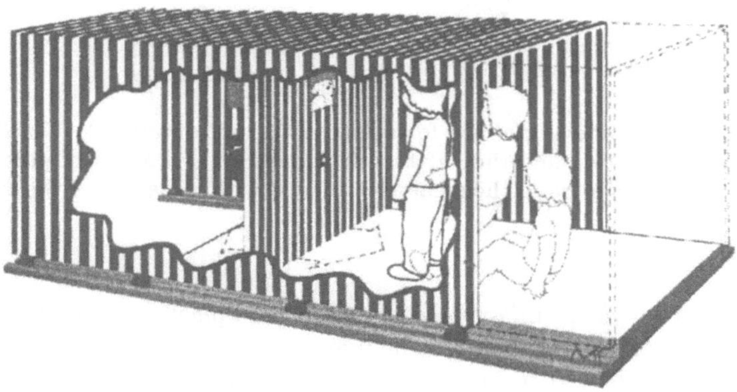

Fig. 1. Schematic drawing of the moving room used as stimulus for sitting infants or standing toddlers. The child portrayed inside the room is falling backward due to perceived optic flow produced by movement of the room. Note that the floor does not move; the subjective perception of self-motion is created by moving the walls and ceiling of the room together.

A forceplate was set in the middle of the floor of the moving room and an infant's bicycle seat was mounted on the center of the forceplate. The forceplate consists of a rigid metal plate suspended from 4 pressure transducers as shown in Fig. 2. The transducers were each sampled at 50 Hz synchronously with the room position and converted to 4 digital signals with an 8 bit A/D converter. The four time series $P1, P2, P3$ and $P4$ from the four transducers were transformed to two center of pressure (COP) time series

$X = \{x_1, x_2, x_3, \ldots x_N\}$ and $Y = \{y_1, y_2, y_3, \ldots y_N\}$ where

$$x_i = \frac{1}{2} D_x \frac{(p1_i + p2_i) - (p3_i + p4_i)}{p1_i + p2_i + p3_i + p4_i}$$

and

$$y_i = \frac{1}{2} D_y \frac{(p1_i + p4_i) - (p2_i + p3_i)}{p1_i + p2_i + p3_i + p4_i} .$$

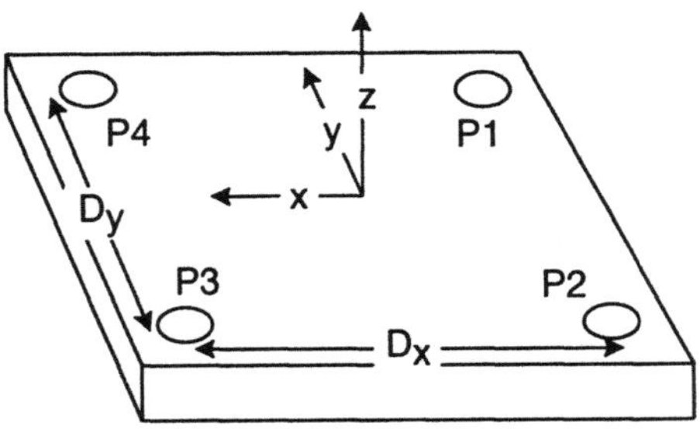

Fig. 2. Schematic drawing of the forceplate. The transducers $P1, P2, P3$ and $P4$ were sampled at 50 Hz and these data transformed to two center of pressure time series X and Y along two orthogonal axes aligned with the edges of the forceplate.

An infant sat in the infant bicycle seat, a rigid plastic seat with a backrest tilted at 80° relative to the floor. The first trial followed a short time in which the infant was allowed to acclimate to the room. At the beginning of each trial, the toy was activated to direct the infant's attention to the front of the room. The walls then oscillated in one of six movement conditions for a period of approximately 12 s during which time the infant's center of pressure was measured. Between successive trials the infant was allowed a brief interval in which to restabilize to the motionless room.

During each trial, the force plate measured center of pressure along two axes: the fore-aft axis (X) aligned with the movement of the room, and the lateral axis (Y) orthogonal to the movement of the room. As the room began to move, the two axes of center of pressure and the unidirectional position of the room were simultaneously sampled at 50 Hz (20 ms sampling interval) for 10.25 s. Thus each trial generated three synchronized time series containing 512 samples: X, Y and R.

Infants were tested in a sequence of 12 trials, two trials for each of six room movement conditions as shown in Table 1. Four conditions consisted of sinusoidal movement with one of two amplitudes (9 cm or 18 cm) combined with one of two frequencies (0.3 Hz or 0.6 Hz). The remaining two conditions

Table 1. The six frequency and amplitude room movement conditions. The baseline condition is labeled as 0 Hz frequency and 0 cm amplitude. "Multi" (for multiple frequencies) refers to a pseudorandom low frequency movement of the room.

	0 Hz	Multi	0.3 Hz	0.6 Hz
0 cm	X			
Multi		X		
9 cm			X	X
18 cm			X	X

were a motionless control condition and a pseudorandom low frequency movement pattern. Each session started with the motionless control trial followed by the moving room trials presented in random order and then followed by the motionless condition again as the last trial.

2.1 Analyses

Three nonlinear techniques will be applied in the analysis of these data: mutual information, surrogate data and false nearest neighbors. Each of these techniques are briefly introduced here and are also covered more completely elsewhere in this volume. These techniques rely on the familiar notion of *time-delay embedding* (Packard et al. 1980, Whitney 1936, Takens 1981, Sauer et al. 1991) in which a multidimensional embedded state space is created from time delayed copies of the time series of interest.

Mutual Information Information theory (Shannon 1949) provides a measure for nonlinear dependence within and between time series. When a sequence of measurements of a variable are taken over a period of time, one can estimate the uncertainty in the prediction of the next measurement given the preceding measurements (see Resnikoff 1989 for an overview). If we have a time series U, the average mutual information $I(\tau)$ about a measurement $u_{t+\tau}$ given a measurement u_t at time t over all N is

$$I(\tau) = \frac{1}{N-\tau} \sum_{t=1}^{N-\tau} p(u_t, u_{t+\tau}) \log_2 \left[\frac{p(u_t, u_{t+\tau})}{p(u_t)p(u_{t+\tau})} \right] , \tag{1}$$

for a time delay of τ.

If we have two time series, U and V, the *average mutual information* between the two time series at a time delay of τ is

$$I_{UV}(\tau) = \frac{1}{N-\tau} \sum_{t}^{N-\tau} p(u_t, v_{t+\tau}) \log_2 \left[\frac{p(u_t, v_{t+\tau})}{p(u_t)p(v_{t+\tau})} \right] . \tag{2}$$

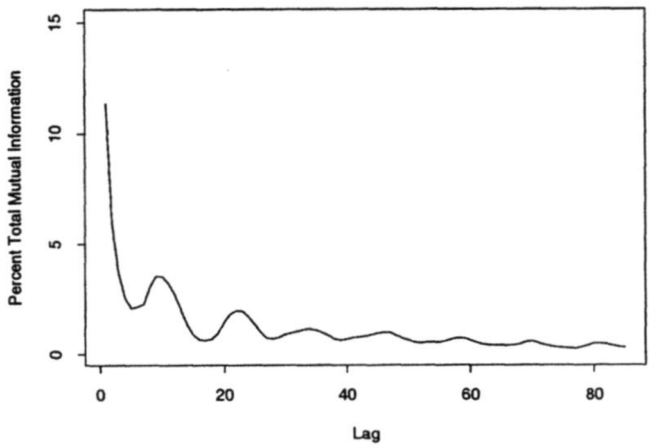

Fig. 3. An example mutual information plot of a time series from the Lorenz system. The X axis plots the number of units of delay, and the Y axis plots the average mutual information as a percentage of the total mutual information for each corresponding delay.

Figure 3 shows an example plot of the average mutual information within a single time series over a range of time delays (or *lags*). Note there is a relatively large amount of information shared between measurements separated by a small time delay, but as the time delay increases the amount of information shared by the measurements decreases. In any continuously varying system it is to be expected that the amount of mutual information between two measurements would increase as the time between the measurements becomes small. However, notice that in Fig. 3 there are some other time delays when the mutual information has a maximum. These maxima are signatures of periodicity in time series.

The mutual information function has proved to be useful in a wide variety of nonlinear analyses, but its calculation can prove to be compute intensive for long time series. However, recent developments by Fraser & Swinney (1986) and Pompe (1993, 1996) have reduced the calculation to manageable proportions. One such approach works with generalized mutual information on the basis of Renyí entropies (Pompe 1996).

False Nearest Neighbors. The method of *false nearest neighbors* is designed to determine how many dimensions are sufficient to embed a particular time series (Kennel et al. 1992). The basic idea behind false nearest neighbors is that points in a state space should be close to each other because their dynamical state is similar, not because they have been projected close to each other as an artifact of constructing the embedding using a dimension that is too low. The time series is embedded in state spaces of increasing dimension and points which are artifactually close to each other are marked and declared to be false nearest neighbors. The resulting percentage of false nearest neighbors for each embedding dimension is then plotted against the corresponding embedding dimension.

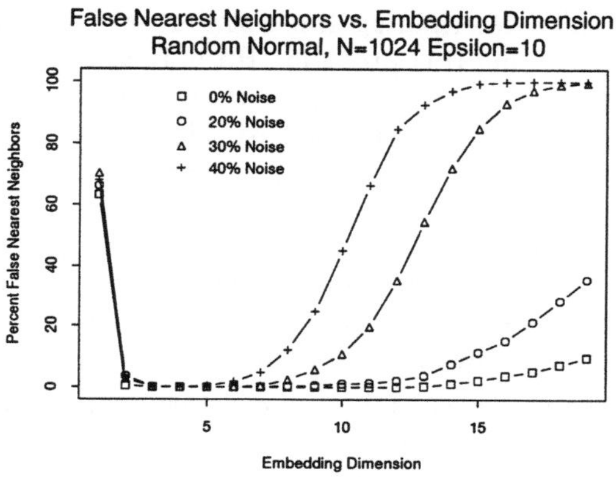

Fig. 4. False nearest neighbors curves for Gaussian noise added to a 1024 point time series of the Lorenz Equation. Note that as the percentage of noise increases, the slope increases for the lines in the right hand portion of the graph.

The effect of additive noise on the false nearest neighbors plot can be exploited to examine the relative amount of additive noise mixed with a nonlinear dynamical signal. The simulation plotted in Fig. 4 exemplifies this phenomenon. A nonlinear signal, a time series of the Lorenz Equation, is mixed with increasing amounts of Gaussian noise. Note that as the amount of noise is increased, the slope increases on the right hand side of the plot. Also note that there is little difference in the lines to the left of the first minimum of the false nearest neighbors curve, the algorithm suggests that a minimum embedding dimension of three is required to represent the dynamical portion of the signal in this series.

Surrogate Data Test for Nonlinearity. Given an arbitrary time series generated by an unknown process, it is important first to test whether any statistical dependencies are present at all, and if this is the case, whether these are completely accounted for by linear pairwise correlations. If there is no evidence for dependencies at all, the data may be considered to be IID, independent identically distributed random numbers. If the only dependencies are linear correlations, there is no point in pursuing a nonlinear dynamical systems model when a linear stochastic model will fit the measured data just as well. In almost every case, the simpler model is to be preferred over the more complex because it contains less bias.

Making these determinations requires a test of significance. Normal statistical theory does help in this case, since calculation of standard errors requires some model of the process which generated the data and in this case the model is not known. A variant on the bootstrap method (Efron 1979a, 1979b) for empirically determining the distribution of a statistic has been used in order to overcome this problem (Horn 1965). Theiler, et al. (1992) have called this the method of ***surrogate data*** and it was independently proposed by Kennel & Isabelle (1992). The basic idea is to generate a population of null hypothesis data sets (surrogates) appropriate to the test of interest and then use the distribution of some nonlinear invariant (such as the fractal dimension of a time delay embedding) of these surrogates to estimate a confidence interval around the mean of the invariant. Then if the nonlinear invariant of the measured data lies outside the confidence interval of the surrogates, the null hypothesis is rejected.

Constructing the surrogate data sets can take many forms and will vary depending on the particular null hypothesis that one desires to test. We are interested in testing whether the postural control data are distinguishable from a linear stochastic system. One method of generating a surrogate data set for this null hypothesis was suggested by Osborne et al. (1986). The Fourier transform of the time series is applied, a uniform random number between 0 and 2π is added to the phase spectrum of the Fourier series, and then the inverse Fourier transform is applied. The effect of this method is to generate a surrogate which shuffles the time ordering of the data while preserving the linear autocorrelations in the time series. The resulting surrogate fits the null hypothesis that the time series is an autocorrelated linear stochastic process (colored noise).

A refinement of this method also starts with a phase randomized sequence. It is then iteratively Fourier filtered and amplitude adjusted to correct for minor anomalies introduced by the phase randomization process (Schreiber & Schmitz 1996). This process is called ***polished surrogates*** and is the method which will be used to generate the null hypotheses for the present work. This process broadens the null hypothesis to include simple nonlinear rescalings of a Gaussian random process, whereas simply phase randomizing the Fourier series includes the restriction that the linear process is Gaussian.

Surrogate data methods have begun to be used in physiological applications (Schiff & Chang, 1992, Collins & De Luca 1993, Collins & De Luca 1994) and have been recently extended to the multivariate case by Prichard & Theiler (1994) and Palus (1995). In this case the surrogates must mimic not only the autocorrelations within each time series, but also all of the cross correlations between the series.

3 Results

Figure 5 plots prototypical time series for the room movement and fore-aft sway for three trials from a single seven month-old infant. The left column of graphs in Fig. 5 plot the room movement for three different stimulus conditions and the right column of graphs plot the corresponding fore-aft center of pressure for the infant during the same 10.25 second interval. In Figs. 5B and 5D the movement of center of pressure for the infant bears little visible resemblance to the corresponding room movement, but Fig. 5F exhibits a noticeable periodicity which appears similar to that of the room movement in Fig. 5E

In order to quickly determine if the infants are responding to the movement of the room, the Fourier spectrum was calculated for the time series resulting from each trial. The Fourier spectra were aggregated for all trials within each condition. Since four of the moving room stimulus conditions represented sinusoidal oscillations, the Fourier spectrum of the moving room time series is represented by one peak at the frequency of the oscillation. It follows that if the infant's behavior was entrained to the linear oscillation of the room, there should be a peak in the mean Fourier spectrum of the infant center of pressure at the corresponding frequency of the four sinusoidal stimulus conditions.

Mean Fourier spectrum plots for the fore-aft center of pressure time series under each condition are shown in Fig. 6. Note that for the 0.6 Hz conditions there is a peak in the distribution corresponding to a frequency of 0.6 Hz. In a similar fashion in the 0.3 Hz, 18 cm condition there is a peak in the distribution at 0.3 Hz. It is evident by inspection that these data exhibit some form of coupling between the visual environment and postural control.

3.1 Surrogate Data Test for Nonlinearity

The time series from two of the experimental conditions were tested to see if a linear stochastic model would be sufficient to describe the center of pressure data. We chose to test the control condition where no room movement occurred and the 0.6 Hz condition in which maximum coupling between the moving room and the infant's center of pressure was expected to occur (Bertenthal et al. 1996). In this way we could examine the question of whether there was evidence of nonlinearity in the center of pressure of a free sitting infant, and if so, whether there was a decrease in nonlinearity when the postural control system was coupled to a sinusoidally oscillating visual stimulus.

For each time series within each selected experimental condition, twenty surrogate time series were generated using "polished surrogates" method (Schreiber & Schmitz 1996). This method generates surrogate time series which match the source time series in mean, variance, distribution of scores and Fourier spectrum. We then calculated the average mutual information $I(\tau)$ for a time-delay of $\tau = 200$ ms for each of the surrogate time series and

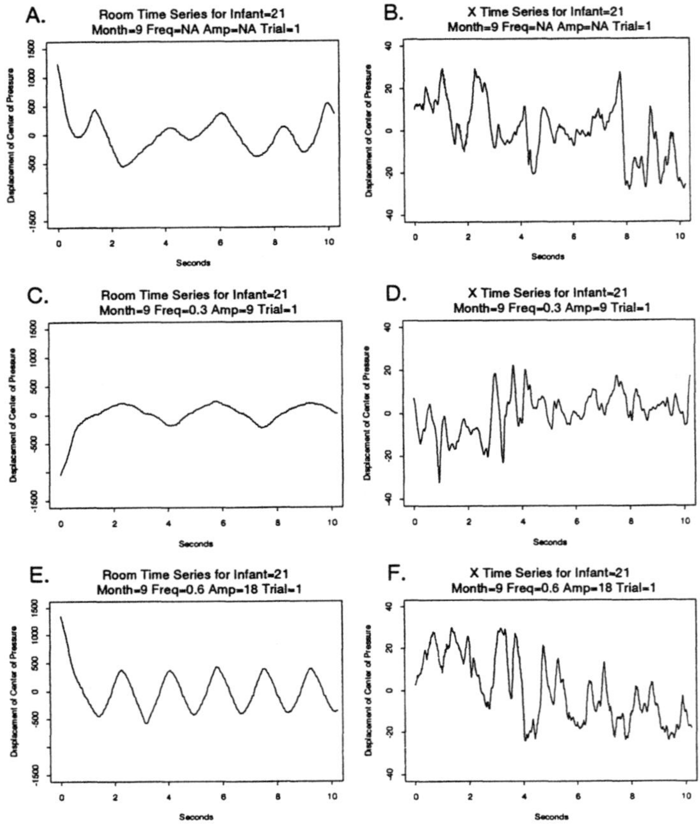

Fig. 5. Example time series for three experimental conditions for infant number 21. In the left column are the time series for the movement of the room. In the right column are the corresponding time series for the movement of the fore-aft center of pressure of the sitting infant in a direction parallel to the movement of the room.

for the source time series. If the mutual information of the source time series was larger than the largest surrogate mutual information, then the null hypothesis of no nonlinearity was rejected. If the null hypothesis is rejected then the hypothesis that a linear model is sufficient is rejected at the ($p = 0.05$) level.

Table 2 summarizes the results of the surrogate data tests for nonlinearity. Recall that at the ($p = .05$) level we should expect that 5% of the null hypotheses conditions should be rejected by chance alone. However, in the free sitting control condition the null hypothesis is rejected 80% of the time. We consider this to be strong evidence of nonlinearity in these data.

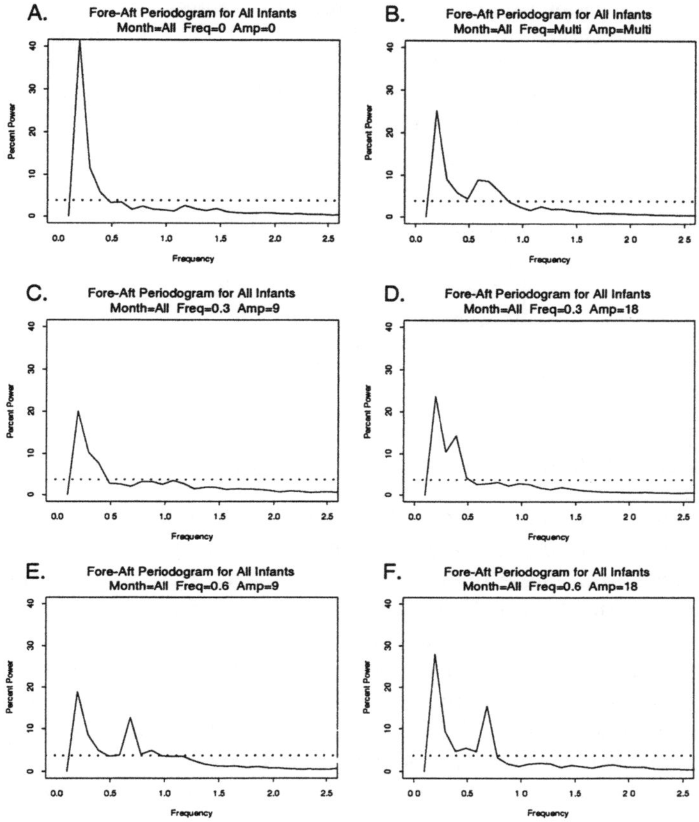

Fig. 6. Mean FFT plots for all trials and all infants for each experimental condition. The 95% confidence interval for distinguishing the frequency components from white noise is indicated by the dotted line in each graph.

Note that in the free sitting condition the evidence for nonlinearity is greatest in the oldest infants. It is also interesting to note that when the center of pressure of the infant is coupled to the 0.6 Hz sinusoidally oscillating room, the percentage of trials in which the null hypothesis is rejected is reduced to 59%. This reduction in the apparent nonlinearity of the center of pressure time series is expected since in this experimental condition the center of pressure is coupled to a linear oscillator.

Table 2. Summary of results of the surrogate data tests for nonlinearity. Each cell lists the percentage of time series for which the null hypothesis of a linear stochastic system was rejected. Thus the hypothesis of a linear stochastic model being sufficient for these data was rejected in 80% of the time series for the free sitting control condition with no room movement.

	0 Hz ($N = 80$)	0.6 Hz ($N = 160$)
All ages	80%	59%
5 Months	75%	65%
7 Months	80%	55%
9 Months	75%	60%
13 Months	90%	55%

3.2 Mutual Information Cross Dependencies

One set of questions that arises with respect to the moving room experiment has to do with the dependency between the room position and the infant center of pressure. These questions are usually addressed using linear measures of cross correlation. Using half the data, five infants at each age, we have analyzed the cross dependencies employing a cross correlation analysis and compared that linear analysis with the results of two methods of calculating the mutual information across the two time series. Figure 7 presents the summary results from these three methods of calculating the mean dependency between the time series. The values within one curve in Fig. 7 can be compared to each other, whereas the values cannot be compared across different curves. What can be learned by these comparisons is how well each of the three methods measure the developmental change occuring during the span of ages covered by the experiment.

Note that each of these methods suggest that between the ages of 5 months and 7 months there is a significant change in the mean cross dependency of the room position and the infant's fore-aft center of pressure. During the interval that separates ages 9 and 13 months, none of the methods detect a significant change in the mean cross dependency. Over these two developmental intervals the three methods provide essentially the same results. However note the difference between the methods when comparing the values at 7 months and the values at 9 months. Due to the larger confidence intervals for the squared correlation, the mutual information methods provide a greater degree of discrimination of the two sample means during this critical developmental time immediately following the onset of self-supported sitting in the infants.

Figure 8 plots the Shannon mutual information of 240 trials of the moving room experiment. The mutual information is calculated for r_t and $x_{t+\tau}$ where r_t is the room position at time t and $x_{t+\tau}$ is the infant's fore-aft center of

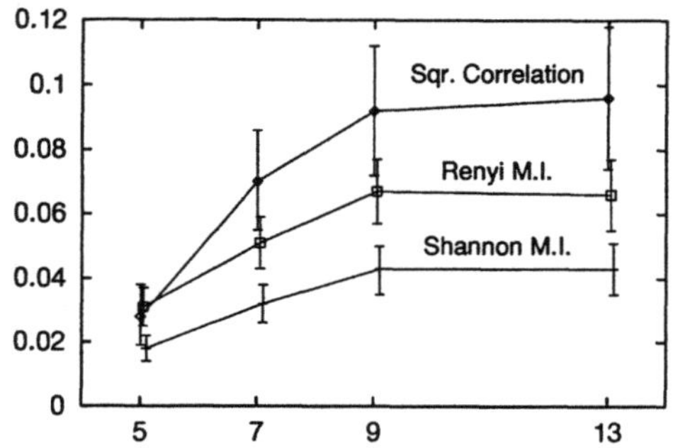

Fig. 7. Three mean cross dependency measures calculated over all experimental conditions. The three dependency measures are the mean over all time delays $-50 \leq \tau \leq 50$ between infant fore-aft center of pressure time series and room position time series. Error bars represent 99% confidence intervals for the associated statistic. Both Shannon and Rényi mutual information are expressed in bits. Note that absolute values of these statistics cannot be compared to each other. However, one can interpret the shape of the three curves with respect to each other.

pressure at time $t+\tau$. The horizontal axis below each graph shows the number of the time series and the horizontal axis label above each graph shows the age of the infant in months.

Overall there is a trend of more dependency with age. This effect is illustrated in Fig. 8A which plots the mean Shannon mutual information for each trial and aggregates and calculates a confidence interval for each age group These mean values and confidence intervals are the values shown in the Shannon M.I. curve in Fig. 7.

Figure 8B plots the value for Shannon mutual information as a grayscale value for each time delay value from $+50$ samples to -50 samples on the vertical axis for each time series on the horizontal axis. There is a horizontally oriented nearly white area at a lag of approximately $+25$ to $+35$ which suggests that there is a minimum of mutual information between the room at time t and the infant's center of pressure at time $t + 30$ over all experimental conditions. Since each sample is 20 ms long, this means that there is a minimum of mutual information between the room and the infant's center of pressure 600 ms later.

There are 12 time series for each infant (6 conditions $\times$ 2 trials). For each infant the experimental conditions are adjacent and always numbered in the same sequence, and the two trials for each condition are adjacent to each other. Thus we can see a vertical striation to Fig. 8B which is due to the ordered nature of the conditions for the time series for each infant. Some conditions cause more entrainment and therefore more dependency than others and therefore some vertical stripes are darker than others.

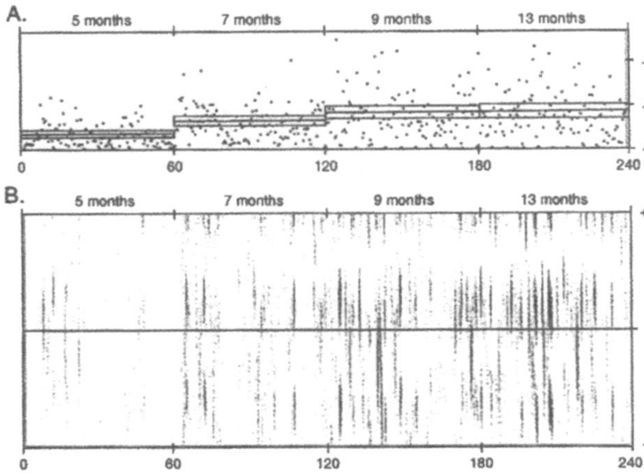

Fig. 8. Shannon mutual information of 240 trials of the moving room experiment. Age in months is shown at the top horizontal axis and time series trial number is shown on the bottom horizontal axis. Figure A plots the mean Shannon entropy in bits for each time series over the range of time delays +50 samples to −50 samples. The elongated rectangles in A represent the mean and 99% confidence intervals for the Shannon mutual information for each age group. Figure B plots the value for Shannon mutual information as a grayscale value for each time delay value from +50 samples to −50 samples on the vertical axis for each time series on the horizontal axis.

It appears that for some trials there is a maximum of mutual information between the room at time t and the infant at time $t+10$, or 200 ms later. Thus for these trials, the room appears to be predicting the infant's position with a lag of about 200 ms. On other trials it appears that there is a maximum of mutual information at a lag of -15. For these trials, the *infant* appears to be predicting the *room* position at a lag of about 300 ms. Practically, this means that the infant's center of pressure is anticipating the room's position in these trials.

3.3 False Nearest Neighbors Analysis

A false nearest neighbors analysis was performed in order to address two questions. First, is there an observable change in the required dimension of an embedding space for the center of pressure time series coincident with the onset of self-supported sitting? If such a dimensional change were observed, it would provide evidence that there may be a qualitative change in the postural control system at the time of the *qualitative* change in sitting behavior. Second, is there an observable developmental change in the amount of noise mixed with the postural control signal in the center of pressure time series? If this developmental change in noise mixture is observed, it would provide evidence that there is a *quantitative* change in the postural control system in which an existing mechanism improves its performance.

The false nearest neighbors analysis was applied to each infant's time series from all experimental conditions. The two trials from each infant were appended into a single time series, the algorithm was applied, and cases of neighbors including the boundary between the pair of trials were excluded from the pool of potential neighbors. Figure 9 plots the false nearest neighbor curves for every infant within each age category for the control stimulus condition.

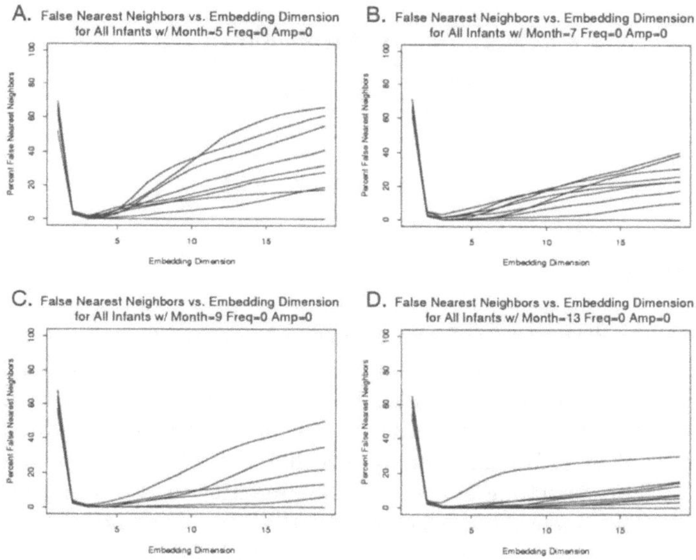

Fig. 9. False nearest neighbors curves for sitting infants for ages 5, 7, 9, and 13 months. Each line represents the False Nearest Neighbors curve for the data from two trials for one infant in the control condition.

Figures 9A, 9B, 9C and 9D are essentially identical for dimensions less than or equal to 3, falling from approximately 60% false nearest neighbors at embedding dimension 1 to approximately 0% false nearest neighbors at embedding dimension 3. Thus no developmental change in the required dimension of an embedding space for these time series is observed using this analysis. This does not rule out a qualitative change in the postural control system at the time of the onset of self supported sitting, but no evidence supporting such a qualitative change was observed.

On the other hand when embedding dimensions greater than 4 are considered, there is an obvious difference in the four graphs in Fig. 9. The slope of a regression line fit to the false nearest neighbors curve as the curve begins to ascend from 0% false nearest neighbors can be evidence for the amount of noise in one time series relative to other time series in the set. In Fig. 9, these slopes are visibly smaller as age increases. A regression line was fit to each of the infant's false nearest neighbors curves for embedding dimensions greater than 4 and the mean values for these slopes are presented in Table 3.

Table 3. Average false nearest neighbors slopes for each infant age group aggregated over all stimulus conditions.

	Age in Months			
	5	7	9	13
Slope	2.09	0.57	0.33	0.34

4 Discussion

There are three main results from the present nonlinear analyses of the moving room experiment. The first result is that the surrogate data analysis exhibits strong evidence that a linear stochastic model of the postural control of infants' sitting behavior is insufficient to capture the dynamical properties of the center of pressure time series. Another way of expressing this conclusion is that there is strong evidence of nonlinearity in the center of pressure time series. This includes the possibility that the dynamics of the control mechanism itself has a nonlinear component. This finding is not what would be predicted by Collins & De Luca (1994), whose analysis of adult center of pressure is derived from data gathered during quiet standing. Two explanations for this inconsistency come to mind: the postural control mechanism underlying infants' self-supported sitting is fundamentally different than the postural control mechanism for adults' quiet standing, or the method which Collins & De Luca used to test for nonlinearity was insufficiently sensitive to detect the difference between their real and surrogate data.

The second main result is that there is evidence for a significant developmental change in the cross dependency between the position of the moving room and the fore-aft center of pressure over the age ranges 5 to 7 months and 7 to 9 months. There is no apparent additional developmental change in the cross dependency over the age range of 9 to 13 months. This result confirms linear analyses performed by Bertenthal et al. (1996) who found a similar pattern of significantly increasing entrainment between infants' center of pressure and the position of the moving room over the age range of 5 to 9 months and no further increase in entrainment in the age range 9 to 13 months.

The third main result derives from the false nearest neighbors analysis of the sitting infants in the control condition, which found that there is a decrease in the slope of the false nearest neighbors curve over the age range of 7 to 9 months with no further decrease in the slope during the age range 9 to 13 months. One way of interpreting this finding is that the decrease in slope of the false nearest neighbors curves is a measure of a decrease in stochastic or high-dimensional noise in the infants' postural control mechanism. The surrogate data tests for nonlinearity in the same time series over the interval 5 to 13 months showed an increase in percentage of trials which rejected the null hypothesis of no nonlinearity. These two findings together suggest that within the center of pressure signal for sitting infants there may be an additive stochastic component mixed with a nonlinear postural control signal for the youngest infants, and as the infants develop, the stochastic component is reduced and is replaced by additional nonlinear dynamical control.

The false nearest neighbors analysis tested for, but did not find evidence for a change in the required minimum embedding dimension for the center of pressure time series. Thus no evidence was found for a qualitative change in postural control coincident with the onset of self-supported sitting. This may be due to the possibility that there is no qualitative change in the postural control mechanism at this critical developmental time, it may be due to insufficient sensitivity of the false nearest neighbors algorithm to dimensional changes in the center of pressure time series, or it may be that whatever qualitative shift occurs does not manifest itself in a change in the dimension of the postural control signal it generates. At the very least, we can report that there is not a large or obvious change in the dimensionality of the center of pressure signal coincident with self-supported sitting behavior.

There are a number of analyses which are planned for these data in order to further understand the development of the coupling of visual perception to postural control. A particularly promising analysis will explore the structure of the lagged nonlinear dependencies within and between the lateral, fore-aft and room signals.

The analyses presented in this chapter highlight the fact that one does not need extremely long time series in order to perform nonlinear analyses. If one is willing to make the assumption that measurements of individuals are representative of a general underlying process, then multiple time series can be used to examine developmental changes in that process. In general, we consider the application of nonlinear techniques to physiological time series to be an active and promising area of research.

References

Bayley, N. (1969): *Manual for the Bayley Scales of Infant Development.* Psychological Corporation, New York.

Bertenthal, B. I., Rose, J. L., Bai, D. L. (1996): Perception-action coupling in the development of the visual control of posture. Journal of Experimental Psychology: Human Perception and Performance (in press).

Bertenthal, B. (1990): Application of biomechanical principles to the study of perception and action. In Bloch, H., Bertenthal, B. I., *Sensory-Motor Organization and Development in Infancy and Early Childhood.* Kluwer, Dordrecht.

Collins, J. J., De Luca, C. J. (1993): Open-loop and closed-loop control of posture: A random-walk analysis of center-of-pressure trajectories. Experimental Brain Research **95**, 308–318.

Collins, J. J., De Luca, C. J. (1994): Random walking during quiet standing. Phys. Rev. Lett. **73**, 764–767.

Collins, J. J. De Luca, C. J. (1995): Upright, correlated random walks: A statistical-biomechanics approach to the human postural control system. Chaos **5**, 57–63.

Efron, B. (1979a): Bootstrap methods: Another look at the jackknife. Ann. Stat. **7**, 1–26.

Efron, B. (1979b): Computers and the theory of statistics: Thinking the unthinkable. SIAM Review **21**, 460–480.

Fraser, A., Swinney, H. (1986): Independent coordinates for strange attractors from mutual information. Phys. Rev. A **33**, 1134–1140.

Horn, J. L. (1965): Fluid and crystallized intelligence. Ph.D. thesis, University of Illinois, Urbana. Unpublished.

Howard, I. P. (1986): The perception of posture, self motion, and the visual vertical. In Boff, K. R., Caufman, L., Thomas, J. P.: *Handbook of Perception and Human Performance: Vol. 1. Sensory Processes and Perception.* Wiley, New York. pp. 1–62.

Kennel, M. B., Brown, R., Abarbanel, H. D. I. (1992): Determining embedding dimension for phase-space reconstruction using a geometrical construction. Phys. Rev. A **45**, 3403–3411.

Kennel, M. B., Isabelle, S. (1992): Method to distinguish possible chaos from colored noise and to determine embedding parameters. Phys. Rev. A **46**, 3111–3118.

Lishman, J. R., Lee, D. N. (1973): The autonomy of visual kinaesthesis. Perception **2**.

Osborne, A. R., Kirwin, A. D., Provenzale, A., Bergamansco, L. (1986): A search for chaotic behavior in large and mesoscale motions in the pacific ocean. Physica D **23**, 75.

Packard, N. H., Crutchfield, J. P., Farmer, J. D., Shaw, R. S. (1980): Geometry from a Time Series. Phys. Rev. Lett. **45**, 712–716.

Paluš, M. (1995): Detecting nonlinearity in multivariate time series. 95–07–059, Santa Fe Institute.

Pompe, B. (1993): Measuring statistical dependences in a time series. J. Stat. Phys. **73**, 587.

Pompe, B. (1996): Ranking and entropy estimation in nonlinear time series snalysis. In Kantz, H., Kurths, J., Mayer-Kress, G., *Nonlinear Techniques in Physiological Times Series Analysis.* Springer, Berlin, Heidelberg.

Prichard, D., Theiler, J. (1994): Generating surrogate data for time series with several simultaneously measured variables. Phys. Rev. Lett. **73**, 951–954.

Resnikoff, H. L. (1989): *The Illusion of Reality*. Springer, Berlin, Heidelberg.

Sauer, T., Yorke, J., Casdagli, M. (1991): Embedology. J. Stat. Phys. **65**, 95–116.

Schiff, S. J., Chang, T. (1992): Differentiation of linearly correlated noise from chaos in a biologic system using surrogate data.. Biol. Cybern. **67**, 387–93.

Schreiber, T., Schmitz, A. (1996): Improved surrogate data for nonlinearity tests. Phys. Rev. Lett. **77**, 635.

Shannon, C. E., Weaver, W. (1949): *The Mathematical Theory of Communication.* The University of Illinois Press, Urbana.

Takens, F. (1981): Detecting Strange Attractors in Turbulence. In Rand, D., Young, L.-S.: *Dynamical systems and turbulence, Warwick 1980: Proceedings of a symposium held at the University of Warwick.* Springer-Verlag, Berlin. pp. 366–381.

Theiler, J., Eubank, S., Longtin, A., Galdrikian, B. (1992): Testing for nonlinearity in time series: the method of surrogate data. Physica D **58**, 77–94.

Whitney, H. (1936): Differentiable Manifolds. Ann. Math. **37**, 645–680.

Symbolic Dynamics of Bimanual Production of Polyrhythms

R. Engbert[1,3], C. Scheffczyk[1,3], R.T. Krampe[2,3], J. Kurths[1,3], and R. Kliegl[2,3]

[1] Institut für Physik,
[2] Institut für Psychologie,
[3] Innovationskolleg "Formale Modelle kognitiver Komplexität", Universität Potsdam, 14415 Potsdam, Germany

Abstract. We analyse time series from a study on bimanual rhythmic movements in which the speed of performance (the external control parameter) was experimentally manipulated. Using symbolic transformations as a visualization technique we observe qualitative changes in the dynamics of the timing patterns. Such phase transitions are quantitatively described by measures of complexity. Using these results we develop an advanced symbolic coding which enables us to detect important dynamical structures. Furthermore, our analysis raises new questions concerning the modelling of the underlying human cognitive-motor system.

1 Introduction

Living systems generally consist of components which are coupled to each other by complicated mechanisms. Both the sub-systems as well as their interactions are mostly nonlinear. Therefore, biological, physiological, or psychological systems can exhibit a wealth of complex behaviour: They possess the "seeds" of deterministic chaos (May 1976). As a consequence of this complexity, data analysts have to deal with the challenging problems of characterization or prediction of these systems (Casdagli et al. 1992). Due to the occurrence of environmental noise or intrinsic fluctuations the detection of underlying deterministic laws of the dynamics becomes even more difficult (Drepper et al. 1994, Engbert and Drepper 1994). Fortunately, the application of new concepts developed in the context of dynamical systems theory has emerged into a promising approach (Haken 1988, Kelso 1995). Using the concepts of complexity (Hao 1991, Kurths et al. 1994, Schwarz et al. 1994, Witt et al. 1994), it may be possible to tackle this problem for a wide class of systems, like physiological (Kurths et al. 1995, Schiek et al., this volume) and cognitive processes (Kelso 1995, Engbert et al. 1997b).

Compared to the physical sciences, typical systems in cognitive psychology cause additional problems, because their key observables are more restrictively prescribed by the experimental procedure and their functional state can be controlled to a much lower extent (Kelso et al. 1993).

In cognitive psychology bimanual coordination is used as a paradigm for the study of movement control. Numerous studies have shown that bimanual coordination is subject to strong performance constraints (Kelso et al. 1979, Swinnen et al. 1988, Keele and Ivry 1991, Treffner and Turvey 1993, Peper et al. 1995). Experimental manipulation of external parameters, e.g. variation of the required speed of performance, permits the systematic study of how the human cognitive system adapts to these external and its own internal constraints.

In this paper we present a symbolic dynamics coding as a technique of nonlinear time series analysis using pilot data from an experimental study on bimanual production of polyrhythms (Krampe et al. 1996). In the experiments subjects had to perform the polyrhythm illustrated in Fig. 1 at different, experimentally controlled tempos. Polyrhythms are especially interesting to investigate because of their conflicting phase relationships between the two hands. The use of performance tempo as an external control parameter in the investigated study enables us to analyse transitions between qualitatively different dynamical regimes. We combine a qualitative description of behavioural dynamics with a quantification of the observed complexity. Using our methods this can be achieved even at the level of individuals.

In the standard approach to human movement timing one has analysed covariance structures among the produced time intervals (Summers et al. 1993, Vorberg and Wing 1996, Krampe et al. 1996). Related methods rest on strong statistical assumptions which are often violated and also require data aggregation to a degree that precludes investigation of interesting qualitative phenomena on the basis of individual performance.

2 Experiments

The 3:4 polyrhythmic task (Fig. 1) was performed on an electronic piano with a weighted keyboard mechanic hooked to a computer which monitored the experiment and recorded time-stamped data with a resolution of 1 ms. Fourteen different metronome tempos ranging from 800 ms per cycle to 8200 ms per cycle were presented in a randomized order. The data reported in this paper came from well-trained amateur pianists (for details cf. Krampe et al. 1996). All subjects tested were right-handed.

In each trial, subjects listened to the exact rhythm generated by the computer as long as they wanted, and then played along (synchronize) with the beat for four cycles after which the computer beat stopped. Participants then had to continue for another 12 cycles during which the time series were recorded. Hence, a single time series consists of 12 bars or cycles. The recorded data are the intervals between successive keystrokes produced by both hands (Fig. 1),

$$\text{left hand (36 values):} \qquad L_1^1,\ L_2^1,\ L_3^1,\ L_1^2,\ \ldots \quad L_2^{12},\ L_3^{12}\,, \tag{1}$$

$$\text{right hand (48 values):} \qquad R_1^1,\ R_2^1,\ R_3^1,\ R_4^1,\ R_1^2,\ \ldots\ R_3^{12},\ R_4^{12}\,, \tag{2}$$

$$\text{combined (72 values):} \qquad I_1^1,\ I_2^1,\ I_3^1,\ I_4^1,\ I_5^1,\ \ldots \quad I_5^{12},\ I_6^{12}\,. \tag{3}$$

3 Data Analysis

Our strategy is to use a coarse-graining of the data in order to explore important structures of the underlying dynamics. This is implemented via a transformation of the recorded time series into symbols. By this procedure a considerable amount of information is discarded, but nevertheless characteristic properties of the dynamics underlying the studied system can be captured by the symbol sequence (Hao 1991, Wackerbauer et al. 1994).

We use symbolic codings in two different contexts. Firstly, we apply a symbolic dynamics as a visualization technique for the detection of qualitative transitions as a function of the speed of performance. Using our symbolic transformation the transition between different regimes of behaviour is transformed into a disorder-order transition in the symbol patterns. This is described quantitatively by different measures of complexity.

Secondly, based on these results we develop an applied symbolic transformation which is highly specific to our problem. As opposed to the more general coding scheme it is more easily interpretable and enables us to give a condensed description of the data. Furthermore, it can be used for a classification of the subjects tested.

3.1 Basic Symbolic Dynamics

Let us denote the realized duration of the kth cycle ($k = 1, 2, 3, ..., 12$) by t^k, defined as the sum of the sub–intervals of the corresponding hand (t_L^k resp. t_R^k). Since we are not mainly interested in fluctuations and trends of the series of realized cycle durations, the corresponding information is discarded. We, therefore, define the relative deviations

$$l_i^k = \frac{3L_i^k - t_L^k}{t_L^k} \quad (i = 1, 2, 3)\,, \qquad r_j^k = \frac{4R_j^k - t_R^k}{t_R^k} \quad (j = 1, 2, 3, 4)\,. \tag{4}$$

These are the deviations from the prescribed rhythm regardless of the accuracy in overall tempo. The motivation for this transformation is as follows: If a subject holds the prescribed tempo within acceptable tolerance, then the relative deviations (4) quantify the precision with which the rhythmic structure is performed. It must be kept in mind, however, that this is not an absolute measure of performance accuracy.

We now further reduce the amount of data by a transformation into symbol sequences. This simplifies our investigation to the analysis of the symbol patterns. In the following we use only two symbols ('0' and '1'). Let us consider the transformed left hand time series (4) as an example. To each value of the relative deviation l_i^k ($i = 1, 2, 3$; $k = 1, 2, ..., 12$) we assign a symbol s_n in the following way:

$$s_n = \begin{cases} 0 & \text{if } l_i^k < 0 \\ 1 & \text{otherwise} \end{cases}, \tag{5}$$

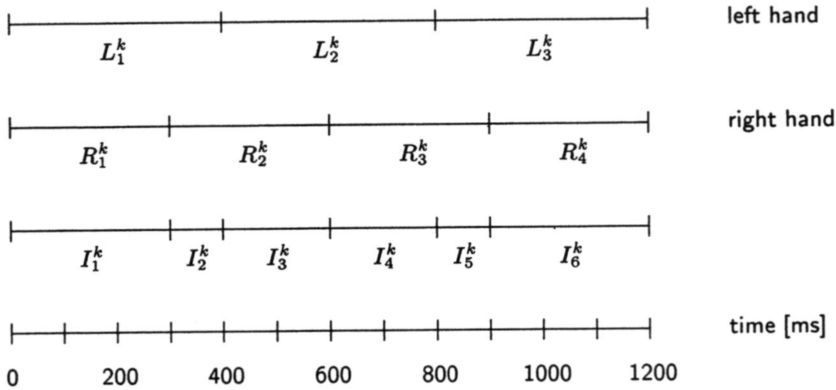

Fig. 1. Schematic illustration of the 3:4 polyrhythmic task used in the study by Krampe et al. (1996); here for a cycle duration of 1200 ms. "R" and "L" in the top panels denote the intervals produced by right and left index fingers, respectively. Each cycle starts with simultaneous strokes of the two hands. Three isochronous intervals, i.e. equidistant strokes in time, in the left hand are performed against four isochronous intervals in the right hand within each cycle. The position of intervals within a certain cycle k is indicated by sub-indices. The two time series can be combined in the series of intervals I_j^k bounded by subsequent strokes.

where $n = 3(k-1)+i = 1, 2, 3, ..., 36$. This coding scheme is called static, since we use a fixed threshold in the conditional part. The symbolic description can be refined more and more by introducing more symbols. The appropriate number of symbols is practically limited by the length of the times series from which the symbol sequence is derived, because the statistical confidence level of the occurrence of the symbols drops down with an increasing number of possible symbols. Furthermore, plots with many different symbols (e.g. more than 5) are much more difficult to interprete.

Besides the fact that transformations into symbol sequences are a basis for measuring complexity, they can already be used as a visualization technique (Fig. 2). Each type of order in Fig. 2a–d indicates a deviation from the prescribed rhythm. Near-perfect performance in this context would yield a completely random pattern, since the relative deviations (4) would be randomly negative or positive with a small absolute value. Using this rather simple technique, we find several transition phenomena in dependence on the cycle duration T. Note especially the one between $T = 1400$ ms and 2000 ms in Fig. 2a at trial number 89. This is an example of a rather sharp transition visible in both hands. As can be seen in Fig. 2, this transition does not always occur in both hands at the same cycle duration. This demonstrates a complex interplay of the hands. The fact that the transitions do not always occur in the fast range of tempos is very important. It implies that the transitions are a consequence of nonlinearity in the human movement control system, instead of a result of "simple" biomechanical constraints. Next, we discuss how these transitions can be quantitatively described using measures of complexity.

3.2 Measures of Complexity

The symbol sequence corresponding to a certain time series consists of 36 (left hand) or 48 (right hand) elements. Due to the fact that the basic rhythmic structure is a cycle, it is appropriate to subdivide the time series into substrings or 'words' of 3 (left hand) or 4 (right hand) symbols and to study the occurrence of these words. Subject A (Fig. 2a) uses almost exclusively the right hand word '0110' for bar durations $T \lesssim 2000$ ms, i.e. the first and last interval of each bar is too short, whereas the two other intervals are too long. From the definition of the relative deviations (4) it is clear that

$$\sum_{i=1}^{3} l_i^k = \sum_{j=1}^{4} r_j^k = 0 \quad (k = 1, 2, ..., 12) ,$$

implying that words consisting entirely of '0's or '1's are impossible. Therefore, we retain $N_w^L = 2^3 - 2 = 6$ words for the left and $N_w^R = 2^4 - 2 = 14$ possible words for the right hand. The relative frequency $p_i = N_i/N_w$ of word i is calculated using all cycles generated during several trials at a certain tempo. The corresponding relative frequency distribution is denoted by p.

As an example, we now analyse such a transition in more detail: The right hand symbol plot of subject A (Fig. 2a), which shows a clear-cut transition. The frequency distribution p (Fig. 3a) exhibits a change with the tempo which is in good agreement with the symbol plot. The fact that the subject uses almost exclusively the word '0110' in the right hand leads to a strongly peaked probability distribution (Fig. 3b).

To distinguish different kinds of probability densities (Beck and Schlögl 1993), we first calculate the well-known Shannon entropy (Shannon and Weaver 1948) of the distribution p,

$$S(p) = c \sum_{i=1}^{N_w} p_i \ln p_i , \tag{6}$$

here normalized with respect to the number of all possible words N_w using $c = 1/\ln N_w$. The qualitative change in the symbol pattern is nicely described by an increase of the Shannon entropy (Fig. 3c). The sharp transition in Fig. 3a is reflected in a sharp transition in the Shannon entropy in Fig. 3c. Thus, this measure characterizes different kinds of transitions. Applying algorithmic complexity (Wackerbauer et al. 1994) to the symbol sequences leads to comparable results.

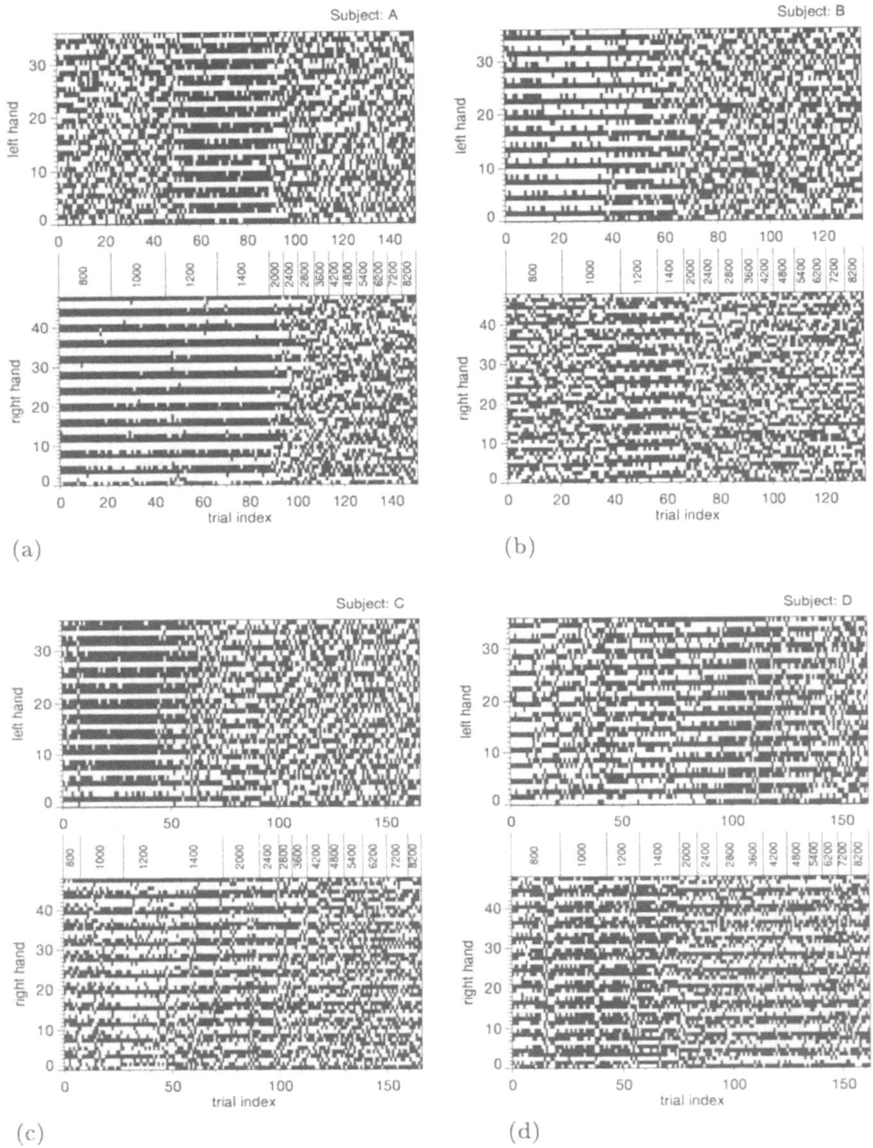

Fig. 2. Symbol sequences of all trials of four individuals (A,B,C,D). The time is increasing on the ordinate, where 36 symbols are plotted for left hand time series and 48 symbols for the right hand. There are several trials for each tempo, as indicated on the abscissa. The bar durations T are indicated by the vertical labels between the plots in milliseconds (ms).

For a direct comparison of probability distributions we apply a χ^2 test (Fig. 3d), which indicates the transition point, i.e. the critical value of the control parameter. It is important to note that traditional measures for fluctuations or accuracy, e.g. covariances, show a much smoother transition which cannot be identified reliably at the level of individuals. The significance of our results has been tested by analysing computer-generated random patterns of the same data length.

3.3 Applied Symbolic Codings

The methods described above were motivated by the general strategy of the application of symbolic dynamics: Theory suggests (Hao 1991) and many examples show (Schwarz et al. 1994, Witt et al. 1994, Kurths et al. 1995, Engbert et al. 1997b) that robust properties of the dynamics can be extracted using a coarse-graining of the time series data.

In our analysis this promise has been kept by the detection of qualitatively different dynamical regimes. A more detailed interpretation of the symbol patterns (Fig. 2) is very complicated. Therefore, we suggest an applied symbolic coding from which the dynamical interpretation can be directly read off.

For a this new symbolic coding we exploit the dominant periodicity or time scale defined by the rhythm task. This idea is also used by (Schiek et al., this volume) in the case of data from paced respiration. The information of a complete cycle is compressed into only one symbol, but as opposed to the first symbolic transformation (5) we use more than two different symbols. Each symbol represents a test for a certain type of dynamics. Let us start with the description of the different regimes of behaviour that we are interested in.

The first criterion which is checked for is what we call 'poly' timing. In the 3:4 polyrhythm (Fig. 1) intervals of three different lengths have to be produced in each cycle k: I_1^k and I_6^k are 1/4 of a cycle, I_3^k and I_4^k are 1/6 of a cycle, and I_2^k and I_5^k represent 1/12 of a cycle. Therefore, a necessary condition for correct timing requires the correct ordering,

$$\text{poly:} \qquad I_1^k, I_6^k > I_3^k, I_4^k > I_2^k, I_5^k \tag{7}$$

within each cycle (Fig. 4a).

In contrast to the ideal polyrhythm we often observe a tendency to relax the rhythmic structure in such a way that all intervals are shifted towards 1/6 of a cycle (Fig. 4b). The corresponding criterion is defined as

$$\text{iso:} \qquad |\tilde{I}_j^k - \frac{1}{6}| < \epsilon_1 \qquad \text{for all } j\ , \tag{8}$$

where $\tilde{I}_j^k$ is the fraction of the cycle duration, i.e.

$$\tilde{I}_j^k = \frac{I_j^k}{t^k}\ , \qquad t^k = \sum_{j=1}^{6} I_j^k\ . \tag{9}$$

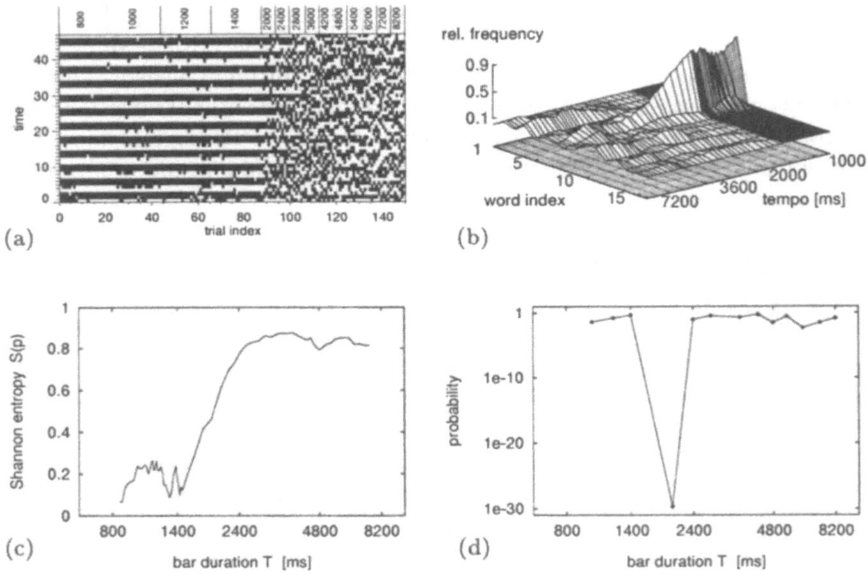

Fig. 3. Frequency distribution (a), statistic of words (b), Shannon entropy (c), and χ^2 probability (d) calculated from the right hand symbol plot of subject A (Fig. 2a).

Another dynamically interesting deviation from ideal timing is the 'locked' mode, where the two shortest intervals I_2^k and I_5^k nearly vanish (Fig. 4c),

$$\text{locked:} \qquad \tilde{I}_2^k,\ \tilde{I}_5^k < \epsilon_2\ . \tag{10}$$

Figure 5 summarizes the corresponding symbol patterns for the same subjects as in Fig. 2. For all plots we have used parameter values $\epsilon_1 = 0.06$ and $\epsilon_2 = 0.075$. It turns out that the results do not sensitively depend on these values. The different criteria (7)–(10) are plotted as grey scales. Black symbols are cycles, where the 'poly' criterion (7) is violated. If this violation is caused by a deviation from the ideal timing in the form of the 'iso' mode (8), then we plot a dark-grey symbol. The 'locked' mode (10) which is allowed in the sense of the 'poly' criterion (7) is observed by a light-grey symbol.

The occurrence of the 'iso' and 'locked' modes have important dynamical interpretations. The production of polyrhythms requires (i) the correct sequence of finger tappings and (ii) the correct timing (7). In most cases the 'iso' mode is observed for the fast tempos ($\lesssim$ 2000 ms cycle duration). Due to the pressure exerted by the fast tempo of performance the ability of timing breaks down (Fig. 5).

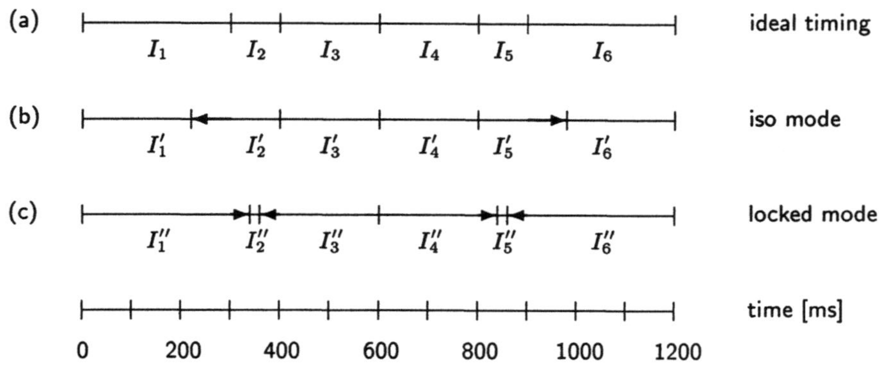

Fig. 4. The ideal 3:4 polyrhythmic timing (a) compared to two important modes of behaviour observed in the experiments. In the 'iso' timing all intervals are shifted towards 1/6 of a cycle (b). The 'locked' mode is a shortening of the interval I_2 and I_5. The strokes which bound these intervals occur nearly simultaneously (c).

The 'locked' mode is typically detected in an intermediate range of tempos ($\gtrsim$ 2000 ms cycle duration). All subject certainly recognize that the intervals I_2^k and I_5^k are the two shortest of the cycle. However, if the mental representation is over-pronounced, then this will yield a deviation from the ideal timing pattern in the form of the 'locked' mode. The fact that this happens in most cases in an intermediate range of tempos is even more interesting, because of the absence of "obvious" no biomechanical or motor constraints in this range of tempos. Therefore, the occurrence of the 'locked' mode is a consequence of nonlinearity in the control system of the brain.

For the problem of characterization of the subjects it is important that the violation of the polyrhythmic timing criterion (7) suggests the definition of a critical tempo for the breakdown of correct timing. Preliminary results on a larger study indicate that the critical tempo is an important parameter which characterizes the performance of the subjects.

4 Discussion

The analysis of natural systems using measures of complexity (Schwarz et al. 1994, Witt et al. 1994, Kurths et al. 1995, Schiek et al., this volume) is a powerful approach. Analysing the dynamics of cognitive processes (Engbert et al. 1996) as an example of such a complex system, we have demonstrated that the application of symbolic codings provides new insights into the cognitive–motor system. In particular, we have found complex transitions in the dynamics of coordination (Kelso 1995, Haken 1996). The existence of such phase

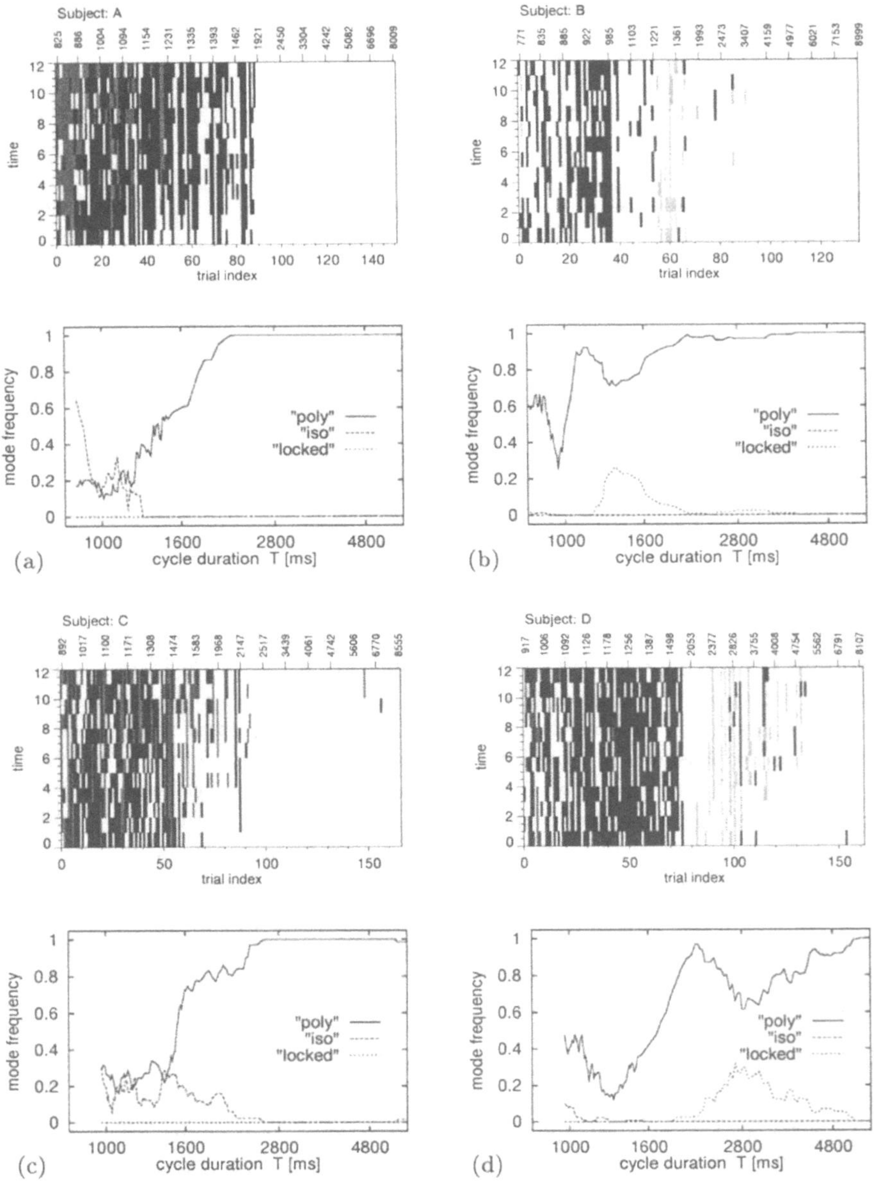

Fig. 5. Symbol sequences of all trials of the same individuals as in Fig. 2 obtained by the applied symbolic coding, where each symbol corresponds to a complete cycle. The bottom panels under the symbol patterns contain the relative frequencies of the symbols. For clarity these curves have been smoothed by a moving average of a width of 10 trials.

transitions in the dynamics of hand movements lends support to nonlinear models of human behaviour (Haken et al. 1985, Engbert et al. 1997a, Scheffczyk et al. 1997), since these transitions cannot occur in linear systems. In this sense our results encourage the comparison of experimental data with nonlinear models on the basis of symbolic codings.

Our approach works even for very short and noisy time series. This is contrasted by standard procedures of data analysis of movement experiments where data aggregation over several individuals is typical. Using symbolic dynamics we are able to describe qualitative changes of the dynamics on the level of individuals.

The applied symbolic transformation (Sec. 3.3) shows that it is possible to extract specific dynamical information from the time series. The characterization of different modes of behaviour may be a unifying concept for the analysis of movement data, since it provides a coarse-grained, qualitative description of the performance of individuals.

Our results are preliminary in the sense that a statistical verification of our findings on a larger group of subjects has to be done. A corresponding experimental study is in progress.

References

Beck C., Schlögl F. (1993): *Thermodynamics of Chaotic Systems.* Cambridge University Press, Cambridge, UK

Casdagli M., Des Jardins D., Eubank S., Farmer J.D., Gibson J., Theiler J., Hunter N. (1992): Nonlinear modeling of chaotic time series: theory and applications. In *Applied Chaos*, J.H. Kim and J. Stringer (Eds). John Wiley, New York

Drepper F.R., Engbert R., Stollenwerk N. (1994): Nonlinear time series analysis of empirical population dynamics. Ecological Modeling **75/76**, 171–181

Engbert R., Drepper F.R. (1994): Chance and chaos in population biology – models of recurrent epidemics and food chain dynamics. Chaos, Solitons & Fractals **4**, 1147-1169

Engbert R., Scheffczyk C., Kurths J., Krampe R.T., Kliegl R. (1996): Nonlinear phenomena in polyrhythmic hand movements. In *Proceedings of the 3rd Experimental Chaos Conference*, R.G. Harrison, W. Lu, W. Ditto, L. Pecora, M. Spano, S. Vohra (Eds). World Scientific, Singapore

Engbert R., Scheffczyk C., Krampe R.T., Rosenblum M., Kurths J., Kliegl R. (1997a): Tempo–Induced Transitions in Polyrhythmic Hand Movements. Physical Review E **56**, 5823–33.

Engbert R., Schiek M., Scheffczyk C., Kurths J., Krampe R., Kliegl R., Drepper F. (1997b): Symbolic dynamics of physiological synchronization: examples from bimanual movements and cardiorespiratory synchronization. Nonlinear Analysis **30**, (in press)

Haken H., Kelso J.A.S., Bunz H. (1985): A theoretical model of phase transitions in human hand movements. Biological Cybernetics **51**, 347-356

Haken H. (1988): *Information and Self-Organization.* Springer, Berlin, Heidelberg

Haken H. (1996): *Principles of Brain Functioning*. Springer, Berlin, Heidelberg
Hao B.-L. (1991): Symbolic dynamics and characterization of complexity. Physica D **51**, 161; and refs. therein
Keele S.W., Ivry R. (1991): Does the cerebellum provide a common computation for diverse tasks? Ann. N.Y. Acad. Sci. **608**, 179
Kelso J.A.S., Southard D.L., Goodman D. (1979): On the nature of human interlimb coordination. J. Exp. Psychol.: Human Perception Performance **5**, 229
Kelso J.A.S., Ding M., Schöner G. (1993): Dynamic pattern formation: A primer. In *A Dynamic Systems Approach to Development*, L.B. Smith and E. Thelen (Eds). MIT Press, Cambridge, MA
Kelso J.A.S. (1995): *Dynamic Patterns*. MIT Press, Cambridge, MA
Krampe R.T., Kliegl R., Mayr U., Engbert R. (1996): The fast and the slow of skilled bimanual movement timing: Parallel and integrated control. (submitted)
Kurths J., Witt A., Atmanspacher H., Feudel F., Scheingraber H., Wackerbauer R. (1994): General remarks on complexity. In *Inside Versus Outside*, H. Atmanspacher and G. Dalenoort (Eds). Springer, Berlin, Heidelberg
Kurths J., Voss A., Saparin P., Witt A., Kleiner H.J., Wessel N. (1995): Quantitative analysis of heart rate variability. Chaos **5**, 88–94
May R. (1976): Simple mathematical models with very complicated dynamics. Nature **261**, 459–67
Peper C.E., Beek P.J., van Wieringen P.C.W. (1995): Frequency-induced phase transitions in bimanual tapping. Biol. Cybern. **73**, 301-309
Scheffczyk C., Zaikin A., Rosenblum M., Engbert R., Krampe R., Kurths J. (1997): Modelling qualitative changes in bimanual movements. Intern. J. Bifurcation and Chaos **7**, (in press)
Schwarz U., Kurths J., Witt A., Benz A.O. (1994): Analysis of solar spike events by means of symbolic dynamics methods. Astron. & Astrophys. **277**, 215–224
Shannon C., Weaver W. (1948): *The Mathematical Theory of Communication*. University of Illinois Press, Urbana
Summers J.J., Rosenbaum D.A., Burns B.D., Ford S.K. (1993): Production of polyrhythms. J. Exp. Psychol.: Human Perception Performance **19**, 416
Swinnen S.P., Walter C.B. (1988): In *Cognition and Action in Skilled Behavior*. North-Holland, Amsterdam
Treffner P.J., Turvey M.T. (1993): Resonance constraints on Rhythmic Movement. J. Exp. Psychol.: Human Perception and Performance **19**, 1221-1237
Vorberg D., Wing A.M. (1996): In *Handbook of perception and action, Vol. 3*, H. Heuer and S.W. Keele (Eds). Academic Press, London
Wackerbauer R., Witt A., Atmanspacher H., Kurths J., Scheingraber H. (1994): A comparative classification of complexity measures. Chaos, Solitons & Fractals **4**, 133; and refs. therein
Witt A., Kurths J., Krause F., Fischer K. (1994): On the validity of a model for the reversals of the Earth's magnetic field. Geophys. Astrophys. Fluid Dynam. **77**, 79–91

Human Postural Control: Force Plate Experiments and Modelling

Michael Rosenblum[1,2], Georgy Firsov[2], Robert Kuuz[3], and Bernd Pompe[4]

[1] Max-Planck-Arbeitsgruppe "Nichtlineare Dynamik" an der Universität Potsdam, Am Neuen Palais 19, PF 601553, D-14415 Potsdam, Germany
[2] Mechanical Engineering Research Institute, Russian Academy of Sciences, 101830 Moscow, Russia
[3] Clinic for Nervous Diseases, I. M. Sechenov Moscow Medical Academy, Rossolimo St. 11, Moscow 119021, Russia
[4] Institut für Physik, Ernst-Moritz-Arndt-Universität Greifswald, Friedrich-Ludwig-Jahn-Str. 16, D-17487 Greifswald, Germany

Abstract. We report the results of time series analysis of human body sway while quiet upright stance. The bivariate records (stabilograms) are measured by means of a force plate. To investigate interrelations between oscillations in anterior–posterior and lateral directions we use several techniques: cross-spectrum analysis, generalized mutual information, and calculation of instantaneous relative phase. We find that the stabilograms can be qualitatively rated into two groups: noisy and oscillatory patterns. Further, we show that oscillatory patterns may demonstrate phase locking. We argue that these patterns are due to stochastic and chaotic dynamics, respectively. We discuss the plausible strategy of postural control and present the model that qualitatively describes transitions from noisy to oscillatory patterns and phase synchronization. The relevance of the results of the time series analysis for the diagnostics of neurological pathologies is discussed.

1 Introduction

An important problem in modern neurology is the development of methods for differential diagnostics of various pathologies of the central nervous system (CNS), both of organic and of psychogenic origin. A manifestation of these pathologies are disturbances of posture and locomotion control. On the one hand, quantitative studies of motor control, and equilibrium maintenance in particular, may provide useful diagnostic information on the functional state of the CNS (Gurfinkel et al. 1965; Terekhov 1976; Cernacek 1980; Furman 1994; Lipp and Longridge 1994). On the other hand, investigations of the strategy of the posture control in humans is very interesting from a dynamical standpoint as an example of control in the multi-degree-of-freedom system.

The force plate technique, also known as stabilography or posturography, has been extensively used since decades (Baron 1983) for the analysis of upright postural control and the evaluation of functional states of the human organism. During the tests small sways of the human body in anterior–posterior and lateral directions are measured simultaneously. These records, called stabilograms, are usually analyzed by means of different statistical techniques, spectral and correlation analysis (Terekhov 1976; Rosenblum and Firsov 1992b; Collins and DeLuca 1994).

In the present paper we describe the results of our force plate investigations of healthy subjects and patients with different neurological pathologies. In our study we concentrate on the joint analysis of both components of the stabilograms. For this purpose we use the standard cross-spectrum analysis technique, the generalized mutual information (Pompe 1993), and the recently found effect of ***phase synchronization*** of coupled self-sustained chaotic oscillators (Rosenblum et al. 1996; Pikovsky et al. 1996). After the discussion of the plausible strategy of the neural regulation of posture we introduce a model of two–dimensional dynamics of the center of gravity of the body.

2 Experiments

The experiments were accomplished in the Clinic for Nervous Diseases of the Moscow Medical Academy using a standard rigid force plate with four tensoelectric transducers. The output of the setup provides current coordinates (x, y) of the center of pressure under the feet of the standing subject. These coordinates are close to that of the center of gravity of the human body. In the following we denote the deviation of the center of pressure in anterior–posterior and lateral direction as x and y, respectively. Every subject was asked to perform three tests of quiet standing with:

EO — eyes opened and stationary visual surrounding,
EC — eyes closed,
AF — eyes opened and additional video–feedback.

In the AF test the current position of the center of pressure was indicated by a light dot on the screen of an oscilloscope. The subject was instructed to watch the screen and to keep the dot within a circle in its center. The AF test can be considered as a simply realized non-invasive method to change the dynamics of the system in order to extract additional information about it. In all tests the plate was fixed, i.e. no artificial mechanical perturbations of the upright posture were used.

The postural sways have been sampled with a frequency of 25 Hz. Every record contains two channels - each of 4096 points ($\approx$ 160 s). About 150 stabilograms have been obtained testing healthy volunteers and neurological patients. By visual inspection we have rejected some trials, for instance that where the subject moved during the test, or where he or she was not able to stay for three minutes. The later turned out to be rather difficult for patients with a neurological pathology. Thus we got 132 bivariate records which can be considered as free of artifacts. A survey of the data is given in Table1.

Table 1. The subjects under study

#	subject	sex	age	state	group	tests
1 – 3	39	f	23	healthy	1	EO, EC, AF
4 – 6	32	f	23	healthy	1	EO, EC, AF
7 – 9	33	f	25	healthy	1	EO, EC, AF
10 – 12	35	f	26	healthy	1	EO, EC, AF
13 – 15	40	f	27	healthy	1	EO, EC, AF
16 – 18	2	f	31	healthy	1	EO, EC, AF
19 – 21	34	f	32	healthy	1	EO, EC, AF
22 – 24	42	f	36	healthy	1	EO, EC, AF
25 – 27	3	f	37	healthy	1	EO, EC, AF
28 – 30	4	f	40	healthy	1	EO, EC, AF
31 – 33	43	m	22	healthy	1	EO, EC, AF
34 – 36	36	m	25	healthy	1	EO, EC, AF
37 – 39	38	m	21	healthy	1	EO, EC, AF
40 – 42	41	m	22	healthy	1	EO, EC, AF
43 – 45	5	m	27	healthy	1	EO, EC, AF
46 – 48	37	m	27	healthy	1	EO, EC, AF
49 – 51	1	m	29	healthy	1	EO, EC, AF
52 – 54	7	f	26	multiple sclerosis	2	EO, EC, AF
55 – 57	6	f	30	multiple sclerosis	2	EO, EC, AF
58 – 60	30	f	30	multiple sclerosis	2	EO, EC, AF
61 – 63	9	f	42	Parkinson disease	2	EO, EC, AF
64 – 66	10	m	53	Parkinson disease	2	EO, EC, AF
67 – 68	11	m	53	Parkinson disease	2	EO, EC
69 – 71	8	f	62	brain tumor	2	EO, EC, AF
72 – 74	44	m	44	atrophy of cerebellum	2	EO, EC, AF
75 – 77	45	m	54	atrophy of cerebellum	2	EO, EC, AF
78 – 80	46	f	56	discircular encephalopathy	2	EO, EC, AF
81 – 82	47	m	58	discircular encephalopathy	2	EO, EC
83 – 85	12	f	36	neurathenia	3	EO, EC, AF
86 – 88	28	f	42	neurathenia	3	EO, EC, AF
89 – 91	15	f	44	neurathenia	3	EO, EC, AF
92 – 94	27	m	52	hysterical hemiparesis	3	EO, EC, AF
95 – 96	14	f	15	hysterical hemiparesis	3	EO, EC
97 – 99	22	f	20	neurotic disorder	3	EO, EC, AF
100 – 102	25	f	28	neurotic disorder	3	EO, EC, AF
103 – 105	18	f	35	neurotic disorder	3	EO, EC, AF
106 – 108	26	f	40	neurotic disorder	3	EO, EC, AF
109 – 111	23	f	41	neurotic disorder	3	EO, EC, AF
112 – 114	21	f	51	neurotic disorder	3	EO, EC, AF
115 – 117	19	f	51	neurotic disorder	3	EO, EC, AF
118 – 120	24	f	62	neurotic disorder	3	EO, EC, AF
121 – 123	20	f	34	functional ataxia	3	EO, EC, AF
124 – 126	13	f	39	functional ataxia	3	EO, EC, AF
127 – 129	17	f	40	functional ataxia	3	EO, EC, AF
130 – 132	16	f	51	functional ataxia	3	EO, EC, AF

The subjects can be divided first of all into three groups:

Group 1: healthy persons,
Group 2: subjects with an organic pathology (tumor, multiple sclerosis, Parkinson disease),
Group 3: subjects with a psychogenic pathology (neurathenia, functional ataxia).

3 Data Analysis

For the analysis of the stabilograms different methods have already been used, including the calculation of various statistical characteristics – the maximal and root mean square displacement in both directions, the length of the trajectory on the plain, the area inside the contour formed by the trajectory, the calculation and approximation of one and two-dimensional distributions, correlation and spectral analysis, calculation of correlation dimension, and random-walk analysis (Terekhov 1976; Dobrynin et al. 1985; Rosenblum et al. 1989; Rosenblum and Firsov 1992b; Firsov et al. 1993; Collins and DeLuca 1994; Yong 1994).

In the present work we concentrate on the joint analysis of two components of stabilograms. The main question we are trying to answer is: Are the oscillations in anterior–posterior and lateral directions independent or not? To our knowledge, this question has not been systematically addressed in the literature. Our previous study (Rosenblum et al. 1989) showed that the x and y components are linearly uncorrelated if the subject is healthy, and some correlation may appear in pathological case. To clarify this point we use the cross-spectrum analysis and the cross generalized mutual information (GMI). Moreover, we look for synchronization phenomena.

Introducing the Data. Stabilograms appear as some random functions of time (Fig. 1). By visual inspection we can conclude that stabilograms are, as a rule, non–stationary. Further, we can distinguish between noisy and oscillatory patterns, although these notation is to some extend arbitrary. Oscillatory patterns appear considerably less frequently – only some few per cent of the records can be identified as oscillatory. We can also conclude that in our data set the oscillatory patterns appear in pathological cases only. The probability distributions of three records (#1: healthy female, 23 years old, EO; #61: Parkinson disease, 42 years old, EO; and #89: neurathenia, female, 44 years old, EO) are shown in Fig. 2. The histograms of healthy subjects are usually close to Gaussian. Oscillatory patterns, and patterns close to them have the bimodal form of the distribution. We note that in the last record the distribution of x is clearly asymmetrical. Our studies show that this asymmetry is a good indicator of a functional pathology.

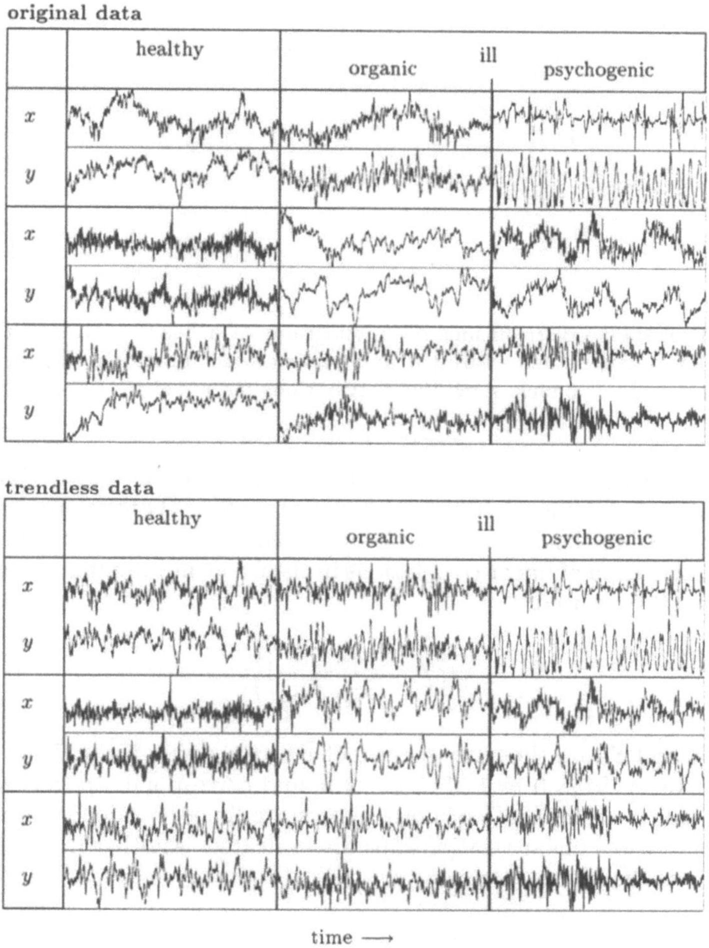

Fig. 1. Typical stabilograms of healthy subjects (records #1, 6, 16, from top to bottom), and patients with organic (records #55, 61, 78) and psychogenic (records #95, 99, 123) diseases. In each signal the time runs from 0 to 164 s

To make the data suitable for our analysis we have removed low-frequency trends. For the cross-spectrum and mutual information analysis it is done by fitting and subtracting a polynomial of order 10. The resulting stationarity of the data was tested with a method proposed elsewhere (Pompe, this volume). The trendless data are shown in the lower panel of Fig. 1. For the calculation of the instantaneous phase the moving average has been subtracted from the data (see below).

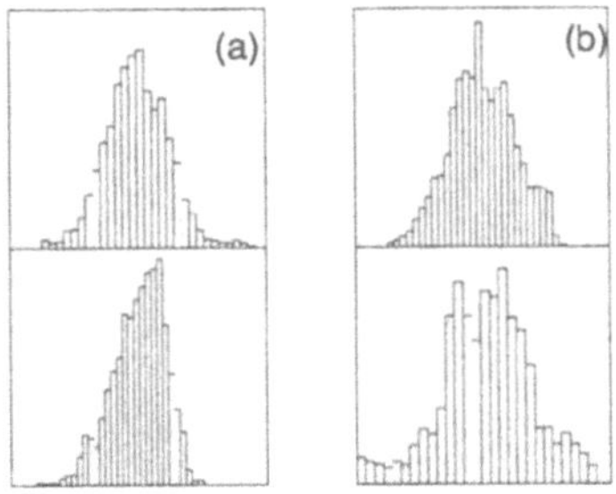

Fig. 2. Probability distributions of stabilograms of a healthy subject, record #1 (a), Parkinsonian patient, #61 (b) and patient with functional ataxia, #89 (c). The upper and low panels show the distributions for x and y

Linear Analysis. A standard technique to analyze the linear relationships between two signals leads to the cross-spectrum $S_{xy}(f)$ of the processes x and y. It is defined as the Fourier transform of their cross-correlation function. For the quantification of linear correlations in the frequency domain the *coherence function* $\gamma^2(f)$ is used:

$$\gamma^2(f) = \frac{|S_{xy}(f)|^2}{S_x(f)S_y(f)} , \tag{1}$$

where S_x and S_y are the auto–spectra of x and y, respectively. The coherence function varies from 0 to 1. The lower bound corresponds to the case where the frequency components of both signals are linearly independent whereas in all other cases linear relationships are indicated. Another important characteristic is the *phase spectrum*

$$\psi(f) = \arg S_{xy}(f) . \tag{2}$$

It represents the phase shift between frequency components, provided they are coherent. Otherwise it has no meaning. In our calculations we use the following parameters: the original record is divided in 12 overlapping samples of 1024 points each and the result is obtained by averaging over 12 periodograms using the Bartlett window.

The power spectra of noisy stabilograms decay monotonically, and the coherence functions γ^2 are considerably less than unity for the whole frequency range (Fig. 3). This indicates the absence of linear correlations between x and y. Although some broad peaks appear in the spectra of the y components in Fig. 3, no significant coherence is seen. In some cases coherence can be observed in the frequency range of the so-called θ-rhythm of corresponding electroencephalograms (4–6 Hz). This may indicate certain intellectual

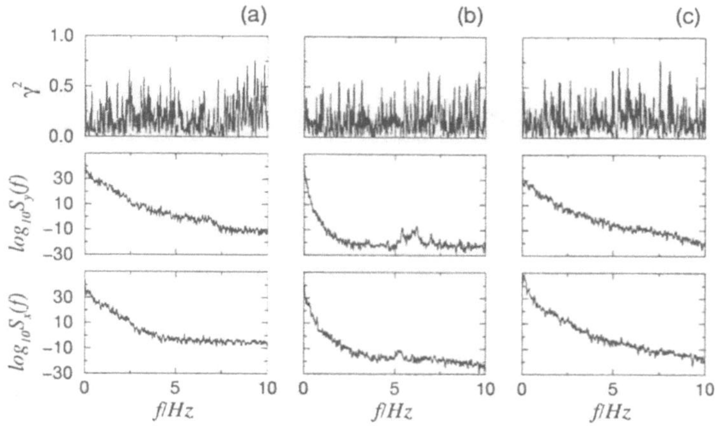

Fig. 3. Typical cross-spectrum of the stabilogram of a healthy subject (record #4) (a) and cross-spectra of Parkinsonian patient (record #61) (b) and the subject suffering from neurathenia (record #89) (c). From bottom to top are shown: the auto-spectra of x and y, and the coherence function γ^2 of the the cross-spectrum. (b): There is some significant increase of the spectral density around 6Hz, but the x and y oscillations are not coherent. Nevertheless, significant dependencies between components of the stabilogram can be revealed by nonlinear techniques (see below)

or emotional stress (Fig. 9c). The results of the linear analysis are summarized in Fig. 4, where the coherence functions are shown for all 132 records as a gray scale picture. The white and full black level corresponds to $\gamma^2 = 0$ and $\gamma^2 = 1$, respectively. We can conclude that with the exception of some pathological records the components of stabilograms are noncoherent.

Analyzing Nonlinear Dependencies in the Data. In order to reveal the presence or absence of dependencies between the postural control signals $y(t)$ and $x(t + \tau)$ we consider the cross mutual information $I_\varepsilon(\tau)$ of them. It represents the mean (over all instants t) information we get from $y(t)$ on $x(t + \tau)$ and vice versa. The parameter $\varepsilon > 0$ denotes the relative level of coarse graining, $\varepsilon \ll 1$. For smaller values of ε more details of the relation between the signals can be detected, however, the reliability of the estimates is decreased in this way. Here we have chosen $\varepsilon = 0.05$ which means that we consider each signal with a precision of 5 % of its total variation range. The time lag τ is taken as the independent variable leading to the mutual information function. It can be considered as a nonlinear analogon to the squared correlation function. In our applications τ runs in an interval around zero (-10 s $< \tau <$ 10 s). For $\tau \to \pm\infty$ the mutual information typically vanishes reflecting the absence of long term relations of the posture control

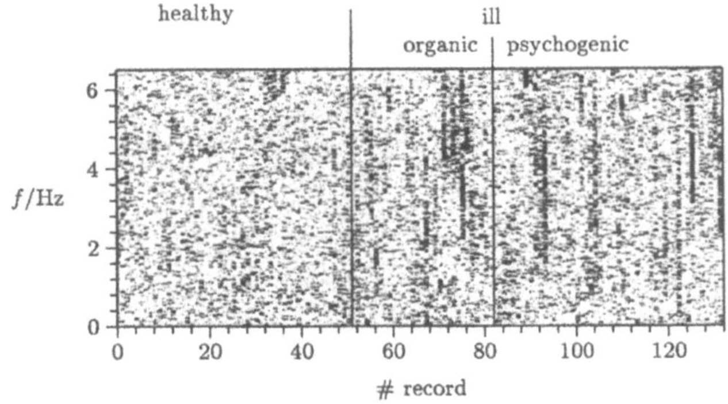

Fig. 4. Coherence functions $\gamma^2(f)$ of of 132 bivariate postural control records $[x(t), y(t)]$ of healthy and ill subjects

data.

We present results for the so-called *generalized mutual information* (GMI). It causes some computational efforts in comparison to the usual Shannonian mutual information. For each τ the GMI fulfills the relations

$$0 \leq I_\varepsilon(\tau) \leq -\log_2 \varepsilon \approx 4.32\,\text{bit} \quad .$$

Their interpretation is as follows: The quantities $y(t)$ and $x(t+\tau)$ can be considered as independent within all accuracy levels $\geq \varepsilon$ if and only if $I_\varepsilon(\tau) = 0$. On the other hand, $x(t+\tau)$ follows uniquely from $y(t)$, within the relative accuracy ε, if and only if $I_\varepsilon(\tau)$ attains its upper bound $-\log_2 \varepsilon$. The larger $I_\varepsilon(\tau)$ the stronger are the linear and nonlinear statistical relations between $y(t)$ and $x(t+\tau)$. Here we work with trendless data – stationarity is crucial for the GMI estimations. A more detailed description of the method is given elsewhere (Pompe, this volume).

We have calculated the GMI function for each of the 132 bivariate records (x, y) of Table 1. In Fig. 5a the functions are encoded by a gray scale and plotted vertically – horizontally the number of the record runs. The GMI functions vary between 0 and 1 bit corresponding to a range of 0...23% of the maximum possible information of about 4.32 bit. In most cases $I_\varepsilon(\tau)$ lies in the range 0.25 ± 0.2 bit indicating that the dependencies are rather week. This is true for healthy as well as for ill subjects. However, in the later case we sometimes observe relatively strong dependencies over a wide range of time lags τ, for instance for the records #61–63 corresponding to a person suffering from the Parkinson disease. (The bivariate record #61 is shown in Fig. 1.) It is important to underline that the cross-spectral analysis shows no coherence for this record (see Fig. 3b) whereas the analysis of the relative phase also shows strong interrelations between x and y.

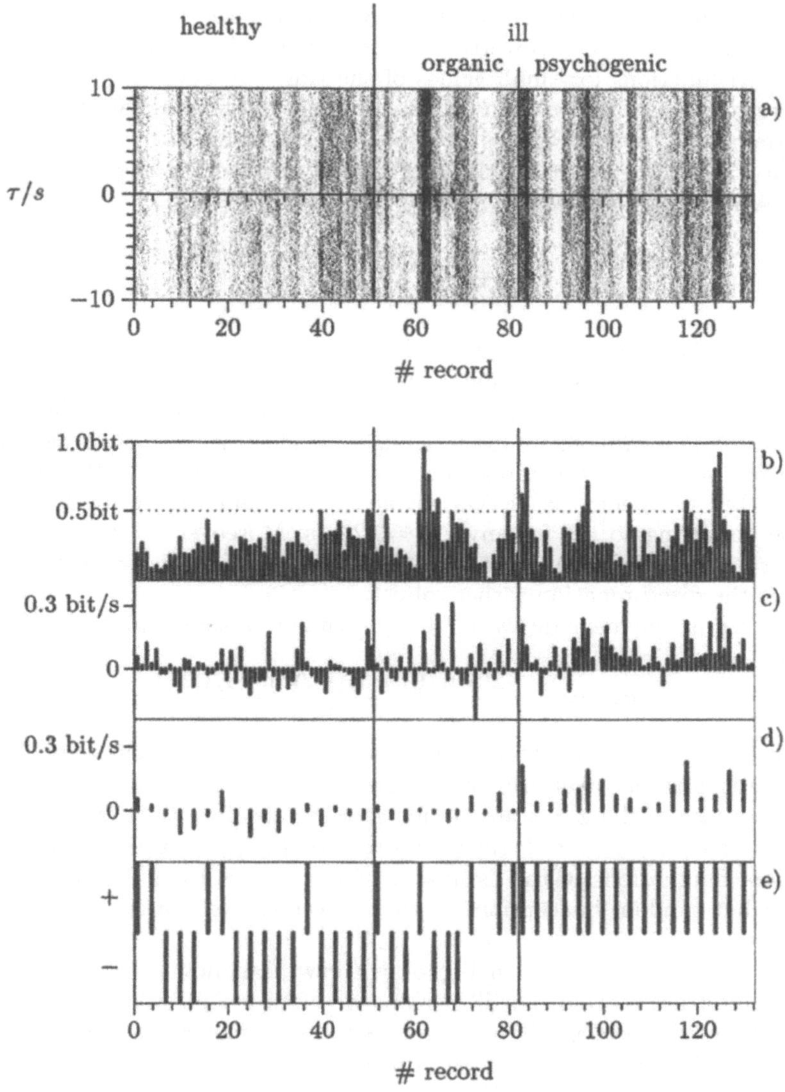

Fig. 5. In (a) the cross mutual information functions $I_\varepsilon(\tau)$ of 132 bivariate postural control records $[x(t), y(t)]$ of healthy and ill subjects are represented. Strong dependencies between $y(t)$ and $x(t+\tau)$ correspond to the more dark regions. The diagrams (b)–(e) represent features extracted from the mutual information functions:

(b) mean of $I_\varepsilon(\tau)$ for $-0.5\text{s} < \tau < 0.5\text{s}$,

(c) decay of $I_\varepsilon(\tau)$ for $\tau = 0 \to 0.5\text{s}$ and $\tau = 0 \to -0.5\text{s}$,

(d) the same as in (c) but only records where the subject had eyes opened,

(e) the same as in (d) but indicating only the sign

From the representation in Fig. 5a we cannot well discriminate between healthy and ill subjects. Nevertheless, the situation improves if we investigate the GMI functions for small values of the time lag, say $-1\ \mathrm{s} \lesssim \tau \lesssim 1\ \mathrm{s}$. In this region we can derive from the functions several features which have some discriminating power. Figures 5b–e give several examples.

In Fig. 5b the mean of the mutual information for small time lags is represented. It is defined as

$$\frac{1}{2\Delta\tau}\int_{-\Delta\tau}^{\Delta\tau} I_\varepsilon(\tau)\,\mathrm{d}\tau \ . \tag{3}$$

For about 15 % of the records of the ill subjects this mean is larger than the corresponding largest value of the healthy subject which are the outliers (records # 40, # 50, and # 51). From the figure we can conclude that the x–y-coupling is more likely to be stronger for ill subjects. If the mean mutual information exceeds the threshold of 0.5bit $\approx 0.11 \times \log_2 \varepsilon^{-1}$ then it is very likely that the person is ill. We get nearly the same figure by deriving the mean according to (3) for any $\Delta\tau$ =0.2–2 s. Moreover, it should be noted that this analysis is done with ranked data (Pompe, this volume). All this makes the discrimination rather robust.

In Fig. 5c the mean decay of $I_\varepsilon(\tau)$ for small time lags is plotted. We define it as

$$\mathrm{feature}_1 \equiv \frac{2 \times I_\varepsilon(0) - I_\varepsilon(-\Delta\tau) - I_\varepsilon(\Delta\tau)}{\Delta\tau} \ . \tag{4}$$

From the figure we conclude that for subjects suffering from a psychogenic disease it is rather likely that this quantity is positive indicating that there is a decrease of the x–y-coupling for growing absolute time lags. If there is no such coupling at $\tau \approx 0$ then the decay of $I_\varepsilon(\tau)$ is mainly determined by random fluctuations of the estimator of $I_\varepsilon(\tau)$. This leads first of all to the somewhat random fluctuations of the corresponding feature of the healthy subjects.

In Fig. 5d the same as in Fig. 5c is shown, but now only the results for the EO test are plotted. For ill persons we often get positive values whereas for healthy subjects negative values seem to be somewhat more likely. For subject with a psychogenic disease this is most striking as it becomes obvious from the plot of the sign in Fig. 5e.

The same investigations were done for the squared cross correlation instead of the GMI functions. The results are presented in Fig. 6. A comparison with Fig. 5 shows that we could expect better distinctive marks from the nonlinear analysis of the data in Fig. 5.

We have done some more attempts to find distinctive marks from the GMI functions. For instance, we also considered higher order cross GMI functions $I_{\varepsilon,\vartheta}(\tau)$ describing relations between $[y(t-\vartheta), y(t)]$ and $x(t+\tau)$, $\vartheta > 0$. In general, the additional knowledge of $y(t-\vartheta)$ cannot decrease the information on $x(t+\tau)$. Indeed, in our numerical experiments we always found an increase

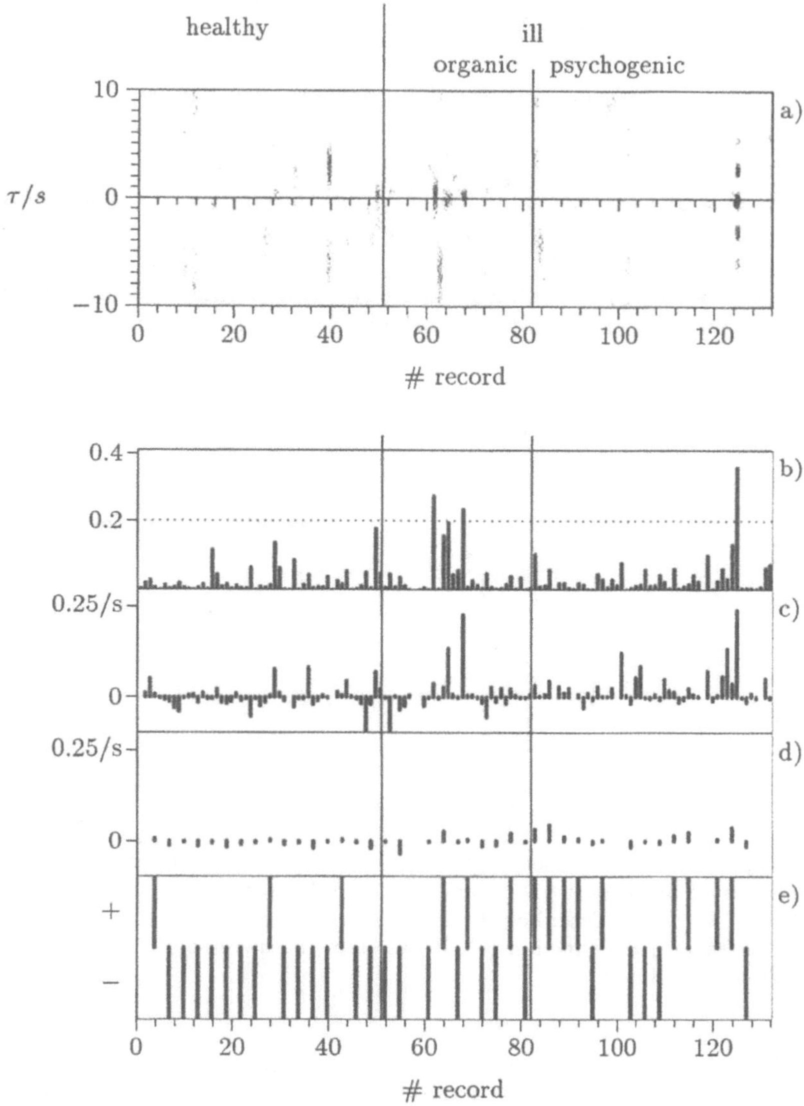

Fig. 6. In (a) the squared cross correlation functions $\varrho^2(\tau)$ of 132 bivariate postural control records $[x(t), y(t)]$ of healthy and ill subjects are represented. Strong correlations between $y(t)$ and $x(t+\tau)$ correspond to the more dark regions. The diagrams (b)–(e) represent features extracted from the squared correlation functions:
b: mean of $\varrho^2(\tau)$ for $-0.5\text{s} < \tau < 0.5\text{s}$,
c: decay of $\varrho^2(\tau)$ for $\tau = 0 \to 0.5\text{s}$ and $\tau = 0 \to -0.5\text{s}$,
d: the same as in (c) but only records where the subject had eyes opened,
e: the same as in (d) but only indicating only the sign

of this information which was somewhat larger for ill subjects. Thus we are inclined to define another feature

$$\text{feature}_2 \equiv \frac{1}{2\Delta\tau}\int_{-\Delta\tau}^{\Delta\tau}\left[I_{\varepsilon,\vartheta}(\tau) - I_{\varepsilon}(\tau)\right]\mathrm{d}\tau \ . \tag{5}$$

Figure 7 represents a plot of feature_2 against feature_1. Almost all feature vectors of healthy subjects are found within the circle whereas that of about 80% of the ill subjects are outside. The discrimination is a bit more striking for subjects suffering from a psychogenic disease.

In our calculations we have chosen $\Delta\tau \approx 0.5\,\text{s}$ and $\vartheta = 40\,\text{ms}$. However, any $\Delta\tau = 0.2$–2 s and $\vartheta = 40$–250 ms would provide a rather similar discrimination of 70...80%.

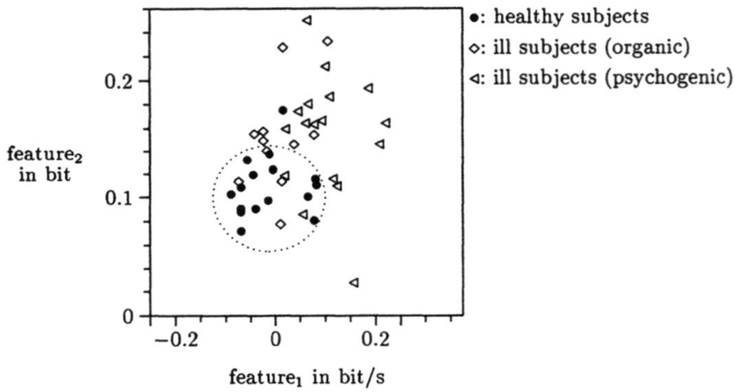

Fig. 7. Plot of two characteristic features of the cross generalized mutual information functions of the postural control data recorded with opened eyes. The feature_1 is defined in (4). It describes the decay of the x–y-coupling for increasing time lag. The feature_2 is defined in (5). It describes the increase of the coupling by an additional coordinate

Our investigations suggest that we could derive some characteristic parameters describing the nonlinear relationship between the postural control data and having some discriminating power between healthy and ill subjects. This might be important for medical diagnosis. However, more data records are needed to get more reliable statements.

Search of Phase Synchronization. To analyze oscillatory patterns we calculate *instantaneous phases* ϕ_1 and ϕ_2 of the signals x and y using the Hilbert transform. An introduction of this technique is given elsewhere (Rosenblum and Kurths, this volume). If the relative phase $\Delta\phi = \phi_1 - \phi_2$ is limited, the presence of phase synchronization of interacting chaotic oscillators may be indicated (Rosenblum et al. 1996; Pikovsky et al. 1996). Possible models of the underlying dynamics are discussed below. We underline that the method is suitable for processing non-stationary data, and it can reveal alternating epochs of qualitatively different behavior. Applications of this method to the analysis of the cardiorespiratory system of a piglet are presented elsewhere (Hoyer et al., this volume).

In order to eliminate low-frequency trends, the moving average computed over the n-point window was subtracted from the original data. The window length n has been chosen by trial to be equal or slightly larger than the characteristic oscillation period. Its variation up to two times does not practically effect the results.

Here we present in detail results of the analysis of several records. For the first example stabilograms of a female subject were investigated (records #124–126: 39 years old, functional ataxia). We can see that in the EO and EC test the patterns are clearly oscillatory (Fig. 8). The difference between these two records is that with eyes opened the oscillations in two directions are not synchronous during approximately the first 110 s, and are phase locked during the last 50 s. In the EC test, the phases of oscillations are perfectly entrained during all the time. In both cases the phase difference fluctuates around zero (the mean value $\langle\Delta\phi\rangle \approx 0.003$). From the power cross-spectra (Fig. 9) we see that, although the low-frequency peaks are clearly seen, the coherence is not very high ($\gamma^2 \approx 0.5$ for the EO test and $\gamma^2 \approx 0.7$ for the EC test), as well as the maximal value of the GMI function ($I_\varepsilon(\tau) < 0.2 \times \log_2 \varepsilon^{-1}$ for the EO test and $I_\varepsilon(\tau) < 0.23 \times \log_2 \varepsilon^{-1}$ for the EC test). The behavior is essentially different in the AF test. The patterns become more noisy and no phase locking or increased coherence in the low-frequency domain is observed. Instead of it, the coherence is increased in a rather broad frequency range ($\approx$ 3–5 Hz) which is close to the frequency of the θ-rhythm in EEG. Such qualitative changes of dynamics (from oscillatory to noisy) was several times observed for psychogenic patients. Further investigations are required in order to find out whether this test can be used as a diagnostic tool.

For the second example we have chosen the stabilogram of a Parkinsonian patient (records #61–63: female, 42 years old). During the EO test both 1:1 and 1:2 synchronous epochs can be found (Fig. 10), although the second one is rather short (about 10 seconds). It is important to notice that in the 1:1 regime the phase difference is significantly non-zero ($\langle\Delta\phi\rangle \approx 0.4$ in the interval 70–100 s and $\langle\Delta\phi\rangle \approx -0.7$ in the interval 100–130 s). During the EC test only an 1:2 phase locking epoch of about 50 s was observed (Fig. 11). No synchronization was found for the AF test.

We underline that the fact that both components of the stabilogram can be rated as oscillatory patterns does not mean the occurrence of the synchro-

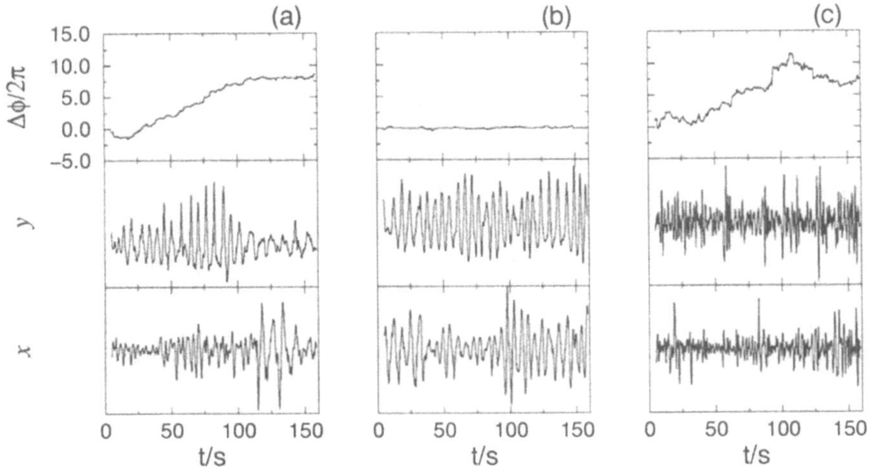

Fig. 8. Stabilograms of an ataxia patient (records #124–126) after trend elimination for EO (a), EC (b), and AF (c) tests. The upper panels show the relative phase between two signals x and y. During the last 50s of the first test and the whole second test the phases are perfectly locked. No phase entrainment is observed in the AF test

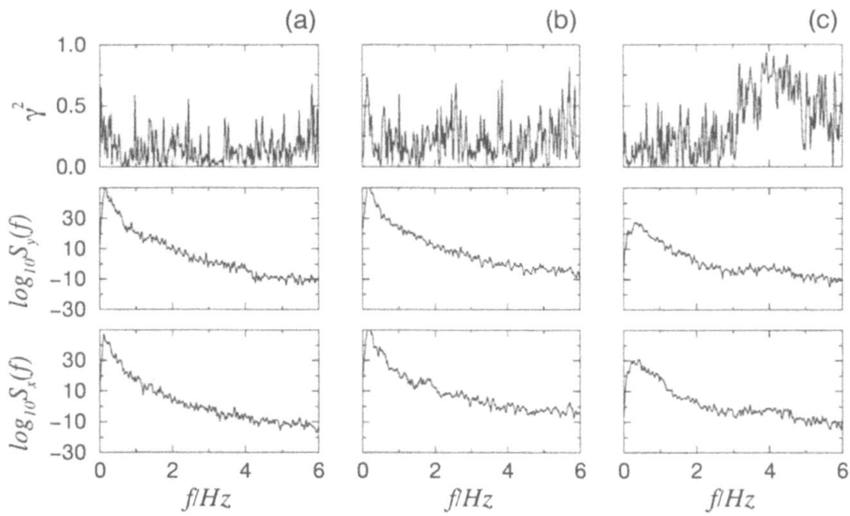

Fig. 9. Auto-spectra and coherence functions for the stabilograms shown in Fig. 8. Although the time series in the EC tests are perfectly phase locked, the spectral analysis shows no significant coherence

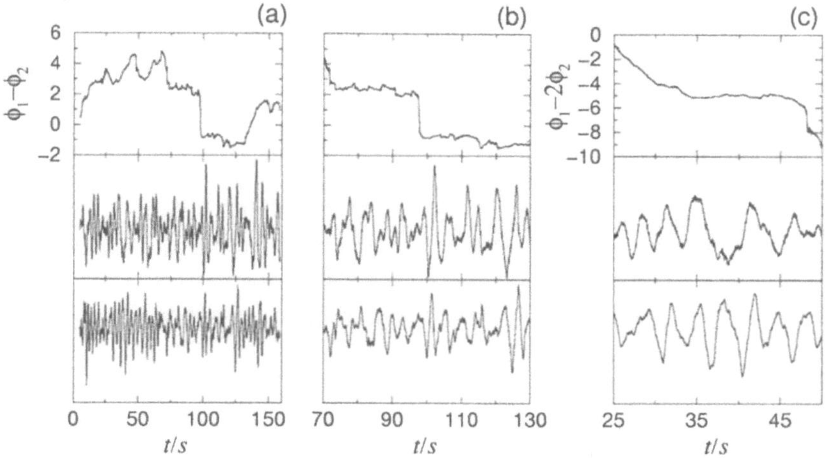

Fig. 10. Stabilogram of a Parkinsonian patient (record #61) after trend elimination for the EO test. The upper panels show the relative phase between two signals x and y. Enlarged parts of (a) show epochs of 1:1 (b) and 1:2 (c) locking

nization. Thus, the patient with the tumor has shown oscillatory patterns in all three tests but no synchronization was found.

4 Model of Postural Control Dynamics

Several mathematical models of posture dynamics have been proposed in the literature. All of them consider only one-dimensional sways of the center of gravity of the human body. The muscle–skeleton subsystem is represented as one-link (Aggashjan and Palcev 1975; Matsushira et al. 1983; Rosenblum et al. 1989) or multi-link inverted pendulum (Rosenblum and Firsov 1992a), or considered as a pinned polymer (Chow and Collins 1995). The crucial point is modelling the control subsystem, i.e. the regulating functions of the CNS. The structure of this system and strategy of the control are highly complicated. Nevertheless, several main principles can be outlined:

- The CNS realizes simultaneously feedforward and feedback control. This provides high reliability of the whole system. The controlled variables are, respectively, stiffness of the joints and elastic torques. These variables are governed by separate cortical systems and adjusted via coactivation and reciprocal activation of muscles (Humprey and Reed 1983).
- The CNS constantly receives information on angles and angular velocities of joints. This information is provided by *proprioceptors*, visual and vestibular analyzers. The main role is played by proprioceptors. Experimental

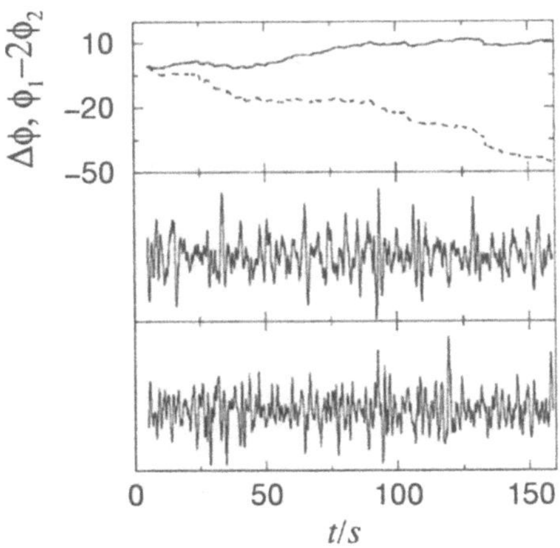

Fig. 11. Stabilogram of an Parkinsonian patient (record #62) after trend elimination for the EC test. The upper panels show the relative phase $\Delta\phi = \phi_1 - \phi_2$ between the signals x and y (solid line) and $\phi_1 - 2\phi_2$ (dashed line). 1:2 phase locking is seen in the time interval 40–90 s.

studies confirm that the nervous system constantly uses this information for the maintenance of posture (Gurfinkel et al. 1982, Litvincev and Tur 1988).

- The characteristics of the proprioceptors (in the muscles and joint spindles, tendon receptors) are essentially nonlinear. Namely, there exist some *sensibility thresholds* which were directly measured in physiological experiments (Gurfinkel et al. 1982). The existence of these thresholds was also indirectly confirmed by results of time series analysis (Collins and De Luca 1994).
- The important property of the feedback loops is time delay caused by the finiteness of the velocity of propagation and processing of the information in the nervous system. The value of the delay is estimated as 0.1–0.8 s (Gurfinkel et al. 1965, Williams 1981).

4.1 Modelling One–Dimensional Sways

The first model of posture dynamics was proposed in (Aggashjan and Palcev 1975) in the form of an one–link inverted pendulum elastically linked to the base

$$\ddot{\varphi} + 2h\dot{\varphi} + \omega^2\varphi + R = \xi(t) \ . \tag{6}$$

$\xi(t)$ is "white" Gaussian noise, and R denotes the regulating action of the CNS. It was assumed that the control is based on position and velocity feedback loops with time delay, $R = c_1\varphi(t-\tau) + c_2\dot{\varphi}(t-\tau)$. Thus, the origin of the body sway while quiet standing was supposed to be a result of fluctuations in the control and mechanical subsystems. A similar model was studied in (Matsushira et al. 1983). To account for the sensibility threshold of proprioceptors, Matsushira et al. have also considered piecewise linearity in the feedback loop. They found that this leads to the excitation of periodic oscillations corrupted by noise.

The importance of the nonlinearity in the feedback loops was demonstrated elsewhere (Rosenblum et al. 1989, Rosenblum and Firsov 1992a). The following model was studied numerically (Fig. 12a):

$$\ddot{\varphi} + 2h\dot{\varphi} + \omega^2\varphi + c_1\mathcal{F}\big(\varphi(t-\tau), \lambda_1\big) + c_2\mathcal{F}\big(\dot{\varphi}(t-\tau), \lambda_2\big) = 0 \ , \qquad (7)$$

where the piecewise linear function $\mathcal{F}$ describes the characteristics of the proprioceptors with sensitivity thresholds $\lambda_{1,2}$,

$$\mathcal{F}(x, x_0) = (x - x_0)\Theta(x - x_0) + (x + x_0)\Theta\big(-(x + x_0)\big) \ , \qquad (8)$$

$\Theta(\cdot)$ is the Heaviside step function. For this model in a broad range of parameters chaotic oscillations arise. It was concluded that oscillations of the center of gravity of the human body may be of deterministic origin.

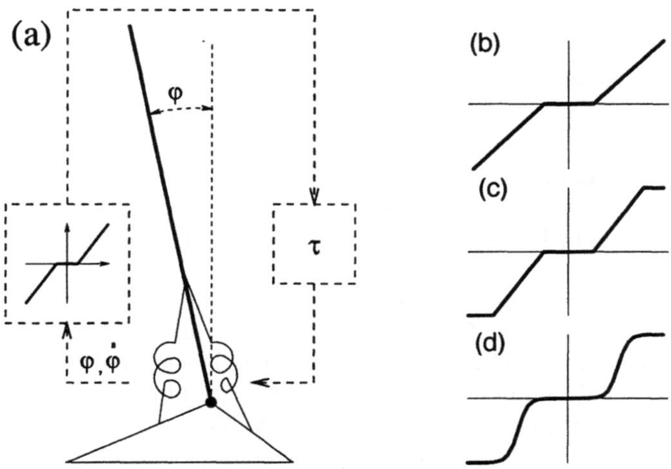

Fig. 12. Scheme of the one-link inverted pendulum model (a). The dashed line denotes the feedback loop with time delay τ. The nonlinear properties of proprioceptors are modelled by a piecewise linear functions (b), (c) or smooth function (c).

Another model (Rosenblum and Firsov 1992a) was proposed on the base of the so-called *equilibrium point hypothesis* (Feldman 1979; Hogan 1985).

According to this hypothesis the muscle–skeleton subsystem is considered as a mass–spring system where the lengths and elasticities of the springs are adjusted by the controller (CNS). Hence, the CNS defines the equilibrium configuration, or state, of the system (in the one-dimensional case we can speak about the equilibrium point). In the process of movement the CNS changes the equilibrium point and the mechanical system goes to this new equilibrium state according to general mechanical laws and in spite of small external perturbations. The control of movements in the presence of rapid external perturbations is mainly achieved by an increase of the stiffness of the joints. The constant posture is maintained by means of shifting the equilibrium point while the stiffness can be considered constant (Humprey and Reed 1983).

Let us denote the coordinate of the center of gravity by x and consider small oscillations in the vicinity of the equilibrium point z,

$$\ddot{x} + 2h\dot{x} + \omega^2(x - z) = 0 \ . \tag{9}$$

The CNS regulates the posture moving the equilibrium point. This regulation is based (a) on the information on x and its derivative that is obtained with some time delay, and (b) information on the current equilibrium state "known" to the controller. Hence, we can write

$$\dot{z} = f\big(z, x(t-\tau_1), \dot{x}(t-\tau_2), \ldots\big) \ .$$

As the first approximation we take

$$\dot{z} = -\tilde{c}_0 z - \tilde{c}_1 \mathcal{F}\big(x(t-\tau), \lambda_1\big) - \tilde{c}_2 \mathcal{F}\big(\dot{x}(t-\tau), \lambda_2\big) \ , \tag{10}$$

where $\mathcal{F}$ is the characteristic of proprioceptors. If $\tilde{c}_0 \gg h$, (9) and (10) reduce to (7) with the feedback coefficients $c_{1,2} = \omega^2 \tilde{c}_{1,2}$. Simulations of the model (9), (10) on the one hand and (7) on the other hand give qualitatively similar results.

The models described above can be divided into two groups: the models of Aggashjan and Palcev 1975, and Matsushira et al. 1983 imply that stabilograms originate from some fluctuations, whereas the models of Rosenblum et al. 1989, and Rosenblum and Firsov 1992a are purely deterministic. The stabilograms are rather short and it is impossible to get considerably longer records even with healthy subject (the tests are rather tiresome). That is why we believe that the identification of the origin of the body sway ("noise versus chaos") on the basis of time series analysis is hardly possible. Nevertheless, the approximately $1/f$ behavior of power spectra of some of the noisy patterns can be considered as a hint (but certainly not as a proof) that the body sways are caused by some fluctuations. Contrary to that, the appearance of oscillatory patterns suggests self-oscillation excitation.[1] More

[1] Although the periodic-like oscillations may appear as a result of filtration of some noise, it seems to be rather unlikely that the control system acts like such a narrow-band filter.

strong evidence of deterministic dynamics, although certainly noisy, is the occurrence of phase synchronization which is a characteristic feature of interaction between self–sustained oscillators. Therefore, we assume that both noise and chaos can be responsible for the observed phenomena. Hence, in modelling we take into account both stochastic and deterministic origin of oscillations under study. Namely, we consider the following model:

$$\ddot{\varphi} + 2h\dot{\varphi} + \omega^2\varphi + c_1\mathcal{F}\big(\varphi(t-\tau),\lambda_1\big) + c_2\mathcal{F}\big(\dot{\varphi}(t-\tau),\lambda_2\big) = \xi(t) \ , \qquad (11)$$

where the characteristics of proprioceptors are described by a piecewise linear or smooth nonlinear function $\mathcal{F}$. $\xi(t)$ is some noise which is taken to be white and Gaussian. We have simulated (11) for three different functions $\mathcal{F}$ (Fig. 12b–d) and have not found essential dependencies of the results on the choice of the function. In further examples we use the function plotted in Fig. 12c. In the purely deterministic case ($\xi(t) = 0$) we found periodic, chaotic, and decaying (transient) solutions depending on the parameter values.[2]

An interesting feature of the system (11) is its response to noisy forcing. Besides "trivial" behavior (random oscillations in the vicinity of the stable equilibrium point that can be considered as a model of noisy patterns Fig. 13a), we have observed the following: Small noise can induce the appearance of a structure in the phase space that is reminiscent to the strange attractor which exists in the phase space of the dynamical system for close parameter values. Alternatively, for the parameter values corresponding to the existence of the limit cycle in the noise-free system, the noise not just corrupts the cycle but makes it very similar to the strange attractor (Fig. 13b, c). As a result, the parameter region corresponding to irregular periodic–like oscillations (that are either noisy chaotic or noisy periodic) is rather broad.

To summarize, the proposed model describes qualitatively the appearance of noisy and oscillatory patterns – forced random oscillations and chaotic self-oscillations disturbed by noise, respectively. The sensitivity thresholds $\lambda_{1,2}$ and the coefficients $c_{1,2}$ may serve as the physiologically relevant bifurcation parameters.

4.2 Modelling Sways in Two Dimensions

From the fact that body sways of healthy persons in anterior–posterior and lateral directions are independent, we can conclude that there exist two separate control systems governing maintainance of the upright posture. We assume that both systems can be described by equations of the form (11).

[2] If the proprioceptor characteristics of Fig. 12b is used, infinitely growing unstable solutions may occur. As we restrict ourselves to modelling small oscillations around the equilibrium, we do not consider these solutions. We do not perform the detailed bifurcation analysis of the model because there are 6 free parameters, and the physiological meaning of two of them, $c_{1,2}$, is not clear.

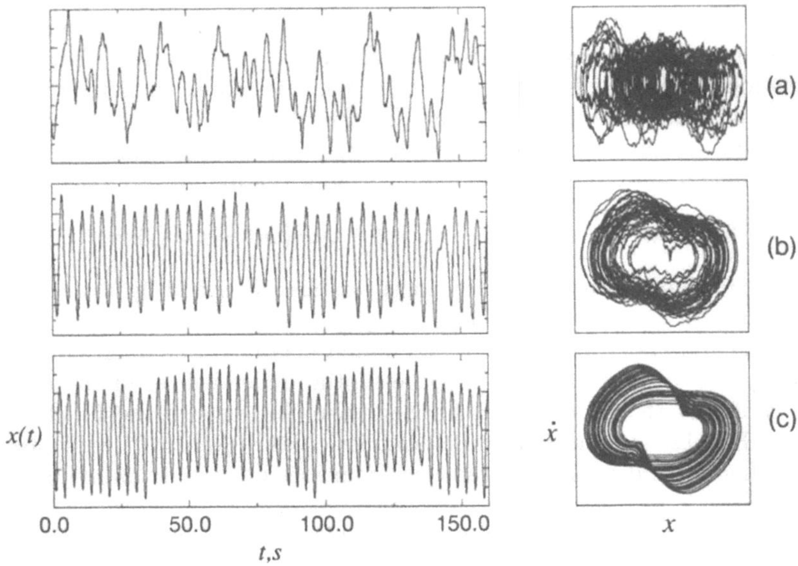

Fig. 13. Simulated time series $x(t)$ (left column) and projections of the corresponding attractors on the plane $x, \dot{x}$ (right column). (a) Simulated noisy pattern ($\lambda_1 = \lambda_2 = 0.01$, $c_1 = 4$, $c_2 = 8$, $\sigma_\xi = 0.1$). (b) Oscillatory pattern ($\lambda_1 = 0.01$, $\lambda_2 = 0.05$, $c_1 = 4$, $c_2 = 8$, $\sigma_\xi = 0.1$). In the absence of noise the attractor of the system is the limit cycle. Influence of noise results in the structure similar to the attractor of the noise–free system for close parameter values (cf. (c)). (c) Chaotic solution ($\lambda_1 = 0.01$, $\lambda_2 = 0.05$, $c_1 = 7$, $c_2 = 8$, $\sigma_\xi = 0$)

Of coarse, the parameters of these equations can differ. In the following we suppose that the time delay τ is the same for both systems because it is determined by the length of the neural fibers and the velocity of the signal propagation. These parameters are likely to be equal for both systems.

As the results of the data analysis show, there are several qualitatively different situations:

1. The x and y are noisy-like. No synchronization was observed in such a case.
2. Both x and y are oscillatory-like and may be either synchronous (with different relation of frequencies) or not. Moreover, synchronous and non-synchronous epochs may alternate within one test.
3. Intermediate situations are also possible – one component is noise-like and another is oscillatory-like. No synchronization is found in this case.

The first and third case can be easily modelled by two equations of the form (11) with two independent noise sources ξ and η. The appearance of the phase

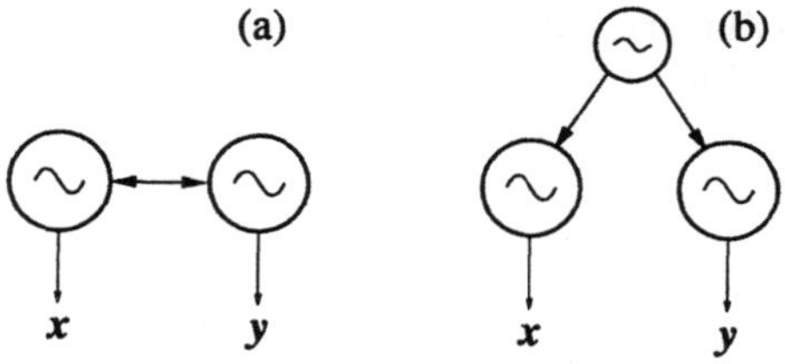

Fig. 14. Two possible oscillatory structures explaining the observed effect of phase synchronization: two coupled self-oscillatory systems (a) and two self-oscillators entrained by the common external driving. In both cases signals x and y may demonstrate phase locking

locking can be explained in different ways:

- Due to some coupling between two chaotic oscillators they can synchronize (Fig.14a). This situation can be described by the following model:

$$\begin{aligned} \ddot{x} + 2h_x\dot{x} + \omega_x^2 x + c_{x,1}\mathcal{F}\big(x(t-\tau),\lambda_{x,1}\big) + c_{x,2}\mathcal{F}\big(\dot{x}(t-\tau),\lambda_{x,2}\big) &= \\ \varepsilon_x(\dot{y}-\dot{x}) + \xi(t) \ , & \\ \ddot{y} + 2h_y\dot{y} + \omega_y^2 y + c_{y,1}\mathcal{F}\big(y(t-\tau),\lambda_{y,1}\big) + c_{y,2}\mathcal{F}\big(\dot{y}(t-\tau),\lambda_{y,2}\big) &= \\ \varepsilon_y(\dot{x}-\dot{y}) + \eta(t) \ , & \end{aligned} \tag{12}$$

  where $\varepsilon_{x,y}$ are the coupling coefficients. The coupling between two control systems may arise, e.g., due to some abnormal function of neural processing making the information from two different channels mutually redundant. The results of the simulation are presented in Fig. 15a.
- If the oscillations excited in two control systems are entrained by some external oscillatory source their phases are also locked (Fig.14b). This external source may appear due to some pathological excitation in the brain, e.g., in the case of Parkinsonian disease. The appropriate model can be written as

$$\begin{aligned} \ddot{x} + 2h_x\dot{x} + \omega_x^2 x + c_{x,1}\mathcal{F}\big(x(t-\tau),\lambda_{x,1}\big) + c_{x,2}\mathcal{F}\big(\dot{x}(t-\tau),\lambda_{x,2}\big) &= \\ \varepsilon_x \sin \Omega t + \xi(t) \ , & \\ \ddot{y} + 2h_y\dot{y} + \omega_y^2 y + c_{y,1}\mathcal{F}\big(y(t-\tau),\lambda_{y,1}\big) + c_{y,2}\mathcal{F}\big(\dot{y}(t-\tau),\lambda_{y,2}\big) &= \\ \varepsilon_y \sin \Omega t + \eta(t) \ . & \end{aligned} \tag{13}$$

  The results of the simulation are presented in Fig. 15b.
- Both control systems are not self-excited and are driven by some external oscillatory force. In this case the phases may be also locked. This explanation seems to be rather unlikely, because in such a case the phases must be always entrained. The transitions between non-synchronous and

synchronous oscillatory patterns, or between different synchronous states would be impossible.

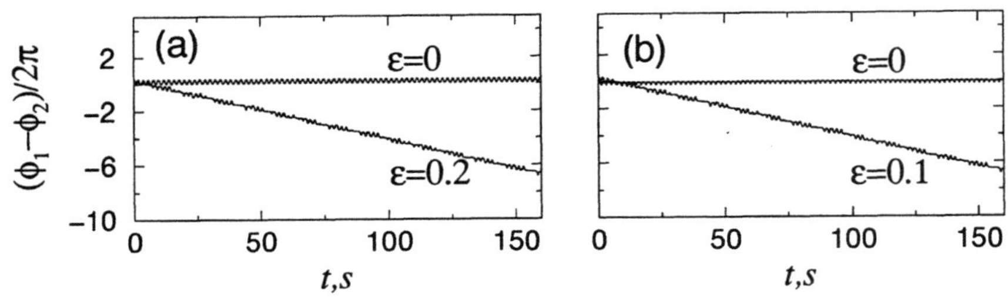

Fig. 15. (a): Mutual synchronization of two postural control systems (see 14a). In the absence of coupling ($\varepsilon_x = \varepsilon_y = \varepsilon = 0$) the phases diverge. Coupling $\varepsilon = 0.2$ leads to phase locking. (b): Synchronization of two control systems by a common external source. The parameters are: $\lambda_{x,1} = \lambda_{y,1} = 0.01$, $\lambda_{x,2} = \lambda_{y,2} = 0.05$, $c_{x,1} = 4$, $c_{x,2} = 8$, $c_{y,1} = 6$, $c_{y,2} = 8$, $\sigma_\xi = \sigma_\eta = 0.1$

We note that, as our goal was to demonstrate the phase synchronization properties of our model, we restricted our simulations to the case of a symmetric coupling only. Certainly, an asymmetric coupling seems to be more realistic. This was confirmed by calculation of higher order GMI functions $I_{\varepsilon,\vartheta}$. Obviously, synchronization can be observed in this case as well.

From the information available we cannot decide which of the described oscillatory structures is responsible for the phase locking observed in experiments. Certainly, different cases might be encountered in different physiological states. The understanding of physiological mechanisms leading to the appearance of the phase locking is a challenge for further investigations.

5 Conclusions

We have studied postural control in humans while quiet standing with open and closed eyes and with additional video–feedback. We have analyzed the interrelations between components of stabilograms using linear and nonlinear techniques. Our investigations demonstrate that in the healthy state the regulation of posture in anterior–posterior and lateral directions x and y can be considered as independent processes. This fact may be expected from the point of view of the control theory because independence of two control loops provides high reliability of a well operation of the whole system.

Further, we demonstrated that the occurrence of certain relationships between the x and y components of stabilograms can be revealed with the help of the generalized mutual information functions and relative phase calculations. We hope that further developments of these techniques might result in the appearance of new diagnostic tools. The comparison of these methods of bivariate data analysis is also interesting in itself.

We have proposed a model of body sways in anterior–posterior and lateral direction. The model qualitatively describes the appearance of noisy and oscillatory patterns in stabilograms, and the arising of phase synchronization. We proposed two plausible oscillatory structures that can explain the observed effect of phase locking between x and y. A very interesting problem is to find out which of these mechanisms (or, perhaps, both) are responsible for the phase synchronization. Further experiments, in particular simultaneous measurement of EEG or/and disturbances of the posture may be helpful.

We believe that further theoretical and experimental studies of postural dynamics can provide better insight in the organization of human motor control, and thus it could help in the development of methods of differential diagnostics.

6 Acknowledgments

We thank P. Tass and A. Pikovsky for helpful comments. M. R. is grateful to the Alexander von Humboldt Foundation for financial support, and B. P. to the Deutsche Forschungsgemeinschaft supporting this work with the project Po442/2–1.

References

Aggashjan R., Pal'cev E. (1975): Reproduction of several special features of upright posture control in humans by means of a mathematical model. Biofizika **20**, 137–142

Baron J. (1983): History of posturography. In Igarashi M., Black F. (Eds.) (1983): *Vestibular and Visual Control of Posture and Locomotor Equilibrium.* Karger, Basel. pp. 54–59

Cernacek J. (1980): Stabilography in neurology. Agressologie **21D**, 25–29

Chow C., Collins J. (1995): A pinned polymer model of posture control. Phys. Rev. E **95**, 907–912

Collins J., De Luca C. (1994): Random walking during quiet standing. Phys. Rev. Lett. **73**, 764–767

Dobrynin S.A., Feldman M.S., Kuuz R.A., Firsov G.I. (1985): Analysis of trajectory of human center of gravity while standing in norm and pathology. In *Solving Mechanical Engineering Problems with Computers.* Nauka, Moscow. pp. 74–81

Feldman A.G. (1979): *Central and Reflex Mechanisms in the Control of Movement.* Nauka, Moscow

Firsov G.I., Rosenblum M.G., Landa P.S. (1993): Deterministic $1/f$ Fluctuations in Biomechanical System. In Handel P.H., Chung A.L. (Eds.) *Noise in Physical Systems and $1/f$ Fluctuations.* AIP Press, New York. pp. 716–719

Furman J. (1994): Posturography: uses and limitations. In: *Baillière's Clinical Neurology* Vol. **3** Baillière Tindall. pp. 501–513

Gurfinkel V., Kots Y., Shik M. (1965): *Regulation of Posture in Humans.* Nauka, Moscow

Gurfinkel V., Lipshits M., Popov K. (1982): Thresholds of kinestatic sensitivity in upright posture. Fiziologija Cheloveka **8**, 981–988

Hogan N. (1985): The mechanisms of multi-joint posture and movement control. Biol. Cybern. **52**, 315–331

Humprey D., Reed D. (1983): Separate cortical systems for control of joint stiffness: reciprocal activation and coactivation of antagonist muscles. In *Motor Control Mechanisms in Health and Disease.* Raven Press, New York. pp. 347–372

Lipp M., Longridge N. (1994): Computerized dynamic posturography: its place in the evaluation of patients with dizziness and imbalance. J. of Otolaryngology **23**, 177–183

Litvincev A., Tur I. (1988): Controlling muscle activity in a joint final position while performing simple movements. Fiziologija Cheloveka **14**, 539–545

Matsushira T., Yamashita K., Adachi H. (1983): A simple model to analyze drifting of the center of gravity. Agressologie **24**, 83–84

Pikovsky A.S., Rosenblum M.G., Kurths J. (1996): Synchronization in a population of globally coupled chaotic oscillators. Europhys. Lett. **34**, 165–170

Pompe B. (1993): Measuring statistical dependencies in a time series. J. Stat. Phys. **73**, 587–610

Rosenblum M.G., Feldman M.S., Firsov G.I., (1989): Investigation of the Biomechanical System. In *Methods and Software of Computer-Aided Experiment in Machine Dynamics.* Nauka, Moscow. pp. 259–277

Rosenblum M.G., Firsov G.I. (1992a): Stochastic self-oscillations in the system of human vertical posture control. I. Strategy of posture control and dynamical model. Biomechanics (Sofia) **24**, 34–41

Rosenblum M.G., Firsov G.I., (1992b): Stochastic self-oscillations in the system of human vertical posture control. II. Simulation and experiment. Biomechanics (Sofia) **25**, 37–43

Rosenblum M.G., Pikovsky A.S., Kurths J. (1996): Phase synchronization of chaotic oscillators. Phys. Rev. Lett. **76**, 1804–1807

Terekhov Y. (1976): Stabilometry as a diagnostic tool in clinical medicine. Canadian Med. J. **115**, 631–633

Williams W. (1981): A system oriented evaluation of the role of joint receptors and other afferents in position and motion sense. CRC Crit. Rev. in Biomed. Eng. **7**, 23–77

Yong J.B. (1994): Contour representation of sway area in posturography and its application. Archives of Physical Medicine and Rehabilitation **75**, 951–956

Delay Induced Patterns of Visually Guided Movements

Peter Tass[1], Michael Rosenblum[2,3], Giovanna Guasti[2], Jürgen Kurths[2], and Harald Hefter[1]

[1] Department of Neurology, Heinrich-Heine-University, Moorenstr. 5, D-40225 Düsseldorf, Germany
[2] Max-Planck-Arbeitsgruppe "Nichtlineare Dynamik", University of Potsdam, Am Neuen Palais 19, PF 601553, D-14415 Potsdam, Germany
[3] A. von Humboldt Fellow. Permanent address: Mechanical Engineering Research Institute, Russian Academy of Sciences, Moscow, Russia

Abstract. We investigated phase dynamics during sinusoidal forearm tracking with delayed visual feedback in healthy subjects. For this reason we introduced two data analysis tools which enabled us to reveal characteristic delay induced tracking movement patterns: (a) determination of instantaneous phase by means of Hilbert transform, (b) symbolic transformation of pointwise relative phase. Our results show that the investigation of the phase dynamics in tracking movements provides us with new insights into the underlying interactions of visual and proprioceptive feedback. In particular we experimentally verified several predictions of a recently presented mathematical model (Tass et al. 1995).

1 Introduction

The control of visually guided movements essentially relies on visual and proprioceptive feedback. The latter is due to proprioception which continuously provides the brain with information concerning the muscles' activity and the resulting biomechanical changes. Many reentrant and feedback loops in the central nervous system are assumed to contribute to this controlling process (Alexander et al. 1990). However, the complex neuronal interactions which realize the matching of proprioceptive and visual information remain unclear so far.

Obviously the control of visually guided movements is influenced by an artificial delay inserted into the visual feedback loop (cf. Glass et al. 1988). Consequently in several tracking experiments an artificial delay was introduced in the visual feedback (for a review see Beuter et al. 1989, Langenberg et al. 1992). Tracking with delay was also used to analyse the oculomotor system alone (Stein and Glickstein 1992) as well as the interaction of skeletomotor and oculomotor control (Vercher and Gauthier 1992).

However, little is known about the tracking movement patterns evolving as a consequence of the artificial time delay. Beuter et al. performed an experiment where healthy subjects had to maintain a constant finger position relative to a stationary baseline (Beuter et al. 1989). Time delays between 40 and 1500 ms were inserted in the visual feedback loop for 100 s. As time delays increased irregular rhythms appeared with short intermittent periods of regular oscillations. The amplitude of these regular oscillations increases with the time delays, whereas the period of the oscillation is about 2 to 4 times the time delay. In particular Beuter et al. did not reveal several characteristic delay dependent movement patterns which qualitatively differ from each other.

In the majority of experiments so far only a limited delay range was tested (cf. Langenberg et al. 1992). The impact of an artificial delay on sinusoidal forearm tracking in humans was analysed systematically for the first time by Langenberg et al. (1992).

Along the lines of a top-down approach and based on a few neurophysiological assumptions Tass et al. (1995) developed a mathematical model for this experiment. The model predicts delay induced transitions between different tracking movement patterns (Tass et al. 1995).

In the present study we analyse sinusoidal forearm tracking with delayed visual feedback with the same experimental set-up as Langenberg et al. (1992). The latter, however, were not able to observe the predicted tracking patterns for two reasons: (a) the recording time in their experiments was too short, (b) they did not use appropriate data analysis techniques. Thus, in the present study we carry out long-term recordings. Moreover we introduce two data analysis tools in order to analyse the phase dynamics of the forearm movement. Therefore we can verify several of the model's predictions.

The comparison between experimental data and the model's behaviour enables us to justify, modify or even reject the model's neurophysiological assumptions. Below it will turn out that our approach also includes analysis of nonlinear interactions between proprioceptive and visual feedback by means of investigating delay induced movement patterns.

The article is organized as follows. In Sect. 2 we present the experimental set-up. The model is presented in Sect. 3. Data analysis is briefly sketched in Sect. 4. In Sect. 5 the experimental data are compared with the model's dynamics. Finally, in Sect. 6, we discuss our results.

2 Experiment

2.1 Subjects

Recordings of sinusoidal forearm tracking with delayed visual feedback were performed in 26 right-handed normal subjects (7 female; 19 male; age range: 19–53 ys.; mean age: 28.4 ys.).

2.2 Experimental Set-Up

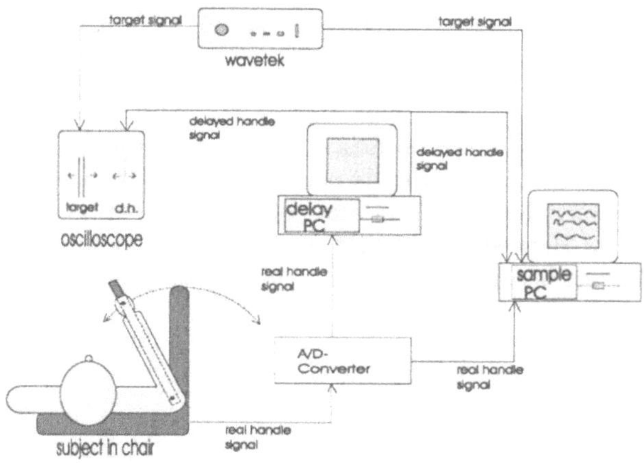

Fig. 1. Experimental set-up

During the experiment subjects are comfortably sitting in a chair in front of an oscilloscope (cf. Fig. 1). Their right elbow is abducted to about 75 degrees and supported by a manipulandum handle of low inertia and low friction. They grasp the manipulandum handle with their dominant hand and perform 15 degrees flexion and extension movements around a 90 degrees elbow angle. The angle of the manipulandum handle is fed into a delay unit and presented with an experimentally controlled delay as a thin line on an oscilloscope. The time delayed displayed manipulandum position is called the *tracking signal.* When the manipulandum is shifted to the right for vanishing delay the tracking signal turns to the right and vice versa.

The target is displayed as a double bar moving sinusoidally with a given frequency across the oscilloscope screen. Subjects have to keep the tracking signal within the double bar, the so-called *target signal.* Target signal, tracking signal, and the manipulandum position signal are digitized each with 250 Hz and stored for off-line analysis.

In contrast to the study by Langenberg et al. (1992) the subject's preferred tracking frequency is used as target frequency which is kept constant throughout the whole experimental session. The preferred tracking frequency is determined in a short preexperimental trial where the oscilloscope is turned off and the subject is asked to perform sinusoidal forearm movements with a most convenient frequency. The preferred target frequency in our group ranges from 0.5 to 0.9 Hz.

Let us introduce the *relative delay* τ_{rel} by putting

$$\tau_{\text{rel}} = \frac{\tau}{T} \ , \tag{1}$$

where τ denotes the artificial delay, and T is the target period. ¿From trial to trial only the relative delay is pseudorandomly varied. Depending on the subject 9 to 15 different delays are tested over a period of 10 to 15 minutes for each delay. Between the single trials pauses had to be made so that the subjects did not become exhausted. The recording session lasted between 2.5 and 4 hours.

3 Model

The model is presented in detail in (Tass et al. 1995). For this reason we will only sketch the main underlying ideas: Our knowledge concerning the neuronal interactions which realize the control of visually guided movements is far from being complete. For this reason along the lines of a *top-down approach* a minimal model was derived for the interactions of the oscillating target signal and the oscillatory forearm movement (Tass et al. 1995). Moreover the model's derivation is based on the notion that phase is an appropriate variable for the description of oscillatory movements (cf. Haken et al. 1985, Haken 1995, Kelso 1995). We first summarize the model's *assumptions*:

1. The amplitude dynamics of the forearm oscillator is neglected in a first approximation.

2. All delays except for the experimentally controlled delay are neglected because under physiological conditions they are assumed to be compensated.

3. The control of the forearm movement relies on nonlinear interactions between target signal, tracking signal, and proprioceptive feedback signal. For vanishing artificial delay the subject is assumed to track without any mistake.

4. We take into account that healthy subjects are able to reproduce the target frequency. For this reason the eigenfrequency of the forearm oscillator equals the target frequency which is denoted by ω.

In order to describe the phase dynamics of the tracking behaviour, we introduce the *phase difference* ϕ between target signal and tracking signal by putting

$$\phi(t) = \theta(t) - \psi(t - \tau) \ , \tag{2}$$

where $\theta(t)$ is the phase of the target signal, i.e. $\dot{\theta} = \omega$. The dot denotes differentiation with respect to time: $\dot{\theta} = d\theta/dt$. $\psi(t - \tau)$ is the phase of the tracking signal, i.e. the phase of the time delayed manipulandum position.

The artificial delay which is inserted into the visual feedback loop is denoted by τ. With these notations the *model equation* reads

$$\dot{\phi}(t) = \underbrace{-\alpha \sin(\phi(t) - \omega\tau)}_{\mathrm{I}} \underbrace{-\beta \sin(\phi(t-\tau))}_{\mathrm{II}} , \tag{3}$$

where α and β are positive real constants. It is important to note that equation (3) has a clear neurophysiological interpretation. Target signal, tracking signal, and proprioceptive feedback contribute to *two matching processes*:

1. *Visual and proprioceptive matching:* The proprioceptive feedback provides the subject with information about the muscles' activity and the resulting biomechanical changes. On the other hand the subject gets information concerning the target position via the visual feedback. As a consequence of nonlinear signal interactions there is a matching between target signal and actual forearm position given by the proprioceptive feedback. This matching process is modeled by term I. α corresponds to the strength of the proprioceptive influence.

2. *Purely visual matching:* The subject has to minimize the angular displacement between target signal and (delayed) tracking signal, both given by the visual feedback. This matching process is modeled by term II. β corresponds to the strength of the visual influence on the signal processing.

According to equation (3) the dynamics of the phase difference ϕ results from the interactions of both matching processes.

¿From a qualitative analysis of this model we get the following main predictions:

1. *Subcritical shift of the fixed point:* Starting with vanishing delay (i.e. $\tau = 0$) and increasing τ causes a shift of the stable fixed point of the phase difference ϕ_0 according to

$$\phi_0 = \frac{\omega\tau}{2} + \arctan\left(\frac{\alpha - \beta}{\alpha + \beta} \tan\frac{\omega\tau}{2}\right) . \tag{4}$$

Thus, as a consequence of the artificial delay τ there is a constant phase difference between target signal and tracking signal.

2. *Oscillations:* When τ exceeds a critical time delay τ_{crit} the system undergoes a Hopf bifurcation giving rise to an oscillation of the phase difference ϕ. Thus, the lead or delay of tracking signal versus target signal is no longer constant: it oscillates periodically.

3. *Running solutions:* For higher delays running solutions occur. This means that the phase difference ϕ is no longer confined to a small region, for instance a cycle. We distinguish two different types of running solutions which differ from each other remarkably as far as the tracking performance is concerned. The parameters (τ, α, β, ω) determine which type occurs.

3.1. *Drift:* The phase difference ϕ decreases or increases rather monotonously. Correspondingly in the experiment we would observe the subject nearly permanently passing the target or vice versa. Thus, drift behaviour is connected with bad tracking performance.

3.2. *Cycle slipping:* Cycle slipping is characterized by low-amplitude oscillations alternating with slipping dynamics. The model parameters (τ, α, β, ω) determine whether the slipping process is combined with an increase or a decrease of the phase difference. For fixed model parameters increasing and decreasing slipping processes may also alternate.

During the low amplitude oscillations the phase difference is typically confined to one phase cycle. Thus, cycle slipping corresponds to a movement pattern where the periods when the tracking signal passes the target (or vice versa) alternate with periods where the phase difference between both signals undergoes small amplitude oscillations. Therefore periods with bad and good performance alternate.

The oscillatory as well as the rather monotonous part of the dynamics may dominate. For this reason there are different types of cycle slipping. Whether a drift or a cycle slipping occurs depends on the model parameters. In particluar, delay induced transitions between different types of running solutions occur.

4. *Chaotic region:* Still increasing the delay the system exhibits chaotic oscillations with amplitudes confined to one or at most two phase cycles.

5. *Windows:* According to the determination of the Ljapunov dimension of the delay dependent sequence of attractors the dynamical behaviour of the model is extremely rich (cf. Fig. 11 in Tass et al. 1995). In particular within the chaotic region and within the region with running solutions there are small delay ranges connected with more simple dynamics, such as limit cycle oscillations or fixed point behaviour.

Note that the model parameters α and β crucially influence the bifurcation route so that even small changes of α and β lead to dramatic modifications of the bifurcation scenario.

4 Data Analysis

In this section we briefly sketch two data analysis techniques which enable us to investigate the phase dynamics of the experimental data and to check the predictions of the model. Both methods are presented in detail by Engbert et al. (in this volume), Rosenblum and Kurths (in this volume). The first method aims at a continuous determination of the phase difference of two signals, whereas the second method is based on a pointwise detection of phase relations.

4.1 Hilbert Transformation

The continuous phase difference between two signals s_1 and s_2 can be defined consistently by means of the analytic signal approach (Panter 1965). This way, we obtain *instantaneous* phase and amplitude of an arbitrary signal $s(t)$. The analytic signal is a complex function of time, $\zeta(t) = s(t) + j\tilde{s}(t) = A(t)e^{j\phi_H(t)}$, where the function $\tilde{s}(t)$ is the Hilbert transform (HT) of $s(t)$

$$\tilde{s}(t) = \pi^{-1}\text{P.V.} \int_{-\infty}^{\infty} \frac{s(\tau)}{t-\tau} d\tau \tag{5}$$

(where P.V. means that the integral is taken in the sense of the Cauchy principal value). The instantaneous phase $\phi_H(t)$ of the signal $s(t)$ is thus defined in a unique way. As the HT can be easily calculated numerically, the phase difference of two signals $s_1(t)$ and $s_2(t)$ can be obtained as

$$\phi_1(t) - \phi_2(t) = \arctan \frac{\tilde{s}_1(t)s_2(t) - s_1(t)\tilde{s}_2(t)}{s_1(t)s_2(t) + \tilde{s}_1(t)\tilde{s}_2(t)} \,. \tag{6}$$

This way, from experimental data we calculate the *normalized phase difference* given by

$$\Delta\phi = (\phi_1 - \phi_2)/2\pi \;, \tag{7}$$

where $\phi_1(t) = \theta(t)$ and $\phi_2(t) = \psi(t-\tau)$ (cf. Eq. 2).

In Sect. 5.1 we will discuss our results revealed by means of the HT method.

4.2 Symbolic Dynamics

Another approach to detect qualitatively different dynamical regimes is provided by symbolic transformation (Hao 1991) of the pointwise relative phase. The latter is determined at discrete time steps given by the peaks of the target signal (cf. Fig 3). Denoting the timing point of the jth maximum of the target signal by $t_j^{(1)}$ and the timing point of the corresponding nearest maximum of the tracking signal by $t_j^{(2)}$, the *pointwise relative phase* φ_j is given by

$$\varphi_j = \frac{t_j^{(2)} - t_j^{(1)}}{T} \;, \tag{8}$$

where T is the period of the target signal. Maximum detection is not problematic because both signals are not obscured by noise. Note that we do not filter the signals.

The relative phases are transformed into a symbolic sequence $s_1, \ldots, s_N$ according to the following rule:

$$s_j = \begin{cases} 0 & \text{if} \quad \varphi_j^r < -a \\ 1 & \text{if} \quad -a < \varphi_j^r < a \\ 2 & \text{if} \quad \varphi_j^r > a \end{cases} \,. \tag{9}$$

where a is a parameter. This one-parameter symbolic transformation reduces the amount of information but emphasizes the robust properties of the dynamics. The symbolic sequences can be visualized by assigning the symbols

0, 1 and 2 to the colors black, grey and white respectively. Grey corresponds to a rather accurate performance of the trial, where the degree of accuracy depends on the value of a. Black and white respectively correspond to the case in which the tracking signal is in advance or behind as compared with the target signal. In Sect. 5.2 we present the results obtained from symbolic transformation of the experimental data.

5 Results

Using the HT method for all subjects we observe characteristic tracking movement patterns corresponding to typical dynamical regimes of the phase difference between target signal and tracking signal: fixed-point behaviour, oscillatory tracking movements, drift, and cycle slipping. Let us first dwell on these characteristic patterns. Next, we will illustrate delay induced transitions between different tracking patterns as obtained from symbolic transformation.

5.1 Tracking Movement Patterns

Figure 2 shows the typical tracking movement patterns revealed by means of determining normalized phase differences according to HT (7).

1. *Fixed point behaviour:* Figure 2 (a) demonstrates a noisy shifted fixed point of $\Delta\phi$ resulting from a small delay. Tracking performance is still quite good. Note that the amplitude of the tracking signal is nearly constant (cf. point 1 of the model's assumptions).

2. *Oscillatory tracking behaviour:* As a result of further increasing the relative delay oscillations of the period difference occur. A typical example is provided by Fig. 2 (b). The values of $\Delta\phi$ remain between -0.1 and 0.4. Thus, the tracking signal does not pass the target signal and vice versa. This means that the phase difference between target signal and tracking signal undergoes low-amplitude oscillations (cf. point 2 of the model's predictions).

3. *Chaos:* A chaos-like oscillatory tracking pattern is shown in Fig. 2 (c). We do not dwell on the question whether the oscillatory tracking patterns rely on chaotic or noisy periodic phase dynamics because we believe that one cannot draw a distinction between these two dynamical states as the data are nonstationary. Transitions between different tracking patterns occuring in several subjects within a single trial (i.e. for fixed delay) clearly indicate that we are not allowed to assume that the data are stationary.

4. *Drift:* Figure 2 (d) exhibits a typical drift behaviour, where the subject passes the target nearly *permanently* (cf. point 3 of the model's predictions). Note that at the end of that trial the subject has passed the target nearly fifty times. Obviously drift dynamics is related with bad tracking performance.

5. *Cycle slipping* as shown in Fig. 2 (e) is *one* dynamical state consisting of low-amplitude oscillations alternating with slipping dynamics. The low-amplitude oscillations are connected with good tracking performance,

whereas slipping dynamics gives rise to bad tracking performance (cf. point 3 of the model's predictions). In Fig. 2 (e) the slipping occurs after 80, 250, 430, and 510 seconds. In the first case the subject passes the target, whereas during the other events the subject is passed by the target. The signals plot in Fig. 2 (e) shows how a subject slips from one cycle into the next one thereby passing the target once. When a subject speeds up the amplitude of the forearm movement typically decreases.

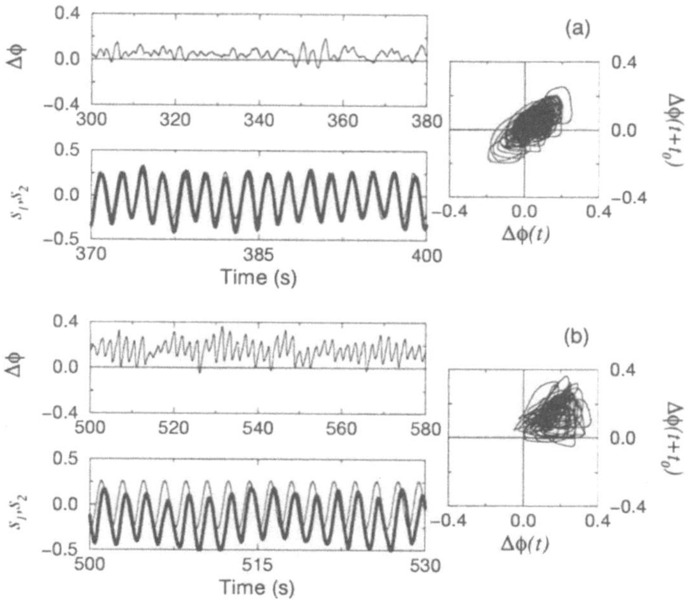

Fig. 2. Typical examples of *theoretically predicted tracking movement patterns* as verified by HT method: Plots show *normalized phase difference* $\Delta\phi$ plotted over time (top) as well as $\Delta\phi$ phase space plot (right, with $t_0 = \tau/4$) and original data (bottom). Target signal s_1 and tracking signal s_2 (bold-face) are both given in radians, where the amplitude of s_1 corresponds to $\pm 14.5°$ manipulandum excursion. (a) Noisy shifted fixed point, where target signal advances tracking signal ($\tau_{\mathrm{rel}} = 0.15$). (b) Oscillatory tracking pattern ($\tau_{\mathrm{rel}} = 0.28$). (c) Chaos-like oscillatory tracking pattern ($\tau_{\mathrm{rel}} = 0.28$). We avoid discussion whether the chaos-like patterns are chaos or noisy periodic oscillations because we believe that we cannot distinguish between these two cases as the data are nonstationary. (d) Drift dynamics, where the subject passes the target nearly permanently ($\tau_{\mathrm{rel}} = 0.9$). (e) Cycle slipping with signals plot showing how the subject passes the target once, thereby slipping from one cycle to the other ($\tau_{\mathrm{rel}} = 0.7$). Target frequency = 0.5 Hz (a), 0.5 Hz (b), 0.5 Hz (c), 0.7 Hz (d), 0.7 Hz (e). Plots (c) to (e) are shown on the next page.

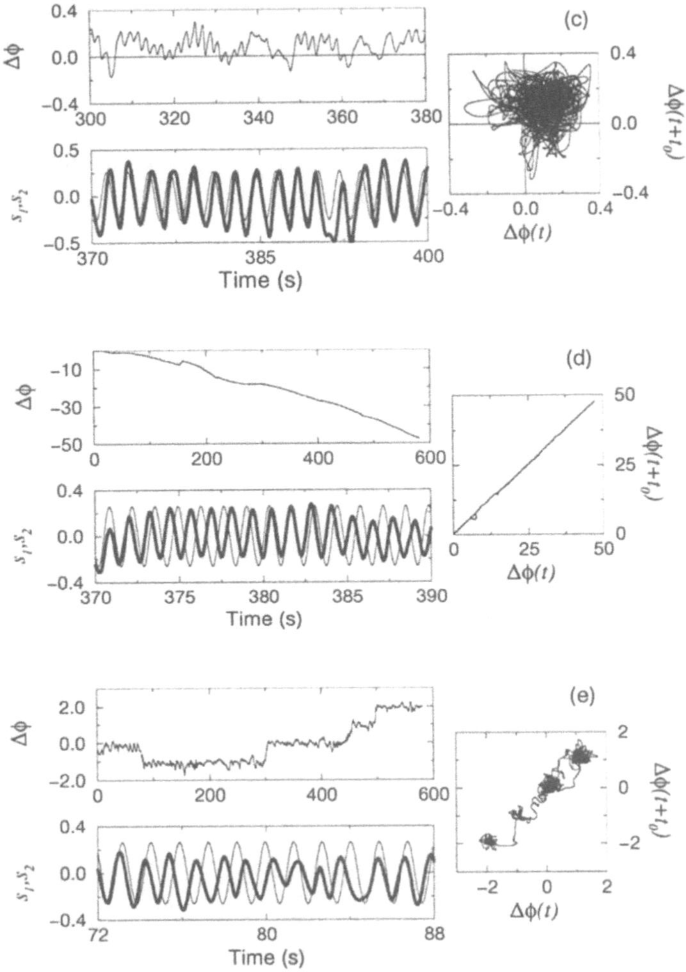

5.2 Delay Induced Transitions

The symbolic patterns of all trials of one subject for a value of $a = 0.025$ for the transformation rule (9) are shown in Fig. 4. This value has been found suitable to reflect the qualitatively different regimes. Already this simple coding rule exhibits four delay dependent dynamical regions all of them verified by means of HT method, too:

1. Fixed point region between $\tau_{\text{rel}} = 0$ and $\tau_{\text{rel}} = 0.075$, where grey dominates corresponding to rather accurate tracking performance.

2. Oscillatory tracking behaviour with worse tracking performance for $\tau_{\text{rel}} = 0.1$ and $\tau_{\text{rel}} = 0.125$.

3. Different types of running solutions between $\tau_{\text{rel}} = 0.2$ and $\tau_{\text{rel}} = 0.9$.

4. Oscillatory tracking behaviour for $\tau_{\text{rel}} = 1$.

In order to quantify transitions between different dynamic regions, measures of complexity have to be applied to the symbolic sequences (Wackerbauer et al. 1994). To this end we compute the Shannon entropy of the symbolic sequences for all values of the control parameter and the relative frequences of all different words of a certain length, in each sequence that can be formed with three symbols. These complexity measures enabled us to significantly quantify transitions between different dynamical regions:

1. between $\tau_{\text{rel}} = 0.2$ and $\tau_{\text{rel}} = 0.3$ (transition from cycle slipping to drift according to HT),

2. between $\tau_{\text{rel}} = 0.5$ and $\tau_{\text{rel}} = 0.6$,

3. between $\tau_{\text{rel}} = 0.6$ and $\tau_{\text{rel}} = 0.7$, where HT method revealed cycle slipping evolving in five ($\tau_{\text{rel}} = 0.5$), two ($\tau_{\text{rel}} = 0.6$) and four ($\tau_{\text{rel}} = 0.7$) cycles.

6 Discussion

In order to investigate the impact of the artificial delay on the interaction of visual and proprioceptive feedback loops, (Tass et al. 1995) presented a mathematical model for the tracking experiment performed by Langenberg et al. (1992). The model predicts characteristic delay induced dynamical states of the phase difference between target signal and delayed handle signal, such as fixed point, oscillations, cycle slipping, and drift. These dynamical states correspond to characteristic tracking movement patterns which were not observed before.

We experimentally verify these theoretically predicted tracking movement patterns in this study which was reported previously in (Tass et al. 1996). To this end we investigated sinusoidal forearm tracking with delayed visual feedback in 26 right handed normal subjects. We used the same experimental set-up as Langenberg et al. (1992). In contrast to them (a) we carried out long-term recordings, (b) we introduced two data analysis techniques which turned out to be appropriate tools for the investigation of the rather complex phase dynamics of tracking movements: The first method aims at a instantaneous determination of the phase difference of two signals by means of the Hilbert Transform. The second method relies on symbolic transformation of pointwise relative phase.

Although the two methods focus on different aspects of the phase dynamics of movements patterns, both enabled us to verify the following predictions of the model:

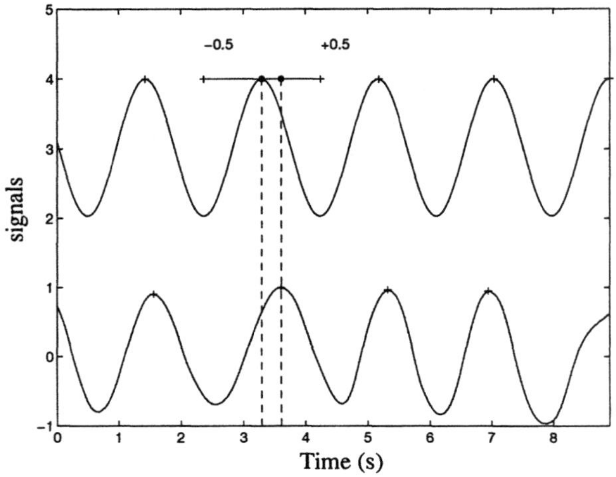

Fig. 3. *Detection of pointwise relative phase:* The plot shows target signal (upper trace) and tracking signal (lower trace). Both are normalized so that target amplitude equals 1. The pointwise relative phase is determined at discrete time steps given by the peaks of the target signal. Denoting the timing point of the jth maximum of the target signal by $t_j^{(1)}$, and the timing point of the corresponding nearest maximum of the tracking signal by $t_j^{(2)}$ the pointwise relative phase is given by $\varphi_j = (t_j^{(2)} - t_j^{(1)})/T$, where T is the period of the target signal. In the example shown in the plot $\varphi_j = 0.17$ holds. Obviously maximum detection is not problematic because both signals are nct obscured by noise. In particular it is not necessary to filter the signals.

1. For small delays we encounter a *fixed point regime* with fixed point shift.

2. With increasing delay *oscillatory tracking patterns* as well as *chaos–like oscillatory tracking patterns* are observed. We avoid discussion whether the oscillatory tracking patterns rely on chaotic or noisy periodic phase dynamics. We believe that one is not able to draw a distinction between these two dynamical states as the data are nonstationary. Transitions between different tracking patterns which occur in several subjects within a single trial (i.e. for fixed delay) clearly indicate that we have to consider the data to be nonstationary.

3. Higher delays give rise to *running solutions*, such as *drift* and *cycle slipping*, which differ from each other as far as the tracking performance is concerned: Drift dynamics is connected with bad tracking perfomance, whereas during cycle slipping periods with good and bad performance alternate.

4. In the majority of trials with large delays we observe running solutions, whereas in the minority of trials with large delays one encounters rather simple dynamics such as fixed point behaviour and low amplitude oscillations.

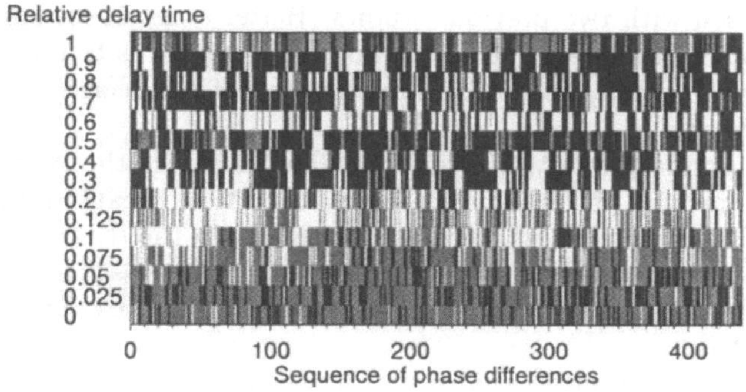

Fig. 4. Delay induced movement patterns revealed by *three symbols coding of the pointwise relative phase* for $a = 0.025$. The sequence of pointwise relative phase φ_j within one trial is plotted over the index j, so that (discrete) time is on the abscissa and τ_{rel} is on the ordinate. The symbol sequences of all trials are plotted starting with $\tau_{\text{rel}} = 0$ at the bottom of the plot.

This corresponds to the *windows* revealed by the determination of the Lyapunov dimension of the delay dependent sequence of attractors in the model (Tass et al. 1995).

5. All 26 subjects investigated in our study showed the above mentioned dynamical phenomena, thereby exhibiting clear *interindividual differences*. For instance, in the delay region connected with running solutions some subjects have a tendency to display cycle slipping, whereas in other subjects drift dynamics predominates. Moreover in some subjects the oscillatory tracking patterns are observed at rather high values of the relative delay, e.g. $\tau_{\text{rel}} = 0.15$, whereas in other subjects oscillatory tracking patterns are already encountered at small delays, e.g. $\tau_{\text{rel}} = 0.03$.

These differences may reflect interindividual differences of prefered tracking strategies (modeled by α and β) as well as interindividual differences of the adaptability to a delay inserted into the visual feedback.

Minimal changes of α and β modify the bifurcation scenario of the model dramatically. Because of the long duration of the experimental sessions, system parameters cannot be considered to be constant in the different trials, e.g. due to fatigue. Therefore we cannot estimate α and β by comparing our experimental data with numerical bifurcation scenarios for fixed α and β. Although we clearly observed the shift of the fixed point, we were not able to verify (4) for two reasons: (a) System parameters vary during the experimental session. (b) The number of trials with fixed point dynamics was too small in all subjects. In order to analyse bifurcation scenarios as well as fixed point shift, in a forthcoming study we will change the delay within a single trial (quasistatically).

In comparison with two previous studies (Hefter et al. 1995, Langenberg et al. 1992) our results clearly point out how important it is to use appropriate data analysis techniques for the investigation of the phase dynamics of tracking movements:

1. In a preliminary study we were not able to detect the characteristic delay induced tracking patterns presented in this article because return maps of the pointwise relative phase were used to quantify the dynamics (Hefter et al. 1995). As a consequence of the elimination of time associated with the return map plots one cannot distinguish between oscillatory dynamics and running solutions.

2. Langenberg et al. (1992) did not observe the characteristic movement patterns for two reasons: (a) They analysed the tracking performance by means of determining the root mean square error of the difference between target signal and tracking signal. Obviously the root mean square error reflects both, phase dynamics and amplitude dynamics. Therefore it is not an appropriate tool for the detection of characteristic phase patterns. (b) Moreover Langenberg et al. (1992) performed short-term recordings where the recording length corresponded to only 20 target cycles. Obviously this is not enough data to analyse the phase dynamics.

The experimental approach presented in this article enables us to induce characteristic tracking movement patterns by inserting an artificial time delay into the visual feedback loop. Both data analysis tools, one based on Hilbert Transform the other based on symbolic transformation, are capable of detecting theoretically predicted delay induced movement patterns.

7 Acknowledgements

This study was supported by grants from the Deutsche Forschungsgemeinschaft (SFB 194, A5). M.R. acknowledges support from Alexander von Humboldt Foundation. G.G. acknowledges support from University of Florence.

References

Alexander G.E., Crutcher M.D., DeLong M.R. (1990): Basal ganglia-thalamocortical circuits: parallel substrates for motor, oculomotor, 'prefrontal' and 'limbic' functions. Progr. Brain Res. **85**, 119–146

Beuter A., Larocque D., Glass L. (1989): Complex oscillations in a human motor system. Journal of Motor Behavior **21**, 277–289

Glass L., Beuter A., Larocque D. (1998): Time delays, oscillations and chaos in physiological control systems. Mathematical Biosciences **90**, 111–125

Haken H. (1997): *Synergetics. An Introduction.* Springer, Berlin, Heidelberg

Haken H. (1995): *Principles of Brain Functioning. A Synergetic Approach to Brain Activity, Behaviour and Cognition.* Springer, Berlin, Heidelberg

Haken H., Kelso J.A.S., Bunz H (1985): A theoretical model of phase transitions in human hand movements. Biol. Cybern. **51**, 347–356

Hao B.-L. (1991): Symbolic dynamics and characterization of complexity. Physica D **51**, 161–176

Hefter H., Tass P., Wunderlin A., Langenberg U., Freund H.-J. (1995): Experimental confirmation of implications from a theoretical model on sinusoidal forearm tracking with delayed visual feedback. In: Elsner N, Menzel R (Eds): *Göttingen Neurobiology Report 1995*. G. Thieme, Stuttgart. p.233

Kelso J.A.S. (1995): *Dynamic Patterns. The Self-Organization of Brain and Behavior*. MIT Press, Cambridge, MA

Kurths J., Voss A., Saparin P., Witt A., Kleiner H. J., Wessel N. (1995): Quantitative analysis of heart rate variability. CHAOS **5**, 88–94

Langenberg U., Kessler K., Hefter H., Cooke J.D., Brown S.H., Freund H.-J. (1992): Effects of delayed visual feedback during sinusoidal visuomotor tracking. Soc. Neurosci. Abstr. Suppl. **5**, 209

Panter P. (1965): *Modulation, Noise, and Spectral Analysis*. McGraw-Hill, New York

Rosenblum M.G., Pikovsky A.S., Kurths J. (1996): Phase synchronization of chaotic oscillators. Phys. Rev. Lett. **76**, 1804 1804–1807

Stein J.F., Glickstein M. (1992): Role of the cerebellum in visual guidance of movement. Physiol. Rev. **72**, 4, 967–1017

Tass P., Wunderlin A., Schanz M. (1995): A theoretical model of sinusoidal forearm tracking with delayed visual feedback. J. Biol. Phys. **21**, 83–112

Tass P., Kurths J., Rosenblum M.G., Guasti G., Hefter H. (1996): Delay induced transitions in visually guided movements. Phys. Rev. E **54**, R2224–R2227

Wackerbauer R., Witt A., Atmanspacher H., Kurths J., Scheingraber H. (1994): A comparative classification of complexity measures. Chaos, Solitons & Fractals **4**, 133–173

Vercher J.-L., Gauthier G.M. (1992): Oculo-Manual Coordination Control: Ocular and manual tracking of visual targets with delayed visual feedback of the hand motion. Exp. Brain Res. **90**, 599–609

V Voice

Detecting Bifurcations in Voice Signals

Hanspeter Herzel[1], Joachim Holzfuss[2], Zbigniew J. Kowalik[3], Bernd Pompe[4], and Robert Reuter[5]

[1] Institute of Theoretical Biology, Humboldt University Berlin, Invalidenstr. 43, D-10115 Berlin, Germany
[2] Institut für Angewandte Physik, Technische Hochschule Darmstadt, Schloßgartenstr. 7, D-64289 Darmstadt, Germany
[3] Institute of Experimental Audiology, University of Münster, Kardinal-Von-Galen-Ring 10, D-48129 Münster, Germany
[4] Institute of Physics, E.-M.-Arndt-University Greifswald, F.-L.-Jahn-Str. 16, D-17487 Greifswald, Germany
[5] Institute of Electronics, Technical University Berlin, Einsteinufer 17, D-10587 Berlin, Germany

Abstract. This chapter is concerned with the detection of bifurcations in voice signals applying several techniques of sliding signal analysis – conventional ones as well as novel methods originating from nonlinear dynamics. The signals come from several models (two-mass and continuum) as well as from an excised larynx experiment and vocalizations of patients with voice disorders. The results of the different techniques were found to be consistent and complementary to each other.

1 The Voice: A Highly Nonlinear Oscillator

The composites of speech production are respiration, phonation, and articulation. The respiratory airflow, typically from the lungs via the bronchi and trachea, serves as the main driving force. Articulation is controlled by the vocal tract which can be regarded as an approximately linear filter (e. g., Fant 1960). The resonance frequencies (termed formants) are governed by the oral and nasal cavities.

In this paper we focus our attention on phonation – the generation of the primary voiced sound within the larynx. The vocal folds are set into vibration by the combined effect of subglottal pressure, the visco-elastic properties of the folds, and the Bernoulli effect (e. g., van den Berg 1958, Titze and Alipour-Haghighi (forthcoming book)). The effective length, mass and tension of the vocal folds are determined by muscle action, and in this way the fundamental frequency ("pitch") and the waveform of the glottal pulses can be controlled.

For sustained vowels the driving lung pressure and the muscle tensions can be regarded as slowly varying parameters since they change on time scales of a few hundred milliseconds whereas vocal fold vibration cycles have periods of only a few milliseconds. More details about the mechanisms of phonation are provided in Sect. 3 in connection with aerodynamic–biomechanical modeling.

Limit cycle oscillations can be considered as a reasonable model of a normal healthy voice, although some small perturbations (of the order of a percent) are always present. However, under certain circumstances much larger irregularities are observed in vocalizations. These are often associated with the term roughness. Also a healthy vocal apparatus can generate a rough voice quality under extreme conditions. Examples are newborn cries (Lind 1965, Mende et al. 1990), simulated creaky voice (Dolansky and Tjernlund 1968, Scherer 1989, Herzel 1993), or Russian lament (Mazo 1994). Bifurcations to subharmonic regimes, toroidal oscillations, and chaos have been reported in these extreme vocalizations.

Of particular interest are vocal instabilities due to organic, neurological, and functional diseases (Kelman 1981, Hammarberg et al. 1986, Hirano 1989, Smith et al. 1992). It has been shown that subharmonics, low-frequency modulations, and biphonation (two independent pitches) are often symptomatic in voice pathology (Herzel and Wendler 1991, Titze et al. 1993, Herzel et al. 1994a). An understanding of the underlying physiological and physical mechanisms will certainly be helpful for diagnosis and treatment of voice disorders.

Our chapter is organized as follows: In the next section five techniques of sliding signal analysis are introduced that are applied afterwards. These methods include conventional tools of voice research (spectrograms, pitch contours) as well as novel nonlinear techniques (generalized mutual information, dimensions, and Lyapunov exponents). Sections 3 and 4 are devoted to modeling. We introduce the two-mass model – the most simple but physiologically reasonable model – and discuss the most sophisticated continuum models. Computer simulations of both models are analyzed with our sliding analysis techniques. In Sect. 5, bifurcations in an excised larynx experiment are discussed. Finally, the methods are applied to vocalizations of patients with certain voice disorders (papilloma and laryngitis).

It turns out that the applied methods provide some insight into transition phenomena in computer-generated and natural voice signals. The results of the different techniques are consistent and complement each other.

2 Sliding Analysis Techniques

In this section we describe signal analysis techniques that are applied in the remainder of this paper. The spectrogram technique – a sliding power spectral analysis – is of widespread use in speech research. It allows the simultaneous characterization of the local (in time) signal properties, e.g. the pitch and the formants, and the long-term changes during utterances. For slowly varying external parameters as subglottal pressure and muscle tension, spectrograms can be regarded as spectral bifurcation diagrams. Such diagrams have been applied also to bifurcation analysis in acoustic cavitation (Lauterborn and Cramer 1981), heterogeneous catalysis (Liauw et al. 1993), and laser models (Merbach et al. 1995).

In order to detect bifurcations in sustained phonation, narrow band spectrograms are required. Typically we use windows of about 200 ms giving a spectral resolution of 5 Hz. Then the (dis)appearance of spectral peaks due to bifurcations of the underlying dynamical system can be monitored. In this way subharmonics (related to period doubling or tripling), side bands (a manifestation of toroidal modulations), and independent frequencies have been identified by spectrograms in voice signals (Mende et al. 1990, Herzel et al. 1994a, Herzel et al. 1995a, Herzel and Reuter 1996).

Instead of using power spectra one can study also correlation measures. For example, short-term autocorrelation functions contain the same information as the corresponding spectra. However, correlation functions detect only linear statistical dependences. Therefore in nonlinear systems the mutual information has been applied frequently (Herzel and Ebeling 1985, Fraser and Swinney 1986, Pompe and Leven 1986). This function vanishes, if and only if, signals are statistically independent (Rényi 1970, Herzel and Große 1995). However, the estimation of the mutual information from finite realizations is problematic (Herzel et al. 1994b). Recently a generalized mutual information has been introduced which is based on correlation integrals (Pompe 1993). Therefore, it can be estimated much more easily and with moderate computational effort. Consequently, also a sliding analysis of signals can be performed which has been termed *miogram* (Pompe and Heilfort 1994). Some more details about this technique are given in figure captions and in another chapter elsewhere (Pompe 1996, this volume).

It has been argued in the preceding section that sustained phonation of normal healthy voices leads to nearly periodic signals. Deviations from periodicity are often symptomatic of voice pathologies. Therefore various "perturbation measures" have been developed to quantify irregularities. Many of these measures are based on "pitch contours", i. e. sequences of consecutive estimations of the local pitch period. Since the pitch originates usually from the larynx, diseases of the voice generating system are effectively studied by pitch contours. In a sense, these contours are comparable to RR-interval series in cardiology. Period-doubling is easily recognized in contours as alternating patterns, and low-frequency modulations lead to oscillating pitch contours (Herzel 1993, Herzel et al. 1994a, Herzel 1996).

Since many voice signals reflect low-dimensional attractors it is natural to estimate dimensions and Lyapunov exponents for some vocalizations (Mende et al. 1990, Herzel and Wendler 1991). However, the well-known nonstationarities on time scales of a few hundred milliseconds prevent precise estimations. In this paper we apply nevertheless related techniques to our data. In order to allow comparison to spectrograms and miograms we calculate local (in time) estimations of dimensions and Lyapunov exponents. We are convinced that these estimates very crudely resemble the actual invariants due to the short windows and nonstationarities. However, they exploit quite different signal properties than spectrograms since they are based on phase space reconstruction and, hence, they reflect attractor properties instead of temporal correlations.

Estimates of the correlation dimension are obtained by averaging local densities (e.g., Grassberger and Procaccia 1983, Holzfuss and Mayer-Kress 1986, Lauterborn and Holzfuss 1986, Kurths and Herzel 1987). In the resulting log–log plot (correlation integral versus distance) scaling regions are detected automatically (Holzfuss and Mayer-Kress 1986). In this way a fast sliding dimension analysis of also rather long voice signals are possible – a signal of 200 000 data points can be processed in a few minutes.

Lyapunov exponents measure the rate of separation of nearby states and are related, therefore, to the predictability of the system. There are several methods to estimate Lyapunov exponents from long stationary time–series (Wolf et al. 1985, Eckmann et al. 1986, Sano and Sawada 1985, Holzfuss and Parlitz 1991). It has been pointed out (Wolff 1992, Kowalik and Elbert 1995, Kowalik et al. 1996, this volume) that often also local stability measures (Herzel et al. 1987, Herzel and Pompe 1987) provide useful information about the dynamics. Deterministic mechanisms of physiological processes allow us to state that at least in the short term, the processes exhibit quasistationary character. The observation of dynamical measures in short time–slices will then characterize their momentary state, and the observation of the temporal evolution of such measures provides information about the development of the physiological system. Combining the local divergence measure with a windowing technique should give us a very useful tool for measurement of global system dynamics.

It will be shown that in view of the intrinsic nonstationarity of voice signals sliding analysis techniques as introduced above are appropriate tools. We demonstrate below that nonstationarities also have an advantage. Slowly varying parameters may induce several bifurcations within a single vocalization. It is not seldom that limit cycles, subharmonic regimes, tori, and chaotic episodes can be detected in a single newborn cry (Lind 1965, Mende et al. 1990) or in a sustained vowel (Herzel et al. 1994a, Herzel and Reuter 1996, Herzel 1996).

3 A Two-Mass Model of the Vocal Folds

Any phonation model has to include aerodynamical and biomechanical components. In principle, the Navier–Stokes equations with time-varying boundaries together with nonlinear and inhomogeneous visco-elastic equations should be solved. However, much insight has been gained with the aid of simplified models, such as a two-mass model (Ishizaka and Flanagan 1972, Ishizaka and Isshiki 1976, Herzel et al. 1991, Smith et al. 1992, Pelorson et al. 1994) or a 16-mass model (Titze 1973). Recently we have reduced the intensively studied Ishizaka–Flanagan model to its very basic features (Herzel

1993, Herzel and Knudsen 1995). Such drastic simplifications allow extensive bifurcation analysis. In the following we present first the symmetric model version in order to demonstrate the underlying physical mechanisms of phonation. Then a time series from the asymmetric version will be analyzed.

In two-mass models both folds are modeled by a lower and upper mass m_1 and m_2, respectively. Their elongations x_i $(i = 1, 2)$ are governed by the usual mechanical equations with spring constants k_i and c_i, damping r_i, and coupling k_c:

$$m_i \ddot{x}_i + r_i \dot{x}_i + k_i x_i + \Theta(-a_i) c_i \left(\frac{a_i}{2l}\right) + k_c(x_i - x_j) = F_i(x_1, x_2) \ . \tag{1}$$

The Heaviside function Θ is related to an additional restoring force during closure of the glottis ($a_i = a_{i0} + 2lx_i < 0$; a_{i0} — rest area; l — length of the glottis). The driving forces F_i can be derived as follows: We assume constant pressure below the glottis (termed subglottal pressure P_s) and above the glottis (vocal tract input pressure $P_i = 0$). Moreover we assume that at the point of minimum area $a_{\min}$ a jet is formed which induces an immediate pressure decay to zero. Consequently, the driving force of the upper mass F_2 is identically zero for all glottal configurations. $F_1 = l\, d_1 P_1$ (d_1 — thickness of the lower mass) is the force exerted by the pressure P_1 on the lower part of the glottis. The corresponding pressure P_1 can be obtained from the Bernoulli law:

$$P_s = \frac{\rho}{2} \left(\frac{U}{a_{\min}}\right)^2 = P_1 + \frac{\rho}{2} \left(\frac{U}{a_1}\right)^2 \ . \tag{2}$$

Here ρ and U denote the air density and the glottal volume flow, respectively. Using

$$U = \sqrt{\frac{2P_s}{\rho}}\, a_{\min} \tag{3}$$

one obtains

$$P_1 = P_s \left[1 - \Theta(a_{\min}) \left(\frac{a_{\min}}{a_1}\right)^2\right] \Theta(a_1) \ . \tag{4}$$

It can be seen that for a convergent glottis ($a_1 > a_{\min} = a_2$) the pressure is reduced by the Bernoulli term. This points to the essential mechanism of vocal fold vibrations: normally the lower mass opens first and, hence, the full subglottal pressure pulls apart the vocal folds. After some delay (about 60° phase shift) the upper mass pair opens leading to an increasing flow. Consequently, during the closing phase the pressure P_1 is reduced according to (4). This pressure asymmetry between opening and closing phase constitutes the main driving force of vocal fold vibration in the chest register. The described interaction between flow and geometry allows the energy transfer from the vertical air flow to the vocal fold motion. The described wave-like motion of the vocal folds is found indeed in stroboscopic and high-speed observations. A more detailed justification of the various simplifying assumptions (laminar flow, no viscous losses, constant subglottal and supraglottal pressures, linear restoring force etc.) can be found elsewhere (Herzel and Knudsen 1995). A representative set of parameters is given in Herzel et al. 1995a.

Bifurcations in the symmetric model version have been studied recently (Herzel 1993, Herzel and Knudsen 1995). As an attempt to model unilateral paralysis we analyzed an asymmetric version of the two-mass model (Steinecke and Herzel 1995). For this purpose the masses and elastic constants of one fold have been scaled as follows using an asymmetry parameter Q:

$$\begin{aligned} k_i^{\text{right}} &= Q\, k_i^{\text{left}} , \\ c_i^{\text{right}} &= Q\, c_i^{\text{left}} , \\ m_i^{\text{right}} &= m_i^{\text{left}}/Q , \\ 0 < Q &< 1 . \end{aligned} \tag{5}$$

In this way the eigenfrequency of the affected fold is reduced by the factor Q. Instabilities have been found for $Q < 0.6$ and subglottal pressure above $P_s = 0.013$ (Steinecke and Herzel 1995). Typically, at the borderline of normal phonation abrupt jumps to subharmonic regimes are observed (Steinecke and Herzel 1995). Figures 1 and 2 show a time series and a miogram for a stepwise decrease of Q from 0.56 to 0.5, respectively. It can be seen that the pitch period increases with decreasing Q.

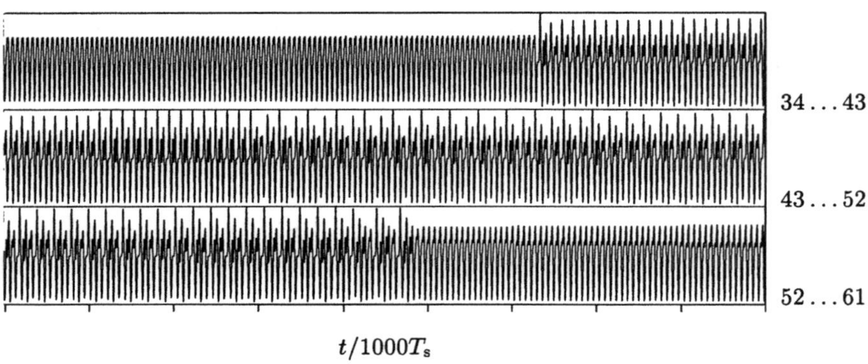

Fig. 1. Smoothed time–series dU/dt for the asymmetric two–mass model. Every 2000 data points (400 ms) the asymmetry parameter Q was decreased by 0.005. In this way bifurcations to subharmonic regimes were induced

At around 40 000 sampling intervals T_s suddenly subharmonics appear at one-half of the pitch. At 46 000 another complicated subharmonic regime is reached. Inspection of the peaks of the elongations x_1^{right} and x_1^{left} reveals that during one cycle (about 5 times the original pitch period) five maxima of x_1^{right} and eight maxima of x_1^{left} occur, i. e. we can interpret this regime as

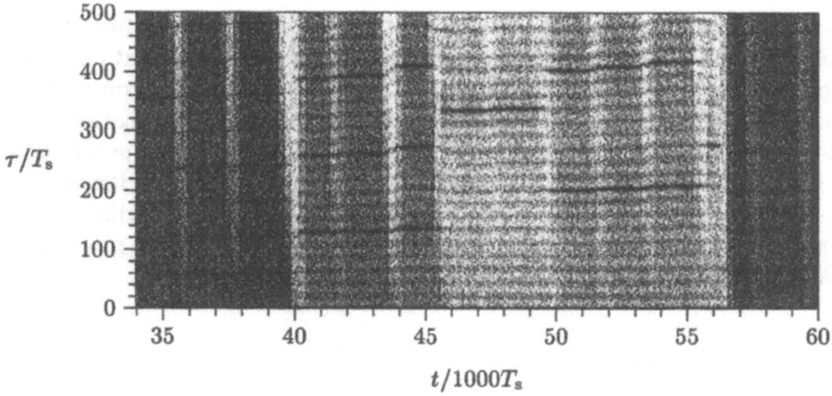

Fig. 2. Miogram of the signal in Fig. 1 (window length — 1000 T_s; window shift — 250 T_s). For each window the generalized mutual information was estimated with a coarse graining of 5 %. The amplitudes are encoded by a gray scale and, hence, dark horizontal bands correspond to peaks of the mutual information and indicate, therefore, periodicities

a 5:8 resonance. From 50 000 subharmonics of one third of the original pitch ("period tripling") dominate corresponding to a 3:5 resonance in the above sense. The sequence of decreasing ratios from 2:3 to 5:8 to 3:5 is reminiscent of "Arnold tongues" in bifurcation diagrams of coupled oscillators. Comprehensive bifurcation diagrams in the P_s-Q parameter plane also reveal the appearance of toroidal and chaotic oscillations (Steinecke and Herzel 1995).

At the beginning of the miogram in Fig. 2 a pitch period T_0 of about 120 data points ($\approx$ 8ms) can be detected. Note that the dark band at about 60 points corresponds to one half of the period, i. e. to the minimum of the autocorrelation function. This reflects the fact that the mutual information is related to the squared autocorrelations (Herzel and Große 1995, Pompe 1996, this volume), and hence we have positive peaks in the mutual information also at time lags where negative peaks appear in the correlation function. Around 40 000 a jump to approximately the double period is visible (note the dark stripes around τ =130, 260, and 390 T_s). Then period five and period three can be detected in the miogram. Similar information can be obtained from the corresponding spectrogram (Herzel et al. 1995a, Fig. 5) and the pitch contour (Herzel 1996).

Despite the qualitative resemblance of our simulations to some observations in unilateral paralysis (Hammarberg et al. 1986, Smith et al. 1992, Herzel et al. 1995a), we have to keep in mind that a 2-mass model is only a crude approximation of the real vocal folds. However, in simulations of a three-dimensional model based on partial differential equations, similar bifurcations to subharmonic regimes and chaos have been found (Titze et al. 1993, Berry et al. 1994). Moreover, the calculation of empirical orthogonal functions from continuum models reveal that the dynamics is often governed by only a few dominant modes. This can be regarded as a justification to study specific aspects of vocal fold vibrations such as left–right asymmetry with appropriate low-order models. An explicit comparison of bifurcations in the two-mass model with results from the continuum models is not possible since the continuum models contain many important features (layered structure of the folds, anterior–posterior modes, prephonatory shape, ...) which have no counterparts in the two-mass model. The incorporation of these physiological details allows better modeling of localized vocal fold lesions.

4 A Continuum Model

The most sophisticated models of the voice have been developed in the last decade at the National Center for Voice and Speech of the USA in Iowa–City (Titze and Talkin 1979, Alipour-Haghighi and Titze 1991, Titze et al. 1993, Berry et al. 1994). The recent models are based on finite element simulations of vocal fold tissue. They incorporate detailed knowledge of the geometry of the vocal folds and allow the modeling of the layered structure (mucosa, ligament, and muscle vocalis). The latest version also includes the numerical solution of the Navier–Stokes equations (Alipour-Haghighi and Titze 1995). However, these simulations are extremely time-consuming and we discuss, therefore, simulations based on the laminar flow hypothesis, i. e., the Bernoulli equation is used as above in the two-mass model. For a simplified geometry of the glottis an analytic treatment of the Navier–Stokes equations was possible (Landa 1996).

The visco-elastic equations are treated in much detail (414 nodal points). These finite element simulations allow a rather realistic modeling of the vibrations. In fact, visualizations of the computer-generated pattern resemble high–speed video observations very closely.

Figure 3 gives an impression of the simulated vocal folds. The model contains nine longitudinal layers. Each layer consists of 32 triangular finite elements. Alltogether, there are 207 nodes per fold free to oscillate, the others are placed on fixed boundaries. Details about the simulations of this continuum model can be found elsewhere (Berry et al. 1994, Herzel et al. 1995b). In these papers we have studied bifurcations due to varying subglottal pressure and muscle tension. It is shown using empirical orthogonal functions that only a few of the 414 potential modes contain most of the variance of the vibrations. More precisely, for normal phonation two modes cover 98 % of the variance, and also in the cases of subharmonic or chaotic vibrations the dominant four modes contain more than 90 % of the variance. It turned out that the empirical modes were quite similar to the modes of the linearized model. This indicates that despite the highly nonlinear excitation via aerodynamical forces the excited modes are governed by the linear visco-elastic properties. The complex vibratory patterns due to decreasing stiffness (Berry et al. 1994)

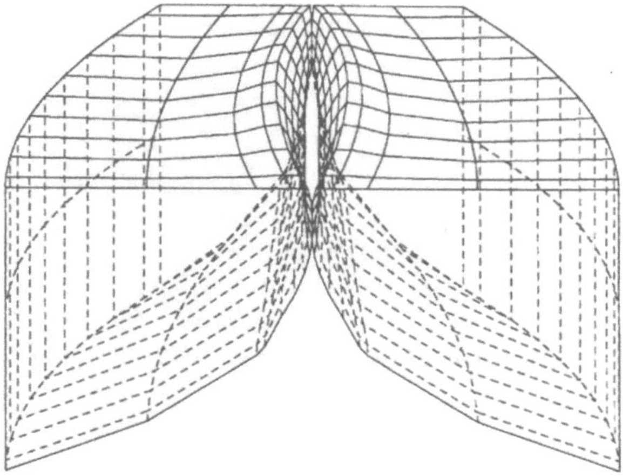

Fig. 3. A view of the continuum model immediately before glottal closure with the posterior edges of the folds in the foreground

can be considered as a model of vocal fry phonation (Scherer 1989, Herzel 1993).

In the following we study the transition to chaos for increasing subglottal pressure. Figure 4 shows a spectrogram of these simulations displaying a transition from regular behavior to irregular oscillations. In the upper graph of Fig. 5 the corresponding time series shows the discussed transition to irregular vibrations. The second and third graph allow a comparison of autocorrelations and mutual information. For each window of 1000 points the mean initial slope (1 to 5 sampling points) was calculated. Both graphs reflect the transition around 90 000 points, but the decay rate of the correlation function decreases whereas the rate for the mutual information increases.

The other graphs show dimensions and Lyapunov exponents of the signal. The estimations of these invariants are consistent with the visual interpretation of the time series and the spectrogram. There is an overall increase of both quantities from left to right and, moreover, there is a stepwise increase of the Lyapunov exponent which is consistent with the increasing irregularity displayed by the spectrogram. Since the decay of the mutual information is related to the Lyapunov exponent the rather similar behavior of the decay rate (3rd graph) and the Lyapunov exponent is understandable.

Of particular interest is the saturation of the dimension estimate around four. This may indicate low-dimensional dynamics and provides, therefore, additional information which cannot deduced from the spectrogram. A low dimensionality is also consistent with the dominance of a few empirical orthogonal functions as discussed above (Berry et al. 1994, Herzel et al. 1995b).

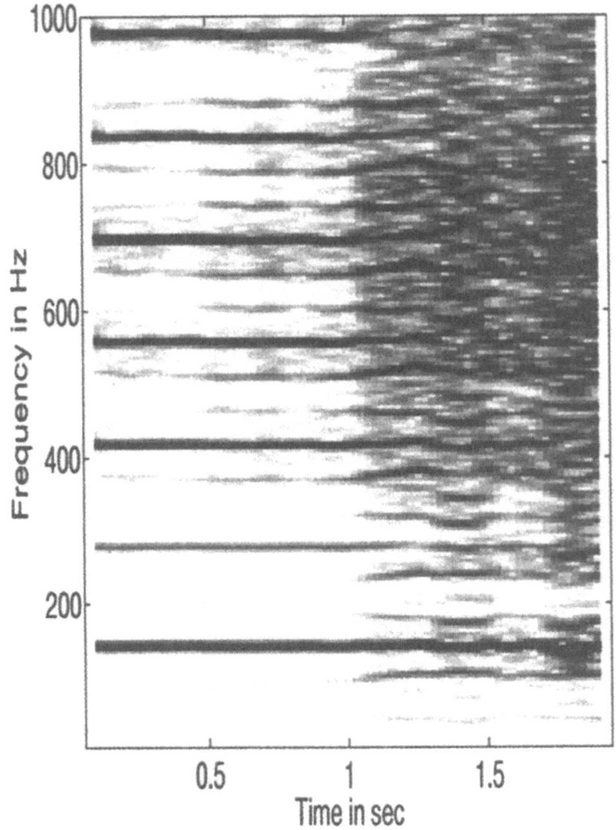

Fig. 4. Spectrogram from the output of the continuum model. The driving subglottal pressure was increased every 8,000 data points (400 ms) from 11.6 cm H_2O to 12.6 cm H_2O. In between the regular oscillations and the highly irregular dynamics some additional spectral peaks indicate subharmonic or quasiperiodic components

5 Excised Larynx Experiments

Experiments with human or animal larynges serve as a link between the human voice source in vivo and computer models (van den Berg 1958, Ishizaka and Isshiki 1976, Baer 1981). They allow controlled and systematic parameter variations and easy observation of vibratory patterns.

We have examined five larynges from large (about 25 kg) mongrel dogs coming from coronary research units at The University of Iowa. The dissected larynges were mounted on an apparatus described in detail elsewhere (Baer 1981, Berry et al. 1996). Heated and humified air was supplied from below as the driving force of the oscillations. The device was attached to several

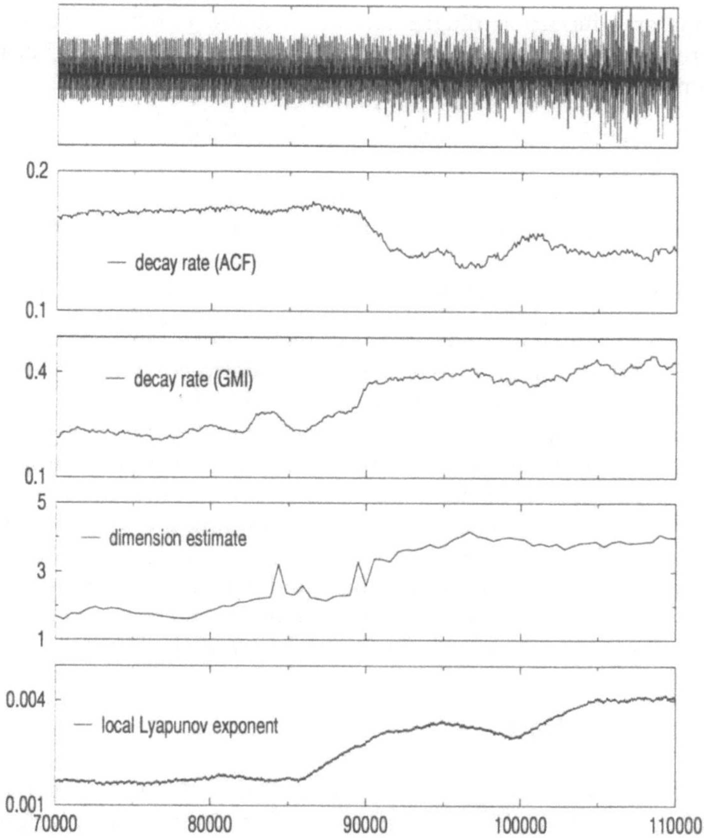

Fig. 5. 1st graph: mouth sound pressure of the continuum model for increasing subglottal pressure (compare Fig. 4). 2nd graph: initial decay of the autocorrelation function for segments of 1000 data points. 3rd graph: decay rate of the mutual information function for segments of 1000 data points. 4th graph: correlation dimension estimated for segments of 4096 data points (embedding dimension — 10; delay–time — 20 points; shift — 512 points). 5th graph: estimations of the maximum Lyapunov exponent for segments of 1024 data points. (The resulting curve has been smoothed by a running average over 512 subsequent estimates.)

micrometers to control the adduction and the elongation of the vocal folds. To facilitate observation of vocal fold movement, a strobe light with adjustable frequency was placed above the glottis. The data were recorded on a color video system and afterwards digitized with 16 bit resolution and a sampling rate of 20 kHz.

In our experiments instabilities have been studied for varying subglottal pressure and for asymmetric adduction and elongation of the vocal folds.

Two-parameter bifurcation diagrams can be found elsewhere (Berry et al. 1996). Here we only summarize briefly the various dynamic regimes which have been observed for overcritical asymmetry and pressure:

- symmetric periodic phonation in head-like and chest-like registers
- whistle-like sound
- periodic vibrations of the lax fold only
- subharmonics, modulations, and irregular vibrations with both folds involved
- aphonia, i. e. vibrations ceased for very strong asymmetric tension.

Typically, the parameter ranges of these regimes overlap, i. e. hysteresis is observed. Sometimes spontaneous transitions between different dynamical regimes appeared without external parameter changes.

The bifurcations discussed below have been induced by slowly increasing the driving pressure. Figure 6 shows the resulting transition from nearly periodic vibrations via period doubling to irregular oscillations with a strong subharmonic component. The smooth period doubling and the sudden frequency jump can be seen very clearly in the corresponding pitch contour. A spectrogram of these bifurcations can be found elsewhere (Berry et al. 1996). Despite the turbulent noise also the estimations of dimensions and Lyapunov exponents exhibit the expected increase in the transition region.

6 Voice Disorders

In the last years bifurcations have been described in vocalizations of patients with nodules, polyps, papilloma, edema, hypofunctional and hyperfunctional dysphonia (Herzel et al. 1994a), chorea Huntington (Kirsch 1995), spasmodic dysphonia (Titze et al. 1993), and unilateral paralysis (Herzel et al. 1995a). To understand the underlying mechanisms of vocal instabilities is the main motivation of computer modeling (Herzel et al. 1991, Berry et al. 1994, Herzel and Knudsen 1995), excised larynx experiments (Berry et al. 1996), and high-speed glottography (Hess et al. 1994).

In this section we study two examples of vocalizations with a rich variety of transition phenomena. Figure 7 shows a miogram of a sustained "i" from a female patient with highly asymmetric papillomas. During most of the time the voice sounds breathy and has a pitch of about 300 Hz. However, intermittently transitions to regimes with another fundamental frequency (about 180 Hz) are found. Around 9000 this frequency dominates whereas around 12 000 and between 25 000 and 35 000 both frequencies are observed. Consequently, the whole signal can be characterized roughly by the following sequence of attractors: limit cycle 1 – limit cycle 2 – decaying toroidal oscillations – limit cycle 1 – torus – limit cycle 1. We note that due to turbulent noise and the obvious non-stationarity only traces of these attractors can be found. However, the analysis of the segment in Fig. 8 is consistent with the

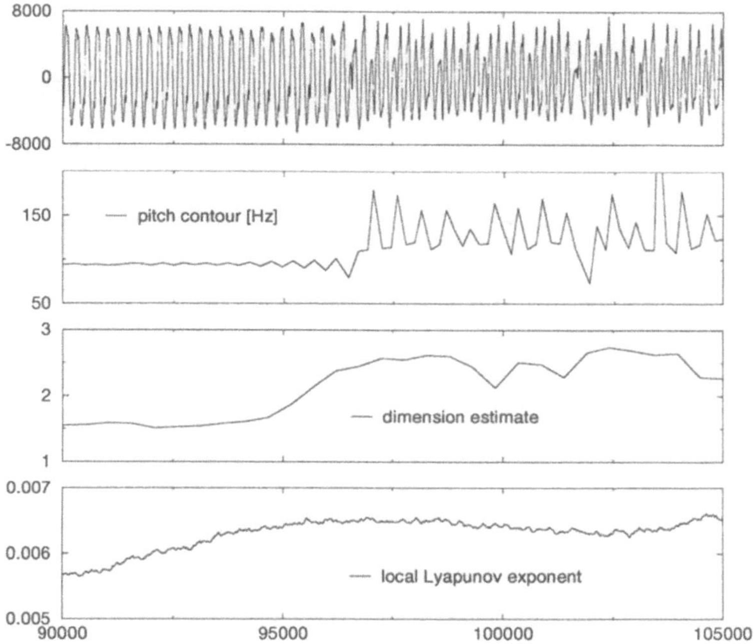

Fig. 6. Bifurcations in an excised larynx experiment for apparantly symmetric folds at large pressure (about 19 cm H_2O). 1st graph: time–series smoothed with a running average over 20 points to reduce the turbulent noise. 2nd graph: pitch contour displaying period–doubling, frequency jump, and subharmonic components (period three). 3rd graph: estimations of the correlation dimension from segments of 4096 data points (embedding dimension — 10; delay — 10; shift — 512 points). 4th graph: smoothed estimates of Lyapunov exponents from segments of 1024 points

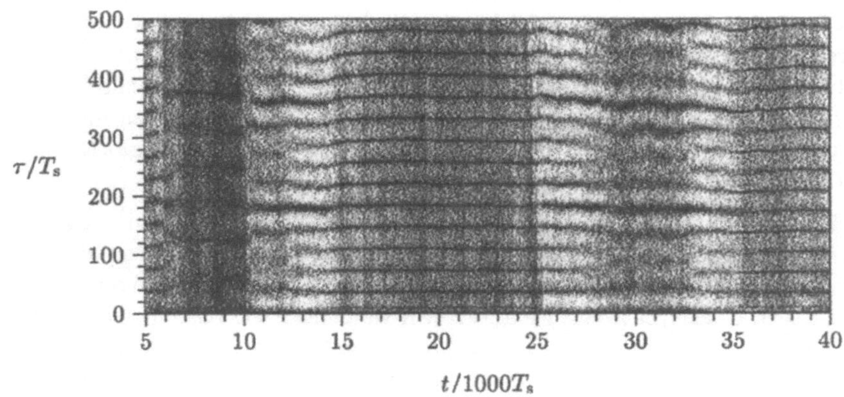

Fig. 7. Miogram of a vowel "i" with frequency jumps and toroidal oscillations (see text and Fig. 8)

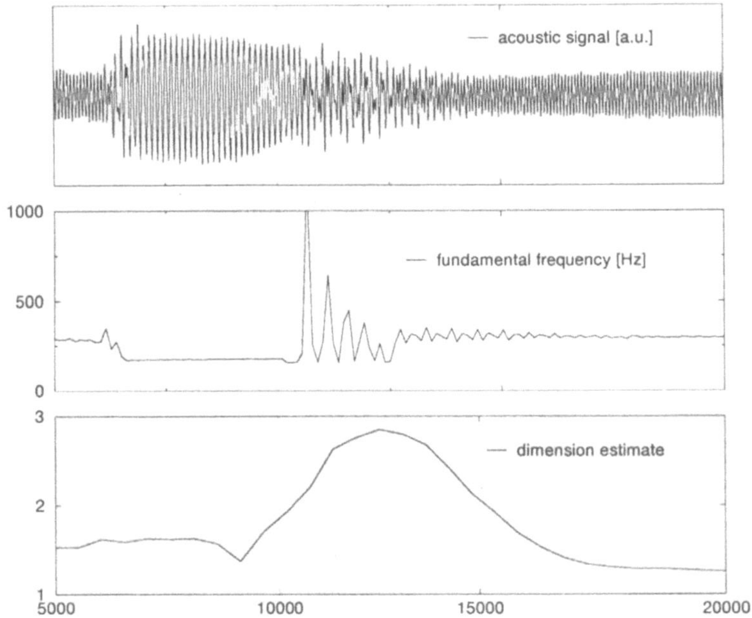

Fig. 8. Upper graph: initial part of the signal analyzed in Fig. 7. Middle graph: pitch contour displaying frequency jumps and toroidal oscillations. Lower graph: correlation dimensions (window length — 4096; shift — 512; embedding dimension — 10; delay — 10)

above characterization. The time series and the pitch contour reveal these episodes very clearly. Since 180 Hz are 3/5 of the other frequency around 12 000 a period five is indicated by the pitch contour. As one would expect there is an increase of the dimension during the irregular segment. It is very likely that the instabilities of this patient are due to the desynchronization of the left and right vocal fold. However, a detailed characterization of the eigenfrequencies and the vibratory modes requires high-speed glottography which was not available during examination of this patient.

The other example is a sustained vowel "a" of a male person with acute laryngitis. Again the spectrogram in Fig. 9 displays a variety of dynamic regimes. First of all, there are two independent frequencies involved. This can be seen clearly between 1 s and 2 s. If we denote the basic frequencies by $f_1 \approx 100$ Hz (increasing in that range) and $f_2 \approx 300$ Hz (almost constant) one can detect the following linear combinations: $f_1 + f_2$, $2f_2$, $2f_2 + f_1$, $3f_2 - f_1$, $2f_2 + 2f_1$, and $3f_2$. Moreover, 1:2 and 1:3 frequency locking can be observed at the beginning and near the end of the signal. In Fig. 10 we present time series of selected segments displaying toroidal oscillations and frequency locking.

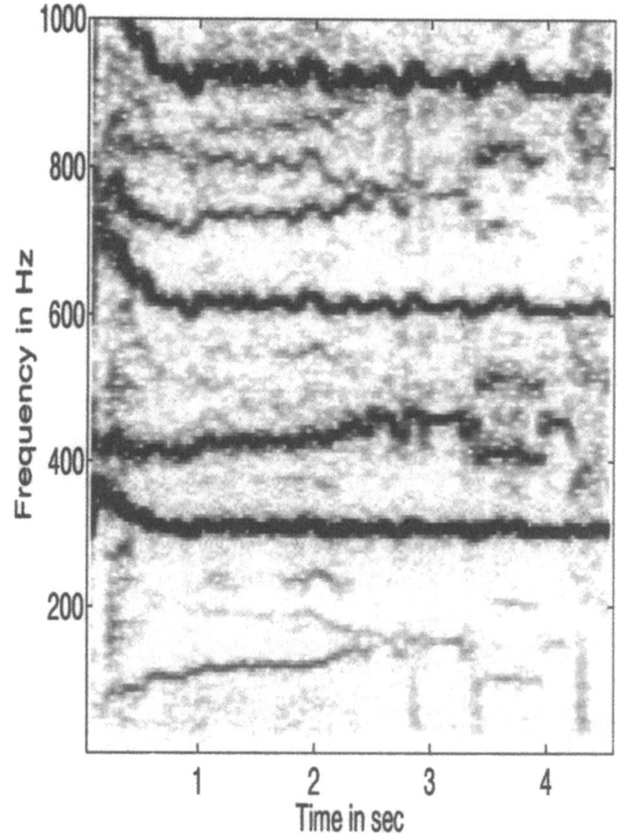

Fig. 9. Narrow–band spectrogram for a rough sounding voice due to acute laryngitis (window size — 4096 points; Hamming window; shift — 512)

7 Summary and Discussion

As demonstrated in earlier papers (Mende et al. 1990, Herzel and Wendler 1991, Herzel 1993, Herzel et al. 1994a) and by our examples in this paper, bifurcations and low-dimensional attractors are frequently observed in voice signals. There is no doubt that the theory of nonlinear dynamics provides the appropriate framework for the analysis of various voice instabilities. A hierarchy of aerodynamic biomechanical computer models exist exhibiting bifurcations as found in voice signals.

In this paper we have analyzed data from a rather simple model and a more sophisticated continuum model of the vocal folds. In addition we studied acoustic signals from an excised larynx experiment and vocalizations of voice patients.

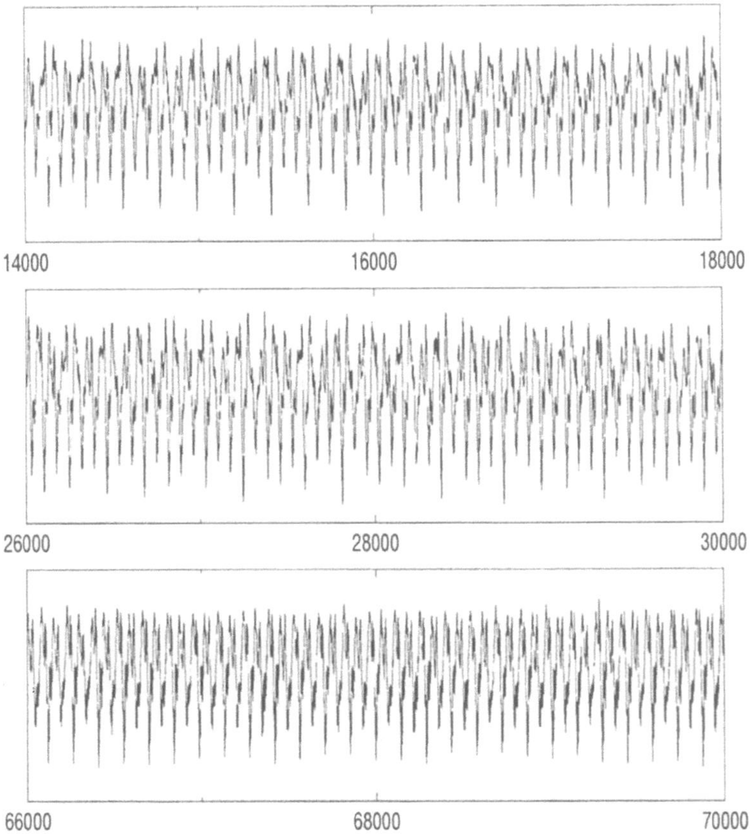

Fig. 10. Representative segments of the signal studied in Fig. 9. Upper graph: 1:3 entrainment around 0.7 seconds. Middle graph: Toroidal oscillations around 1.3 seconds. Lower graph: 1:2 entrainment around 3.1 seconds

The central idea of this paper was the comparative application of five analysis techniques to all these data sets. Since natural voice data are characterized by slow variations of external parameters we have varied slowly the parameters in our computer models as well. In view of these nonstationarities the analysis techniques have been applied using sliding windows of typically 200 ms. In the following we comment on the insight obtained by the different methods.

Interestingly, conventional methods such as spectrograms and pitch contours proved to be very useful in the detection and characterization of bifurcations. Compared with global linear models (e.g., autoregressive processes) spectrograms allow a much more detailed analysis of signals. Local spectra are in a sense state-dependent quantities and, hence, spectrograms can be regarded as a first step towards intrinsically phase-space based techniques. The other conventional tool in voice analysis – the pitch contour – is also closely related to phase space analysis. The detection of events which define the borderlines of pitch periods (maxima, zero crossing, ...) are just special versions of Poincaré sections. Consequently, contours are intimately related to Poincaré maps of continuous systems.

The mutual information is a more powerful indicator of statistical dependences than correlation functions and spectra. In our examples, on long time scales spectrograms reveal essentially the same information about the signals as miograms. However, as demonstrated in Fig. 5 linear correlations measures such as the autocorrelation decay may behave quite differently. Consequently, in our examples, on short time scales miograms contain more information comparable to Lyapunov exponents or metric entropies (sum of all positive Lyapunov exponents).

It is now widely realized that the estimation of dimensions and Lyapunov exponents from physiological data is quite complicated. However, these quantities provide information about the underlying attractors which cannot be obtained by spectral analysis. One should not expect precise estimations from relatively short segments of the data but a sliding analysis allows the comparison of the estimates for different dynamic regimes. For the simulations of the continuum model and the excised larynx data there was indeed the expected increase of dimensions and Lyapunov exponents due to bifurcations. Although the absolute values of dimension estimates from only 4096 data points should be interpreted with caution, the calculated dimensions are consistent with expectations: for noisy limit cycles the estimates are between 1 and 2 and more irregular segments lead to values between 2 and 4.

In summary, the analysis of voice signals has to be adapted to the specific characteristics of the voice source. For example, pitch contours remove effects of the vocal tract filter and constitute, therefore, a valuable variant of a Poincaré section. Known parameter variations on time–scales of a few hundred milliseconds require a sliding application of any measure based on stationarity assumptions. However, slowly varying parameters have the advantage that bifurcations of the underlying dynamical system can be detected in single voice signals.

Acknowledgments

Many of the analyzed data are the output of a fruitful cooperation with colleagues from Iowa City (Ingo Titze, David Berry, and Brad Story) and Berlin (Jürgen Wendler and Ina Steinecke). Support was provided by the Deutsche Forschungsgemeinschaft.

References

Alipour-Haghighi F., and Titze I. R. (1991): Elastic models of vocal fold tissues. J. Acoust. Soc. Am. **90**, 1326–1331

Alipour-Haghighi F., and Titze I. R. (1995): Combined Simulation of two-dimensional airflow and vocal fold vibration. In *Status and Progress Report National Center for Voice and Speech* Vol. 8. pp. 9–14

Baer T. (1981): Investigation of the Phonatory Mechanism. ASHA Reports **11**, 38–47

Berry D. A., Herzel H., Titze I. R., Krischer K. (1994): Interpretation of biomechanical simulations of normal and chaotic vocal fold oscillations with empirical eigenfunctions. J. Acoust. Soc. Am. **95**, 3595–3604

Berry D. A., Herzel H., Titze I. R., Story B. (1996): Bifurcations in excised larynx experiments. J. Voice **10**, 129–138

Dolansky L., Tjernlund P. (1968): On certain irregularities of voiced-speech waveforms. IEEE Trans. **AU–16**, 51–56

Eckmann J.–P., Kamphorst S., Ruelle D., Ciliberto S. (1986): Liapunov exponents from time series. Phys. Rev. A **34**, 4971–4979

Fant G. (1960): *Acoustic Theory of Speech Production.* Mouton, The Hague

Fraser A. M., Swinney H. L. (1986): Using mutual information to find independent coordinates for strange attractors. Phys. Rev. A **33**, 1134–1140

Grassberger P., Procaccia I. (1983): Measuring the strangeness of strange attraktors. Physica D **9**, 189–208

Hammarberg B., Fritzell B., Gauffin J., Sundberg J. (1986): Acoustic and perceptual analysis of vocal dysfunction. J. of Phonetics **14**, 533–547

Herzel H., Ebeling W. (1985): The decay of correlations in chaotic maps. Phys. Lett. A **111**, 1–4

Herzel H., Ebeling W., Schulmeister Th. (1985): Nonuniform chaotic dynamics and effects of noise in biochemical systems. Zeitschr. Naturf. **42a**, 136–142

Herzel H., Pompe B. (1987): Effects of noise on a nonuniform chaotic map. Phys. Lett. A **122**, 121–125

Herzel H., Steinecke I., Mende W., Wermke K. (1991): Chaos and bifurcations during voiced speech. In Mosekilde E., Mosekilde L. (1991): *Complexity, Chaos and Biological Evolution.* Plenum, New York. pp. 41–50

Herzel H., Wendler J. (1991): Evidence of chaos in phonatory samples. In *EUROSPEECH* ESCA, Genova, 1991, pp. 263–266

Herzel H. (1993): Bifurcations and chaos in voice signals. Appl. Mech. Rev. **46**, 399–413

Herzel H., Berry D. A., Titze I. R., Saleh M. (1994a): Analysis of vocal disorders with methods from nonlinear dynamics. J. Speech Hearing Res. **37**, 1008–1019

Herzel H., Schmitt A., Ebeling W. (1994b): Finite sample effects in sequence analysis. Chaos, Solitons, and Fractals **4**, 97–113

Herzel H., Große I. (1995): Measuring correlations in symbolic sequences. Physica A **216**, 518–542

Herzel H., Knudsen C. (1995): Bifurcations in a vocal fold model. Nonlinear Dynamics **7**, 53–64

Herzel H., Berry D., Titze I., Steinecke I. (1995a): Nonlinear dynamics of the voice: signal analysis and biomechanical modeling. CHAOS **5**, 30–34

Herzel H., Krischer K., Berry D., Titze I. R. (1995b): Analysis of spatio-temporal pattern by means of empirical orthogonal functions. In Cladis P. E., Palffy-Muhoray P. (1995): *Spatio-Temporal Patterns in Nonequilibrium Complex Systems* Addison-Wesley, Santa Fe. pp. 505–518

Herzel H. (1996): Possible Mechanisms of Vocal Instabilities. In: Davis P. J., Fletcher N. H. (1996): *Vocal Fold Physiology: Controlling Complexity & Chaos.*, Singular Publ. Group, San Diego (in press)

Herzel H., Reuter R. (1996): Biphonation in voice signals. In Katz R. A., Frison T. W., Kadtke J. B., Bulsara A. R. (1996): *Nonlinear, Chaotic, and Advanced Signal Processing Methods For Engineers and Scientists.* American Institute of Physics, Woodbury

Hess M. M., Gross M., Herzel H. (1994): Hochgeschwindigkeitsaufnahmen von Schwingungsmoden der Stimmlippen. Otorhinolaryngologia NOVA **4**, 307–312

Hirano M. (1989): Objective evaluation of the human voice: clinical aspects. Folia phoniat. **41**, 89–144

Holzfuss J., Mayer-Kress G. (1986): An approach to error-estimation in the application of dimension algorithms. In Mayer-Kress G. (1986): *Dimensions and Entropies in Chaotic Systems.* Springer, Berlin, Heidelberg. pp. 114–122

Holzfuss J., Parlitz U. (1991): Lyapunov exponents from time series. In Arnold L., Crauel H., Eckmann J.–P. (1991): *Lyapunov Exponents* Lecture Notes in Mathematics, vol. 1486, Springer, Berlin, Heidelberg. pp. 263–270)

Ishizaka K., Flanagan J. L. (1972): Synthesis of voiced sounds from a two-mass model of the vocal cords. Bell Syst. Techn. J. **51**, 1233–1268

Ishizaka K., Isshiki N. (1976): Computer simulation of pathological vocal-cord vibrations. J. Acoust. Soc. Am. **60**, 1194–1198

Kelman A. W. (1981): Vibratory pattern of the vocal folds. Folia phoniat. **33**, 73–99

Kirsch R. (1995): *Statistische Analyse akustischer Signale.* Diplomarbeit, Humbolt Universität zu Berlin, Berlin

Kowalik Z. J., Elbert T. (1995): A practical method for the measurements of the chaoticity of electric and magnetic brain activity. Int. J. Bif. Chaos **5**, 475–490

Kurths J., Herzel H. (1987): An attractor in a solar time series. Physica D **25**, 165–172

Landa P. S. (1996): *Nonlinear oscillations and waves in dynamical systems.* Kluwer, Amsterdam

Lauterborn W., Cramer E. (1981): Subharmonic routes to chaos observed in acoustics. Phys. Rev. Lett. **47**, 1445–1448

Lauterborn W., Holzfuss J. (1986): Evidence for a low-dimensional strange attractor in acoustic turbulence. Phys. Lett. A **115**, 369–372

Liauw M. A., Koblitz K., Jaeger N. I., Plath P. (1993): Periodic perturbation of a drifting heterogeneous catalytic system. J. Phys. Chem. **97**, 11724–11730

Lind J. (Ed.) (1965): *Newborn Infant Cry.* Almquist and Wiksells Boktrycken, Uppsala

Mazo M. (1994): Lament made visible: a study of paramusical elements in russian lament. In Yung B., Lam J. S. C. (1994): *Themes and Variations.* Harvard University, Cambridge. pp. 164–211

Mende W., Herzel H., and Wermke K. (1990): Bifurcations and chaos in newborn cries. Phys. Lett. A **145**, 418–424

Merbach D., Hess O., Herzel H., Schöll E. (1995): Injection-induced bifurcations of transverse spatio-temporal patterns in semiconductor laser arrays. Phys. Rev. E **52**, 1571–1578

Pelorson X., Hirschberg A., van Hassel R., Wijnands A., Auregan Y. (1994): Theoretical and experimental study of quasisteady-flow separation within the glottis during phonation. Application to a modified two-mass model. J. Acoust. Soc. Am. **96**, 3416–3431

Pompe B., Leven R. W. (1986): Transinformation of chaotic systems. Physica Scripta **34**, 8–13

Pompe B. (1993): Measuring Statistical Dependencies in a Time Series. J. Stat. Phys. **73**, 587–610

Pompe B., Heilfort M. (1994): Miogram for short-time signal analysis with an application to speech. In Gruber R., Tomassini M. (1994): *Proc. 6. Joint EPS–APS Int. Conf. on Physics Computing.* European Physical Society, Boite Postale 69, CH–1213 Petit–Lancy 2, Geneva, Switzerland.

Rényi A. (1970): *Probability Theory.* North–Holland, Amsterdam

Sano M., Sawada Y. (1985): Measurement of the lyapunov spectrum from a chaotic time series. Phys. Rev. Lett. **55**, 1082–1085

Scherer R. (1989): Physiology of creaky voice and vocal fry. J. Acoust. Soc. Am. **86**, (S1), S25(A)

Smith M. E., Berke G. S., Gerrat B. R., Kreimann J. (1992): Laryngeal paralyses: theoretical considerations and effects on laryngeal vibration. J. Speech Hear. Res. **35**, 545–554

Steinecke I.,Herzel H. (1995): Bifurcations in an asymmetric vocal fold model. J. Acoust. Soc. Am. **97**, 1874–1884

Titze I. R. (1973): The human vocal cords: a mathematical model, part I. Phonetica **28**, 129–170

Titze I. R., Talkin D. T. (1979): A theoretical study of the effects of various laryngeal configurations on the acoustics of phonation. J. Acoust. Soc. Am. **66**, 60–74

Titze I. R., Baken R., Herzel H. (1993): Evidence of chaos in vocal fold vibration. In Titze I. R. (ed.) (1993): *Vocal Fold Physiology: Frontiers in Basic Science.* Singular Publishing Group, San Diego. pp. 143–188)

Titze I. R., Alipour-Haghighi F. (forthcoming book): *Myoelastic Aerodynamic Theory of Phonation*

van den Berg J. (1958): Myoelastic-aerodynamic theory of voice production. J. Speech Hearing Res. **1**, 227–244

Wolf A., Swift J., Swinney H., Vastano J. (1985): Determining lyapunov exponents from a time series. Physica D **16**, 285–317

Wolff R. C. L. (1992): Local lyapunov exponents: looking closely at chaos. J. R. Stat. Soc. B **54**, 353–371

Printing: Mercedesdruck, Berlin
Binding: Buchbinderei Lüderitz & Bauer, Berlin